Lecture Notes in Physics

Volume 835

For further volumes:
http://www.springer.com/series/5304

The Lecture Notes in Physics

The series Lecture Notes in Physics (LNP), founded in 1969, reports new developments in physics research and teaching—quickly and informally, but with a high quality and the explicit aim to summarize and communicate current knowledge in an accessible way. Books published in this series are conceived as bridging material between advanced graduate textbooks and the forefront of research and to serve three purposes:

- to be a compact and modern up-to-date source of reference on a well-defined topic
- to serve as an accessible introduction to the field to postgraduate students and nonspecialist researchers from related areas
- to be a source of advanced teaching material for specialized seminars, courses and schools

Both monographs and multi-author volumes will be considered for publication. Edited volumes should, however, consist of a very limited number of contributions only. Proceedings will not be considered for LNP.

Volumes published in LNP are disseminated both in print and in electronic formats, the electronic archive being available at springerlink.com. The series content is indexed, abstracted and referenced by many abstracting and information services, bibliographic networks, subscription agencies, library networks, and consortia.

Proposals should be sent to a member of the Editorial Board, or directly to the managing editor at Springer:

Christian Caron
Springer Heidelberg
Physics Editorial Department I
Tiergartenstrasse 17
69121 Heidelberg/Germany
christian.caron@springer.com

Martin Bojowald

Quantum Cosmology

A Fundamental Description of the Universe

 Springer

Prof. Dr. Martin Bojowald
Institute for Gravitation and the Cosmos
Pennsylvania State University
104 Davey Laboratory
University Park, PA 16802-6300
USA
e-mail: bojowald@gravity.psu.edu

ISSN 0075-8450
ISBN 978-1-4614-3017-9
DOI 10.1007/978-1-4419-8276-6
Springer New York Dordrecht Heidelberg London

ISSN 1616-6361
ISBN 978-1-4419-8276-6 (eBook)

Cover design: eStudio Calamar, Berlin/Figueres

Printed on acid-free paper

Springer is part of Springer Science+Business Media (www.springer.com)

Contents

Part II   Effective Descriptions

Part III   Beyond Isotropic Models

Chapter 1
Introduction

Ay! There are times when the great universe
Like cloth in some unskilful dyers' vat
Shrivels into a hand's-breadth, and perchance
That time is now! Well! Let that time be now.
Let this mean room be as that mighty stage
Whereon kings die, and our ignoble lives
Become the stakes God plays for.

Oscar Wilde: A Florentine Tragedy

The universe, ultimately, is to be described by quantum theory. Quantum aspects of all there is, including space and time, may not be significant for many purposes, but are crucial for some. And so a quantum description of cosmology is required for a complete and consistent worldview. At any rate, even if we were not directly interested in regimes where quantum cosmology plays a role, a complete physical description could not stop at a stage before the whole universe is reached. Quantum theory is essential in the microphysics of particles, atoms, molecules, solids, white dwarfs and neutron stars. Why should one expect this ladder of scales to end at a certain size? If regimes are sufficiently violent and energetic, quantum effects are non-negligible even on scales of the whole cosmos; this is realized at least once in the history of the universe: at the big bang where the classical theory of general relativity would make energy densities diverge.

One might ask a quantum theory of *what* should be considered. The classical universe is described by general relativity, which may be quantized on its own if its degree of freedom, space–time geometry, is seen as fundamental. Alternatively, general relativity might be seen as ultimately being a phenomenological continuum theory, much as in hydrodynamics. By itself, it would not reveal what the fundamental, microscopic degrees of freedom should be. Nonetheless, general relativity serves as a crucial guideline in constructing a quantum theory of gravity, for it is to be reproduced as the semiclassical limit on a certain range of scales, including those on which we currently probe the universe. This by itself is challenging enough a task owing to the existence of many mathematical consistency conditions. Most

M. Bojowald, *Quantum Cosmology*, Lecture Notes in Physics 835,
DOI: 10.1007/978-1-4419-8276-6_1, © Springer Science+Business Media, LLC 2011

current theories indeed point to the presence of new microscopic entities, and they provide insights into some of their properties—be they described as strings, loops or something else. Irrespective of what exactly is realized, general relativity must be extended for it is singular; and quantum theory must play a role.

Quantum gravity applies to many situations, most importantly early-universe cosmology and black holes. Cosmology complicates and simplifies considerations at the same time. It comes with severe conceptual problems of how to interpret the wave function of the whole universe, with all observers having to be situated within the system. Despite many activities for several decades, a proper understanding of this situation remains a challenge. But there is also an advantage in this context: the cosmological principle, which states the assumption of homogeneity on large scales and is by now well justified by extensive galaxy maps, reduces the number of degrees of freedom. One obtains a technically simpler framework, which is helpful for testing existing general theories but also provides possible physical insights.

Much of quantum gravity is part of mathematical physics due to the heavy tools required. But one should keep in mind that the objective is quite different from usual mathematical edifices: quantum gravity at present is not constructed on firm ground; it rather grows toward a certain, vaguely formulated aim. Some principles must certainly exist and be used, but there are no generally accepted axioms from which one could start, stepping ahead theorem by theorem. This situation often makes developments, even crucial ones, appear fuzzy. Nevertheless, progress is clearly visible by models becoming more and more controlled and realistic and, put the other way, by several developments having been ruled out with later progress.

Accordingly, the focus in this book will not be so much on specific cases, unless they illustrate key features, but rather on a general framework which, at the current stage in time and in the author's personal opinion, summarizes distinctive properties of quantum cosmology. This guide through the scaffolding should provide readers interested in working on those problems with a quick route to the construction site, discussing tools and stating open issues to provide an entrance into this rather messy field. Keep in mind that, while the final building is likely to follow the shape and height already indicated, the scaffolding itself eventually will have to be torn down. But before the building stands, the reader is advised not to pay too much attention to all kinds of details: one should not measure carpets before the walls are set down.

There are two main instances in which quantum physics arises: quantum dynamics and quantum geometry. Quantum dynamics includes the usual conceptual problems of measurements, observables and the role of quantum variables such as fluctuations and correlations. Especially in quantum cosmology it also, ultimately, requires one to understand the meaning and the arrow of time. As general issues, all this is rather insensitive to the specific realization of quantum space–time, and can thus be found and analyzed already in the Wheeler–DeWitt formulation of quantum cosmology started in the 1960s. Quantum geometry, on the other hand, depends more sensitively on the quantization framework used. Here, the structure of space and time on their smallest, possibly atomic scales and their refinement in the course of time is crucial. The most specific such realization so far has been made within the framework of loop quantum gravity, which will be introduced and used throughout this book. But all general aspects, whenever applicable, will be discussed with as small a number of ingredients from a loop quantization as possible.

We will begin the exposition with a rather detailed discussion of quantum theory in the context of cosmology. This introduction will show why an atomic understading of space–time structure is relevant, a specific form of which is then provided using the methods of loop quantum gravity. In this canonical quantization one starts with a "kinematical" quantization of spatial geometry, already illustrating the discrete nature. Dynamics then shows how such atomic structures change in time according to a quantum Hamiltonian. At this stage, control over atomic quantum space–times will be gained.

An analysis of dynamical equations in general is complicated, especially if gravity is involved which requires self-interaction and non-linearity. Part II of this book will introduce the key tool to a manageable analysis: effective descriptions. They will first be applied to quantum cosmology in the Wheeler–DeWitt form, and then to loop quantum cosmology which introduces additional non-linearities. At this stage, we will see the first intuitive mechanism of resolving singularities by repulsive quantum forces. At the same time it will become clear that quantum dynamics, not just quantum geometry, is highly relevant for understanding the big bang. With these results, a discussion of what the actual meaning of resolving singularities might be will be given.

The third part extends constructions and results from isotropic models to several less symmetric cases, first to anisotropic ones which also include models of the Schwarzschild black-hole interior. Here we will see the first applications to black hole dynamics. (There are other important applications of quantum gravity to black holes, mainly in the context of black-hole thermodynamics. They are not part of this book since those methods differ considerably from what one uses in quantum cosmology. This line of research has so far provided scant insight about the dynamics.) General black-hole models, including a phase of gravitational collapse and possibly one of Hawking evaporation, require inhomogeneous geometries. Spherically symmetric ones are the simplest among those and will be discussed in detail. Also here, results about singularity resolution are available, but still inconclusive. By similar constructions one can describe models such as Einstein–Rosen waves or those of Gowdy type, which do have local gravitational degrees of freedom. They provide further interesting examples of singularities and possible resolutions, but investigations have only just begun. Inhomogeneities can also be introduced as perturbations on a homogeneous background space–time, which brings us back to applications in cosmology.

Part IV is a discussion of the typical mathematical issues involved in quantum cosmology: properties of difference equations, the derivation and use of physical Hilbert spaces, and general aspects of effective descriptions.[1]

---

[1]   Some of the material in this book is based upon work supported by the National Science Foundation under Grant No. 0748336. Any opinions, findings, and conclusions or recommendations expressed in this material are those of the author and do not necessarily reflect the views of the National Science Foundation.

Part I
Quantizing the Whole Universe

Curved space-times in general and cosmology in particular require special tools to address their quantum physics. Canonical quantization methods are most highly controlled, which will first be seen as the basis of Wheeler–DeWitt quantization in the simplest, isotropic models. Wheeler–DeWitt methods do not apply to full general relativity beyond formal definitions, but those of loop quantum gravity do. They do not only allow rigorous mathematical constructions but also have their own physical implications. The brief introduction in this part leads up to the cosmological realization of loop quantum gravity.

Chapter 2
Cosmology and Quantum Theory

Next to flat Minkowski space, the simplest solutions of general relativity are given by isotropic Friedmann–Lemaître–Robertson–Walker (FLRW) universe models. Thanks to spatial isotropy, there is just a single parameter, the scale factor $a(t)$ determining the spatial scale via the line element

$$ds^2 = -N(t)^2 dt^2 + a(t)^2 d\sigma_k^2 \tag{2.1}$$

with

$$d\sigma_k^2 = \frac{dr^2}{1 - kr^2} + r^2(d\vartheta^2 + \sin^2\vartheta\, d\varphi^2). \tag{2.2}$$

As usual, $k = 0, \pm 1$ distinguishes a three-space of constant curvature k. Although there is a second coefficient in (2.1) in addition to $a(t)$, the lapse function $N(t)$, it can be removed by rescaling the time parameter such that $d\tau = N(t)dt$, where τ is proper time as measured by co-moving observers. The scale factor, on the other hand, cannot be eliminated completely unless it is time-independent, and so it describes the changing spatial scales of an evolving isotropic universe. Its dynamics is given by specializing Einstein's equation to isotropic metrics (2.1): the Friedmann equation

$$\left(\frac{\dot{a}}{Na}\right)^2 + \frac{k}{a^2} = \frac{8\pi G}{3}\rho \tag{2.3}$$

where the dot is a derivative by the original t and ρ is the energy density of matter, and the second-order Raychaudhuri equation

$$\frac{(\dot{a}/N)^\bullet}{aN} = -\frac{4\pi G}{3}(\rho + 3P) \tag{2.4}$$

in which also the pressure P of matter appears. The latter equation is not independent but can be derived from the Friedmann equation using the continuity equation

$$\dot{\rho} + 3\frac{\dot{a}}{a}(\rho + P) = 0 \tag{2.5}$$

of matter.

M. Bojowald, *Quantum Cosmology*, Lecture Notes in Physics 835,
DOI: 10.1007/978-1-4419-8276-6_2, © Springer Science+Business Media, LLC 2011

2.1 Scaling

In all these equations, the rescaling freedom of the time variable does not matter since it is always $N(t)dt$ that enters; the equations are thus time-reparameterization invariant. While the scale factor cannot be removed completely by a change of coordinates, it can, in the spatially flat case, be rescaled by a constant: a changes to a/λ when coordinates are rescaled by $r \mapsto \lambda r$. (For non-vanishing spatial curvature, fixing $k = \pm 1$ removes any rescaling freedom.[1] For $k = 0$ no natural normalization is available.) Thus, a does not have absolute meaning; only ratios such as the Hubble parameter $\dot{a}/Na$ or the deceleration parameter $q = -aN(\dot{a}/N)^\bullet/\dot{a}^2$ are unambiguous. For $k = +1$, one can give meaning to the volume $a^3 \mathscr{V}_{\text{unit}} = \int_\Sigma \sqrt{\det h} d^3 x$ with $\mathscr{V}_{\text{unit}} = 2\pi^2$ the volume of the three-dimensional unit sphere, as the measure given by the spatial metric h_{ab} integrated over the whole compact space. However, even in this case a itself can hardly be considered a relevant physical parameter because determining its value would require one to measure the size of the universe. In principle, this might be possible, for instance if one can detect repeated patterns in the sky from light traversing the compact space more than once [1]. But no such patterns have been found so far with certainty, and so the full information in $a^3 \mathscr{V}_{\text{unit}}$ cannot play a role in current cosmology.

Since general relativity is generally covariant and fully independent of the choice of space-time coordinates, any equations of motion it provides automatically have the invariance property under changes of coordinates, or under those changes that respect the symmetry of a given model. Nevertheless, such scalings can sometimes have a subtle flavor especially in canonical formulations or their quantizations: the Hubble parameter is a scaling-independent measure for the rate of change of a, which can provide a canonical momentum, but there is no scaling-independent measure for the size itself unless one chooses and fixes extra structures such as a distinguished spatial region. While equations are then coordinate-scaling independent, they might depend on the region chosen. Such issues will play a central (though somewhat formal) part in the fundamental set-up of quantum cosmology, and their treatment will, rather surprisingly, turn out to be deeply related to the atomic nature of space and time.

> In full general relativity, coordinate changes are implemented as gauge transformations generated by constraints. The Friedmann equation (2.3) constitutes the isotropic form of the Hamiltonian constraint, generating time reparameterizations. From this perspective, one would expect rescalings of a, multiplying only spatial coordinates in the line element, to be generated by the diffeomorphism constraint. However, the diffeomorphism constraint, depending on gradients of the fields, vanishes identically when evaluated in isotropic configurations, and thus cannot contribute to isotropic models. Moreover, a non-trivial diffeomorphism (or other) constraint would remove one pair of canonical degrees of freedom—and there is only one in isotropic models. Even before considering the dynamics, all gravitational degrees of freedom would be removed. Rescaling a thus cannot be a gauge transformation; it rather corresponds to transformations between different reductions to isotropy, between different models rather than within one model. This observation already indicates that the rather innocent-looking rescaling freedom is closely related to symmetry reduction.

---

[1] Rescaling freedom in the closed model is discussed in more detail in Sect. 3.2.1.

2.2 Wheeler–DeWitt Quantization

Wheeler–DeWitt models [2, 3] provided the first quantizations of cosmological models. Here, indeed, the scale factor itself was taken as one of the canonical variables, conjugate to

$$p_a = -\frac{3}{4\pi G}\frac{a\dot{a}}{N} \tag{2.6}$$

according to the Einstein–Hilbert action reduced to isotropic metrics. Following the principles of basic quantum mechanics, wave functions are thus of the form $\psi(a)$, supported on the positive real line $a > 0$.

> The quantization of a phase space with a configuration variable a restricted to positive values requires some care. If one introduces the usual operator $-i\hbar d/da$ to represent $\hat{p}_a$ such that it obeys the required commutation relation with the multiplication operator $\hat{a}$, it is not self-adjoint with respect to an inner product based on the integration $\int_0^\infty da$. One can see this in several ways: The derivative operator has an imaginary eigenvalue with eigenfunction $\exp(-a)$, which is normalizable on the positive half-line. Moreover, the operator, when exponentiated, generates translations of wave functions along a, which is not unitary because it moves the wave function out of the integration range of the inner product. With a non-self-adjoint $\hat{p}_a$, the construction of a self-adjoint Hamiltonian will be complicated.
>
> One solution to the problem is to begin with a different, non-canonical algebra of phase-space variables a together with $D := ap_a$, such that $\{a, D\} = a$ is still closed. Moreover, a and D generate canonical transformations that are complete in the sense that they define a transitive and free group action leaving the phase space fixed. In particular, the factor of a in D ensures that the Hamiltonian vector field $X_D = \{\cdot, D\}$ is tangent to the boundary at $a = 0$, unlike X_{p_a}. These are the requirements of group-theoretical quantization [4], which proceeds by constructing a quantum theory from unitary representations of the group generated. The representation space then provides a Hilbert space on which $\hat{a}$ and $\hat{D}$ act as self-adjoint operators. For quantum gravity in general terms, this program has been adopted in the context of affine quantum gravity [5, 6]. In our treatment we will be led to replace a with a variable taking both signs (thanks to orientation) so that we will not have to deal with this problem.

Quantization in cosmology introduces all the usual quantum properties such as fluctuations and quantum uncertainty for the size of the universe itself. However, expectation values are not easy to compute since it is not a priori clear what inner product should be taken: There is, first, the possibility of non-normalizable "scattering" states if a universe is to expand forever and to reach arbitrarily large values of a; secondly, the Friedmann equation only depends on the canonical variables a and p_a (and those for matter), but not on their time derivatives. Canonically, it is not an evolution equation, which could become some kind of Schrödinger equation upon quantization, but a constraint

$$C = -\left(\frac{4\pi G}{3}p_a\right)^2 a^{-1} + ka + \frac{8\pi G}{3}a^3\rho = 0. \tag{2.7}$$

We may directly quantize it by inserting operators,

$$\frac{4\pi G\hbar}{3\sqrt{a}}\frac{\partial}{\partial a}\left(\frac{4\pi G\hbar}{3\sqrt{a}}\frac{\partial}{\partial a}\psi\right) + ka\psi = -\frac{8\pi G}{3}\hat{H}_{\text{matter}}\psi \qquad (2.8)$$

with some matter Hamiltonian $\hat{H}_{\text{matter}}$. In the coefficients, the Planck length $\ell_P = \sqrt{G\hbar}$ appears. Unlike in quantum mechanics, the ordering of factors is not unique.

2.3 Evolution

Only states annihilated by the quantized constraint $\hat{C}$ can be considered physical, forming the so-called physical Hilbert space.[2] As the next problem we thus see the need of having to understand the solution space of the Friedmann constraint operator, and being able to endow it with an inner product. Only expectation values and fluctuations or other moments of a *physical* state can correspond to observable quantities.

Traditionally, the focus in quantum cosmology has been on finding solutions for states in a diverse set of models, mainly with a semiclassical interpretation in mind. The form of the Wheeler–DeWitt equation, quadratic in the momentum p_a, then suggests that states may be normalized by a Klein–Gordon type inner product (and in some models one is exactly dealing with the Klein–Gordon equation; see Example 5.1). For instance, if one treats a as a measure of time and includes a scalar matter degree of freedom as well, one obtains a hyperbolic differential equation [7] with an initial-value problem that suggests to pose an initial wave function at a fixed value of a. In some cases, however, such an interpretation is problematic, such as in a closed model where the same value of a may be reached twice, at times that would be considered far apart in the classical picture. It is sometimes suggested that the notion of time changes considerably in quantum cosmology, requiring a reversal of the arrow of time in a collapsing phase [8].

Complete constructions of the physical Hilbert space are possible [9], but cannot often be performed explicitly; usually a successful implementation hinges on special forms of matter ingredients that can serve the purpose of an internal time, such as a free, massless scalar or dust.

Example 2.1 (Dust) The Hamiltonian formulation of dust-like matter with vanishing pressure, as developed in [10], reduces in isotropic models to a simple Hamiltonian $H_{\text{matter}} = p_T$ linear in the momentum conjugate to proper time T of dust particles. Upon quantization of $\hat{p}_T = -i\hbar\partial/\partial T$, the constraint equation (2.8), ignoring questions of self-adjointness, then takes the form of a Schrödinger equation with time T. If we transform the variables (a,T) to $V := (2\pi G)^{-1}a^{3/2}$ and $\lambda := 3(4\pi G)^{-1}T$, the equation takes the simple form

---

[2] Alternatively to this Dirac quantization of constraints, one may quantize the reduced phase space of observables $\mathscr{O}$ invariant under gauge transformations $\delta_\epsilon\mathscr{O} = \{\mathscr{O}, \epsilon C\}$ generated by (2.7). Again, it is difficult to see an evolution picture because in general there is no obvious time parameter among the observables; see Chaps. 12 and 13.

$$i\hbar \frac{\partial \psi}{\partial T} = \frac{\hbar^2}{2} \frac{\partial^2 \psi}{\partial V^2} + K V^{2/3} \psi \tag{2.9}$$

of a non-relativistic particle at "position" V in a potential $W(V) = -K V^{2/3}$ with a constant $K = \frac{1}{2}k(2\pi G)^{2/3}$. (Time T must be reversed in order to have agreement for all the signs in this interpretation.)

The flat model corresponds to a free particle with stationary states $\psi_p(V) = \exp(ipV/\hbar)$. We can write a general solution for a wave packet as

$$\Psi(V, T) = \int C(p)\psi_p(V) \exp(-iE(p)T/\hbar) \mathrm{d}p \tag{2.10}$$

with $E(p) = -p^2/2m$. The coefficients $C(p)$ are determined by an initial wave function at $T=0$, $\Psi(V, 0) = \int C(p)\psi_p(V)\mathrm{d}p$, via Fourier transformation. For the free particle, the Fourier transformation and its inversion can be done explicitly, providing the general time-dependent solution of travelling and spreading wave packets. If the initial momentum expectation value is negative, corresponding to a collapsing universe, the wave packet will reach the boundary $V = 0$ and disappear from the configuration space, implying a loss of unitarity of the evolution. This problem is a consequence of the self-adjointness issue mentioned in Sect. 2.2, which can be dealt with by specifying appropriate boundary conditions for wave functions. For these wave functions, the dynamics will then differ from the free particle [11, 12]. Alternatively, one can use the operator $\hat{D}$ quantizing ap_a, requiring extra factors of a^{-1} in the Wheeler–DeWitt equation which again lead to deviations from the free particle; see also Sect. 5.3.1.

2.4 Bohmian Viewpoint

The Bohmian interpretation [13–16] is a reformulation of quantum mechanics which has the consequence of amending the classical equations of motion by correction terms, summarized as a quantum potential. The quantum potential depends on the wave function and thus introduces non-classical degrees of freedom. Solving the equation would require one to solve for the expectation values and the wave function at the same time, which usually can be done only in approximations and truncations. (Starting from a different viewpoint, canonical effective equations, introduced later in this book and used throughout, can be understood as a systematic way of organizing the quantum corrections in terms of parameters characterizing the state, the moments subject to their own evolution.) The Bohmian interpretation has been used in (loop) quantum cosmology for instance in [17–20].

In addition to the Schrödinger equation for the wave function, the Bohmian viewpoint postulates the guiding equation

$$\dot{V} = \frac{\hbar}{2i} \frac{\psi^* \partial \psi/\partial V - (\partial \psi^*/\partial V)\psi}{\psi^* \psi} \tag{2.11}$$

for a variable V of classical type. One can interpret this equation as relating the classical velocity of V to the quantum-mechanical probability current, or the V-derivative of the phase of the wave function.

Splitting the wave function $\psi(V, T) = R(V, T) \exp(iS(V, T)/\hbar)$ into norm R and phase S implies a pair of equations equivalent to the Schrödinger equation: the continuity equation

$$\frac{\partial R^2}{\partial T} - \frac{\partial(R^2 \partial S/\partial V)}{\partial V} = 0 \tag{2.12}$$

for the density $|\psi|^2 = R^2$, which follows from the imaginary part of the Schrödinger equation, and the Hamilton–Jacobi equation

$$\frac{\partial S}{\partial T} - \frac{1}{2}\left(\frac{\partial S}{\partial V}\right)^2 - W(V) + \frac{1}{2}\hbar^2 \frac{\partial^2 R}{\partial V^2} = 0 \tag{2.13}$$

with the quantum potential $-\frac{1}{2}\hbar^2 \partial^2 R/\partial V^2$.

The Bohmian viewpoint is sometimes used in quantum cosmology because it provides equations for the scale-factor or volume variable, rather than just wave functions which would be difficult to interpret. Another interpretational scheme often claimed to avoid conceptual difficulties in quantum cosmology is the consistent-histories approach [21–24].

2.5 WKB Approximation

Similarly to the derivation of the quantum potential of the Bohmian interpretation, the WKB approximation to quantum mechanics begins by making an exponential ansatz for the wave function, now as an $\hbar$-expansion $\psi(V) = \exp(i\hbar^{-1} \sum_{n=0}^{\infty} \hbar^n S_n(V))$ with $\hbar$-independent functions $S_n(V)$. Leading orders in $\hbar$ for small n should then give semiclassical physics and the first quantum corrections. As usual for $\hbar$-expansions, the WKB series is an asymptotic expansion, not a converging series.

In the reformulated Wheeler–DeWitt equation of the form of a non-relativistic particle, (2.9), we use the WKB ansatz for stationary states with time dependence $\exp(iET/\hbar)$. The second-order derivative of the wave function by V is

$$\hbar^2 \frac{d^2\psi}{dV^2} = \left(-\left(\frac{dS_0}{dV}\right)^2 + \hbar\left(i\frac{d^2 S_0}{dV^2} - 2\frac{dS_0}{dV}\frac{dS_1}{dV}\right) + O\left(\hbar^2\right)\right)\psi.$$

If we include the potential $W(V)$ and the constant E, the first two orders in $\hbar$ of the stationary Wheeler–DeWitt equation result in the Hamilton–Jacobi equation

$$\frac{1}{2}\left(\frac{dS_0}{dV}\right)^2 + W(V) = E$$

solved by

$$S_0(V) = \int \sqrt{2(E - W(V))}\,dV \tag{2.14}$$

and the potential-independent equation

$$i\frac{d^2 S_0}{dV^2} - 2\frac{dS_0}{dV}\frac{dS_1}{dV} = 0$$

which, using the equation for dS_0/dV, we solve by

$$S_1(V) = \frac{i}{2}\log\frac{dS_0}{dV} = i\log\left((2(E - W(V)))^{1/4}\right).$$

To first order in $\hbar$, the wave function thus takes the form

$$\psi_E(V) = \frac{C}{\sqrt[4]{E - W(V)}}\exp(iS_0(V)/\hbar) \tag{2.15}$$

with a solution $S_0(V)$ of the Hamilton–Jacobi equation and a constant C.

The next order provides an equation

$$i\frac{d^2 S_1}{dV^2} - \left(2\frac{dS_0}{dV}\frac{dS_2}{dV} + \left(\frac{dS_1}{dV}\right)^2\right) = 0$$

which can be expressed as a condition relating dS_2/dV to the first and second derivative of the potential devided by dS_0/dV squared. Since the leading order of the WKB approximation assumes that S_2 is negligible or vanishes, the momentum dS_0/dV squared must be much larger than the first two derivatives of the potential. Thus, the leading terms of the WKB approximation are valid as long as the potential is sufficiently slowly varying compared to the wave length of the wave function. In particular, the approximation breaks down at turning points of the classical motion, where the momentum vanishes. (Notice that this breakdown is unrelated to the problem of time since turning points of V, not T are relevant).

In quantum cosmology, the WKB approximation is useful in order to see the behavior of wave functions as mathematical solutions. However, it also presents several problems: First, it requires a choice of time variable and does not easily show how much of the results depends on the choice. Moreover, it usually produces solutions of plane-wave form rather than travelling semiclassical wave packets. Even after finding approximate wave functions, one must still compute expectation values, fluctuations, correlations or other quantum variables in order to derive predictions for observations, a task which is often complicated, or even impossible if one does not have control over the physical inner product. A less well-known disadvantage is that its results do not agree with those obtained by low-energy effective actions [25], adding to the problem that its semiclassical interpretation is not all that clear. In later chapters of this book we will mainly use the framework of effective canonical dynamics [26, 27] which can overcome all these problems.

2.6 General Problems

Non-semiclassical properties of states at small volume, near the classical singularity, are of interest as well. In the Wheeler–DeWitt setting, this regime has mainly been addressed in the form of posing boundary conditions on boundaries of minisuperspace, primarily at $a = 0$. However, the form of possible conditions turned out not to be very much restricted, and so rather different candidates are equally viable depending on which arguments one makes to evaluate them. Also the singularity issue remains largely unaddressed in the Wheeler–DeWitt quantization; only further ingredients such as those from loop quantum cosmology provide sufficient information about quantum geometry at small volume to address this important problem. In addition to loop quantum cosmology, whose properties are laid out in detail in this book, further extensions of the typical models of quantum cosmology are being considered. First, it is important in any setting to include inhomogeneities, for instance by a mode expansion [28]. Back-reaction terms from inhomogeneity then contribute to the usual Wheeler–DeWitt equation. But inhomogeneity does not just imply additional terms in the homogeneous equation; it also brings in more constraints. A difficult anomaly problem arises to make sure that this set of constraints is consistent. In the Wheeler–DeWitt approach, this problem has not been tackled; we will later approach it with ingredients from loop quantum gravity.

In addition to bringing in new degrees of freedom, several features suggested by full quantum theories of gravity can be combined with Wheeler–DeWitt techniques. These ingredients include supersymmetry in quantum cosmology [29, 30] (or just fermionic matter, which can change some of the usual properties [31]), higher dimensions or higher-curvature terms, and finally the effects from loop quantum gravity. Loop quantum cosmology is the approach to quantum cosmology that has brought in the most characteristic features. Finally, quantum cosmology has consistently proven a fertile ground for (even) more exotic explorations. Among those are speculations about a possible turn-around of the arrow of time in a collapsing universe [8, 32–34] and multiverse considerations [35].

In what follows, we focus on issues closely related to the singularity problem on the one hand, and microscopic, potentially observable effects on the other. For this, the specifics of quantum geometry are required, provided in the most detailed form suitable for quantum cosmology by a loop quantization. But before getting to these issues, which constitute some of the traditional problems of quantum cosmology and will play a role later, there is the basic problem of wave functions depending on the ambiguous scale factor a. One must ensure that the classical scaling invariance finds an analog as a transformation on wave functions. In this context, it is useful to have a closer look at quantum geometry and draw the analogy between material atomic systems and, figuratively speaking, space-time atoms. A material body is composed of many atoms, and its total energy can be changed in two ways: we can add further atoms to enlarge the body, or we can excite the atoms already there. The analog of energy in a general relativistic or cosmological situation is volume, and so, if geometry is quantized, we expect that quantum cosmology allows an atomic universe to grow in two different ways: by generating new spatial atoms or by exciting those already present. How exactly these processes are realized will have to be determined from a fundamental theory of quantum geometry and its dynamics, but applied in the cosmological situation it already shows a key property: there is not just a single function like the classical $a(t)$ to determine the growth of a universe; there must be at least two functions to keep track of the number of spatial atoms as well as their excitation levels or individual sizes.

If we choose some finite region in space, measured in coordinates by a size $\mathcal{V}$ independent of time, its geometrical size determined via the line element (2.1) is $a(t)^3 \mathcal{V}$. The atomic picture of a homogeneous universe, on the other hand, gives an expression $\mathcal{N}(t)v(t)$ simply as the product of the individual atomic volume $v(t)$ (which is the quantity changing by excitations) with their number $\mathcal{N}(t)$. Thus, the key identity

$$V(t) = \mathcal{N}(t)v(t) = \mathcal{V}a(t)^3 \qquad (2.16)$$

directly shows how the role of a single classical time-dependent function is now replaced by the interplay of two. Moreover, since coordinates are not quantized but only geometry is, the coordinate volume $\mathcal{V}$ does not enter in the atomic expression. That already shows that a proper implementation of this picture in quantum cosmology will lead to a description invariant under the classical rescaling-freedom of the scale factor—only the invariant $v(t)$ enters or, if a given spatial region is specifically referred to, also $\mathcal{N}(t)$, but never only $a(t)$. To make this precise at least in a sufficiently general class of models, we need to understand the dynamics of quantum geometry and its atomic structure, which will then tell us what the possible behaviors of $v(t)$ and $\mathcal{N}(t)$ are, how they determine the behavior of wave functions, and how this affects observable or conceptual issues.

References

1. Lachieze-Rey, M., Luminet, J.P.: Phys. Rept. **254**,135 (1995), gr-qc/9605010
2. DeWitt, B.S.: Phys. Rev. **160**(5), 1113 (1967)
3. Wiltshire, D.L.: In: Robson B., Visvanathan N., Woolcock W.S. (eds.) Cosmology: The Physics of the Universe, pp. 473–531. World Scientific, Singapore (1996). gr-qc/0101003
4. Isham C.J.: In: DeWitt, B.S., Stora, R. (eds.) Relativity, Groups and Topology II. Lectures Given at the 1983 Les Houches Summer School on Relativity, Groups and Topology, Elsevier Science Publishing Company (1986)
5. Klauder, J.: Int. J. Mod. Phys. D **12**, 1769 (2003), gr-qc/0305067
6. Klauder, J.: Int. J. Geom. Meth. Mod. Phys. **3**, 81 (2006), gr-qc/0507113
7. Giulini, D.: Phys. Rev. D **51**(10), 5630 (1995)
8. Kiefer, C., Zeh, H.D.: Phys. Rev. D **51**, 4145 (1995), gr-qc/9402036
9. Blyth, W.F., Isham, C.J.: Phys. Rev. D **11**, 768 (1975)
10. Brown, J.D., Kuchař, K.V.: Phys. Rev. D **51**, 5600 (1995)
11. Alvarenga, F.G., Fabris, J.C., Lemos, N.A., Monerat, G.A.: Gen. Rel. Grav. **34**, 651 (2002), gr-qc/0106051

12. Acacio de Barros, J., Pinto-Neto, N., Sagiaro-Leal, M.A.: Phys. Lett. A. **241**, 229 (1998), gr-qc/9710084
13. Bohm, D.: Phys. Rev. **85**, 166 (1952)
14. Bohm, D.: Phys. Rev. **85**, 180 (1952)
15. Bohm, D.: Phys. Rev. **89**, 458 (1953)
16. Goldstein, S.: Stanford Encyclopedia of Philosophy. http://plato.stanford.edu/entries/qm-bohm/
17. Colistete, R. Jr, Fabris, J.C., Pinto-Neto, N.: Phys. Rev. D **62**, 083507 (2000), gr-qc/0005013
18. Falciano, F.T., Pinto-Neto, N., Santini, E.S.: Phys. Rev. D **76**, 083521 (2007), arXiv:0707.1088
19. Falciano, F.T., Pinto-Neto, N.: Phys. Rev. D **79**, 023507 (2009), arXiv:0810.3542
20. Shojai, A., Shojai, F.: Europhys. Lett. **71**, 886 (2005), gr-qc/0409020
21. Griffiths, R.B.: Phys. Rev. Lett. **70**, 2201 (1993)
22. Omnés, R.: Rev. Mod. Phys. **64**, 339 (1992)
23. Gell-Mann, M., Hartle, J.B.: Phys. Rev. D **47**, 3345 (1993), gr-qc/9210010
24. Dowker, F., Kent, A.: J. Statist. Phys. **82**, 1575 (1996), gr-qc/9412067
25. Dias, N.C., Mikovic, A, Prata, J.N.: J. Math. Phys. **47**, 082101 (2006), hep-th/0507255
26. Bojowald, M., Skirzewski, A.: Rev. Math. Phys. **18**, 713 (2006), math-ph/0511043
27. Bojowald, M., Sandhöfer, B., Skirzewski, A., Tsobanjan, A.: Rev. Math. Phys. **21**, 111 (2009), arXiv:0804.3365
28. Halliwell, J.J., Hawking, S.W.: Phys. Rev. D **31**(8), 1777 (1985)
29. D'Eath, P.D.: Supersymmetric Quantum Cosmology. Cambridge University Press, Cambridge (2005)
30. Moniz, P.V.: Quantum Cosmology—The Supersymmetric Perspective. Springer, Berlin(2010)
31. D'Eath, P.D., Halliwell, J.J.: Phys. Rev. D **35**, 1100 (1987)
32. Gold, T.: Am. J. Phys. **30**, 403 (1962)
33. Hawking, S.W.: Phys. Rev. D **32**, 2489 (1985)
34. Page, D.N.: Phys. Rev. D **32**, 2496 (1985)
35. Carr, B.(ed.): Universe or Multiverse? Cambridge University Press, Cambridge (2007)

Chapter 3
Kinematics: Spatial Atoms

Quantum geometry determines properties of quantized space–time structures, which can be interpreted as providing an atomic understanding of space–time. A view results which is fascinating not only in its physical implications but also in its rich combination of aspects of geometry and quantum theory. Many relevant features can already be seen by analogy with quantized particles, then borne out by rigorous constructions of quantum space–time.

3.1 Quantized Particles

Different aspects seen already in single-particle quantum mechanics are important in the context of quantum space–time as well. First, we consider an ordinary free and non-relativistic particle. Its well-known solutions show that the wave function in general spreads out in time even if the particle is not moving. Clearly, there is more freedom in quantum compared to classical dynamics: quantum variables such as fluctuations usually change independently of what one classically considers as the degrees of freedom; even a particle which classically would stay at rest can have non-trivial quantum dynamics. The degree of spreading can easily be determined by solving an equation of motion for the position fluctuation: With the Hamiltonian $\hat{H} = \hat{p}^2/2m$ and the general identity

$$\frac{\mathrm{d}}{\mathrm{d}t}\langle \hat{O} \rangle = \frac{\langle [\hat{O}, \hat{H}] \rangle}{i\hbar} \tag{3.1}$$

for an arbitrary operator $\hat{O}$, we have the equation

$$\frac{\mathrm{d}}{\mathrm{d}t}\left(\langle \hat{q}^2 \rangle - \langle \hat{q} \rangle^2\right) = \frac{\langle [\hat{q}^2, \hat{H}] \rangle - 2\langle \hat{q} \rangle \langle [\hat{q}, \hat{H}] \rangle}{i\hbar} = \frac{1}{m}\langle \hat{q}\hat{p} + \hat{p}\hat{q} \rangle - \frac{2}{m}\langle \hat{q} \rangle \langle \hat{p} \rangle = \frac{2}{m}C_{qp}. \tag{3.2}$$

The fluctuation $(\Delta q)^2 = \langle \hat{q}^2 \rangle - \langle \hat{q} \rangle^2$ is not guaranteed to remain constant in time; more precisely, its spreading is controlled by the covariance

M. Bojowald, *Quantum Cosmology*, Lecture Notes in Physics 835,
DOI: 10.1007/978-1-4419-8276-6_3, © Springer Science+Business Media, LLC 2011

$C_{qp} = \frac{1}{2}\langle \hat{q}\hat{p} + \hat{p}\hat{q}\rangle - \langle \hat{q}\rangle\langle \hat{p}\rangle$ of the state. A usual unsqueezed Gaussian state, for instance, which is often used for an initial profile and has the form $\psi(q) \propto \exp(-q^2/4\sigma^2)$ with a real variance σ, has vanishing covariance. Such an initial state would ensure that the initial spreading does not change momentarily. However, as a function of time the covariance must satisfy another equation of motion:

$$\frac{d}{dt}C_{qp} = \frac{2}{m}(\Delta p)^2 \qquad (3.3)$$

again derived using (3.1). The covariance could be constant only if the momentum fluctuation Δp vanishes, which cannot be the case for a normalizable state. On the other hand, the equation of motion for Δp itself, derived by the same methods, tells us that it is a constant in time. We can thus solve (3.3) for $C_{qp}(t) = 2(\Delta p)^2 t/m + C_{qp}^{(0)}$ in terms of its initial value $C_{qp}^{(0)}$. This solution, in (3.2), gives

$$\Delta q(t) = \sqrt{\frac{2}{m^2}(\Delta p)^2 t^2 + \frac{2}{m}C_{qp}^{(0)}t + \Delta q^{(0)}} \qquad (3.4)$$

as the well-known result showing the spreading of a free-particle state in time.

We have discussed this familiar example, already encountered in Sect. 2.3, at some length because it illustrates useful methods which we will come back to later, and because it provides important lessons. As seen clearly in this simple example, while one is always free to choose an initial state and make it as simple as possible, quantum dynamics in general changes its properties as time goes on. Here, we have seen that a vanishing covariance cannot be maintained in time; states tend to get "squeezed" and develop non-vanishing correlations. This is true even in situations which one would consider as semiclassical, and here correlations are even an integral part of decoherence scenarios [1].

To visualize the meaning of correlations, we use the second-order moments $(\Delta q)^2$, C_{qp} and $(\Delta p)^2$ of a state to define the family of ellipses

$$q^2(\Delta p)^2 + 2qpC_{qp} + p^2(\Delta q)^2 = \text{const} \qquad (3.5)$$

around the origin in the $q - p$-plane. These ellipses demonstrate the amount of quantum fluctuations: for $C_{qp} = 0$, for instance, we have an ellipse of semimajor axes Δq along the q-axis and Δp along the p-axis. For $C_{qp} \neq 0$, these ellipses are rotated such that certain linear combinations of q and p show the maximal and minimal fluctuations. A distribution function with these properties can be computed from the wave function: the Wigner function

$$W(q, p) = \frac{1}{2\pi\hbar} \int\limits_{-\infty}^{\infty} \psi^*\left(q + \frac{1}{2}\alpha\right)\psi\left(q - \frac{1}{2}\alpha\right)e^{-ip\alpha/\hbar}d\alpha. \qquad (3.6)$$

The factor of $1/2\pi\hbar$ ensures that $W(q, p)$ and the marginal distributions it provides by integrating over q or p, respectively, are normalized: $\int_{-\infty}^{\infty}\int_{-\infty}^{\infty} W(q, p)dqdp = 1$.

(A probability distribution in a strict sense, that is a non-negative function, is obtained if and only if $\psi(q)$ is Gaussian.)

For a Gaussian state

$$\psi(q) = \exp\left(-\tfrac{1}{4}(\sigma_R^{-2} + i\sigma_I^{-2})q^2\right) \tag{3.7}$$

of arbitrary squeezing, we obtain the Wigner function

$$W(q,p) = \frac{1}{\pi\hbar}\exp\left(-\tfrac{1}{2}\left(\sigma_R^{-2} + \sigma_R^2/\sigma_I^4\right)q^2 + 2\sigma_R^2\sigma_I^{-2}qp/\hbar - 2\sigma_R^2p^2/\hbar^2\right). \tag{3.8}$$

In terms of fluctuations and the covariance, related to $\sigma_{R/I}$ via $(\Delta q)^2 = \sigma_R^2$, $(\Delta p)^2 = \hbar^2/4\sigma_R^2 + \hbar^2\sigma_R^2/4\sigma_I^4$ and $C_{qp} = -\hbar\sigma_R^2/2\sigma_I^2$ (see for instance (13.6)), we can write the exponent as

$$E := -2\hbar^{-2}\left(q^2(\Delta p)^2 + 2qpC_{qp} + p^2(\Delta q)^2\right). \tag{3.9}$$

Constant-level lines of the Gaussian Wigner function in phase space are thus ellipses. In order to determine their proportions, we choose reference values q_0 and p_0 with the dimensions of q and p. respectively, and work with dimensionless ratios q/q_0 and p/p_0. In the absence of a ground state or other specific features of states, no distinguished values for q_0 or p_0 can be provided (unlike, for instance, if one could refer to the harmonic-oscillator ground state with fixed fluctuations of the correct dimensions). After dividing E by $q_0^2 p_0^2$, we express all terms by dimensionless variables, in which we find the major axis of the ellipse rotated against the q-axis by an amount

$$\tan(2\alpha) = \frac{q_0 p_0 C_{qp}}{q_0^2(\Delta p)^2 - p_0^2(\Delta q)^2}. \tag{3.10}$$

An interpretation of the covariance is thus as the rotation of the likelihood ellipse in phase space. The axes lengths of the ellipse are

$$\frac{p_0^2(\Delta q)^2 + q_0^2(\Delta p)^2 \pm \sqrt{(p_0^2(\Delta q)^2 - q_0^2(\Delta p)^2)^2 + 4q_0^2 p_0^2 C_{qp}^2}}{2q_0^2 p_0^2}$$

which thanks to $(\Delta q)^2(\Delta p)^2 - C_{qp}^2 = \hbar^2/4$ (a Gaussian state saturates the uncertainty relation) can be written as

$$\frac{p_0^2(\Delta q)^2 + q_0^2(\Delta p)^2 \pm \sqrt{(p_0^2(\Delta q)^2 + q_0^2(\Delta p)^2)^2 - q_0^2 p_0^2 \hbar^2}}{2q_0^2 p_0^2}.$$

Since there are no distinguished values for q_0 and p_0 in general, the only available meaning of the squeezing of states is by non-vanishing correlations, rotating the likelihood ellipse. Changing position and momentum fluctuations while keeping the uncertainty relation saturated at vanishing correlations provides a meaningful sense of squeezing only if one can refer to a distinguished state, such as the harmonic-oscillator ground state. In quantum cosmology, no ground state is available to define squeezing in the absence of correlations. From now on, we will use only the general sense of squeezing as determined by $C_{qp} \neq 0$.

If $(\Delta p)^2$ is constant in time (as for the free particle), $(\Delta q)^2$ must be large for large C_{qp}. The major axis of the ellipse then has a length approximately given by Δq, while the minor axis is very small. The ellipse is stretched out in one direction, along a linear combination of q and p determined by the covariance or the angle α in (3.10). While this linear combination has large fluctuations, the orthogonal one has very small ones and thus behaves rather classical. In this way the emergence of a classical degree of freedom via decoherence can be seen, whose precise form is determined by the underlying dynamics. Squeezed states with large covariance automatically arise in the process, and play an important role for the nearly classical behavior. In general quantum systems deviations from Gaussian form more general than squeezing arise.

As far as semiclassical regimes are concerned, we arrive at our

First Principle State dynamics is important and to be derived. In particular, the form of appropriate semiclassical states cannot always be guessed, or assumed to be unsqueezed Gaussians.

From elementary particle physics, for which perturbations around the free vacuum state are often sufficient, one is used to Gaussian states to play a central role. But the form of the vacuum is a dynamical question, and general situations in quantum cosmology may not even allow a distinguished vacuum or ground state. More general classes of states and methods to deal with them must be used. In Sect. 5.4.1.3 we will see an example of states rapidly moving away from Gaussian form, then settling into a new, better preserved shape as determined by its moments [2]. Other example systems have dynamical coherent states of exactly preserved shape, but these systems, such as the harmonic oscillator, are very special and rarely realistic. More generally, states of "stable" shape exist [3], but this is a mathematical rather than physical generalization of the desired properties of dynamical coherent states. In fact, in these more general states the shape does change: As time goes on, the entire state evolution is determined by the change of as many parameters as there are classical degrees of freedom; however, unlike in the harmonic-oscillator case, these parameters are not in one-to-one correspondence with expectation values. They partially affect fluctuations and other moments of the state as well, and thus the state's shape evolves. In particular, semiclassicality may be lost as the states evolve.

Considerations of the free particle in quantum mechanics offer another observation. If the particle is very massive or macroscopic with large m in (3.4), it takes a long time for the wave function to spread out from some tightly peaked initial state. In a naive interpretation, this would suggest that macroscopic bodies do not show quantum effects at all. This conclusion can, of course, not be true since there are important properties even in macroscopic situations, such as conductivity, which rely on quantum aspects of their constituents. At this stage, the composite, atomic nature of matter becomes important: microscopic building blocks are much smaller, and they almost always behave very quantum. This is an obvious statement for condensed-matter physics, but it shows that quantum cosmology, where the dominant view is usually one of a large macroscopic and homogeneous universe, must properly take into account the underlying atomic nature of space–time if it is to describe all quantum phases of the universe reliably.

Second Principle Microscopic physics is important. In cosmology, even homogeneous models must include the proper small-scale quantum behavior. They constitute many-body systems when seen in quantum gravity; the large "number of particles", corresponding to $\mathcal{N}(t)$ of the preceding chapter, may lead to characteristic effects.

For instance regarding the singularity problem, a massive homogeneous "blob" universe, which contains all its matter smeared-out, may be non-singular in some models. For instance, examples exist in which effective violations of energy conditions can trigger a "bounce" where the isotropic universe volume is minimal [4]. But if the microscopic dynamics of its quantum building blocks remains singular, which is a question more complicated to address, the theory is still in danger of breaking down.

For the dynamics of many-particle systems at high energies, quantum field theory rather than particle quantum mechanics is required. Here, one starts with fields on a given space–time and applies quantization techniques. For gravity and cosmology, however, it is the space–time itself that is to be quantized. Familiar techniques, which always implicitly assume the availability of a background space–time, then fail. Canonical quantization of a scalar field on Minkowski space–time, for instance, might make use of a mode expansion

$$\phi(x) = \int \frac{d^3k}{\sqrt{2\omega_k}} \left(\hat{a}_k e^{ik\cdot x} + \hat{a}_k^\dagger e^{-ik\cdot x} \right). \tag{3.11}$$

to define annihilation operators $\hat{a}_k$ and distinguish the vacuum state $|0\rangle$ as the state annihilated by all $\hat{a}_k$. Many-particle states are obtained by acting with creation operators, the adjoints of annihilation operators, on the vacuum:

$$|k_1, n_1; \ldots; k_i, n_i\rangle = (\hat{a}_{k_1}^\dagger)^{n_1} \cdots (\hat{a}_{k_i}^\dagger)^{n_i} |0\rangle. \tag{3.12}$$

In such states, the total normal-ordered energy as the eigenvalue in

$$\hat{E}|k_1, n_1; \ldots; k_i, n_i\rangle = \sum_{j=1}^{i} \hbar n_j \omega(k_j) |k_1, n_1; \ldots; k_i, n_i\rangle \tag{3.13}$$

is non-zero.

The mode decomposition, however, requires space–time equipped with a background metric to be defined: at least the integration measure $d^3x\sqrt{\det h}$ must be known in order to integrate the original field and obtain its modes (or use preferred Cartesian coordinates in which the spatial metric is $h_{ab} = \delta_{ab}$). In (3.11), the scalar product $k \cdot x$ refers to a flat background. Without the modes, we cannot even define the vacuum state. This consideration finally provides the third principle:

Third Principle Tools of quantum field theory must be appropriately adapted to deal with quantum geometry in a background-independent way. While the simplest cosmological models are homogeneous and of finitely many degrees of freedom, allowing straightforward quantizations of geometrical variables, significant changes in the quantization methods due to the generally covariant nature of the underlying field theory must also be reflected in quantum cosmology.

Quantum cosmology must take into account the lessons learned in attempted constructions of quantum gravity. Constructing quantum gravity or even deriving quantum cosmology from it remains a formidable challenge, but important features can nevertheless be implemented and explored in sufficiently general formulations of quantum cosmological models. Sufficient generality is important for reliable conclusions and for stringent tests of the full framework, even if it comes at the expense of additional ambiguities.

All three principles will be revisited throughout this book; they are important for conceptual properties, for instance regarding singularities, and observational ones. We will begin by reviewing the reformulation proposed by loop quantum gravity to incorporate the Third Principle. The first two principles will have to be faced once we deal with the dynamics.

3.2 Quantized Space–Time

In general relativity, the dynamical object is the space–time metric, now to be quantized. As mentioned in the introduction, we do not require a viewpoint of general relativity being fundamental, but rather take a more general one: even if there is a more fundamental underlying theory, which may eventually be arrived at by the quantization procedure, there must be a consistent way of endowing the metric with fluctuations and uncertainty. There is a tried-and-true traditional method to unravel quantum properties of hitherto classical theories: canonical quantization, a procedure that takes a classical phase space and returns a non-commuting algebra of observables, such as pairs of basic operators on whose representation quantization can be built.

Coordinates are often used to describe space–times, but this is only superficially related to expressing point-particle dynamics by coordinates that become operators $\hat{q}$. Unlike the positions of point particles, space–time coordinates are not measurable; they can play no role in the final algebra of quantum observables. What can be measured is only the geometry and dynamics of space–time, which requires extended objects. Fully coordinate-independent observables, on the other hand, are infamously difficult to construct: only very few general expressions are known, and even approximate constructions become very tedious in anything but the simplest models. Instead of trying to quantize classical observables, a two-step procedure looks more promising: one first considers just spatial quantum geometry at a fixed time (referring to objects such as lengths, areas, volumes which for given regions are coordinate independent), and then imposes additional constraints to make sure that kinematical objects are combined suitably to space–time observables. Once complete, the results will show the quantum dynamics of space–time; but even before all constraints are implemented, spatial quantum geometry already provides interesting results.

3.2.1 Scaling

Kinematically, we consider objects such as the volume $V_R = \int_R d^3x \sqrt{\det h}$ of some spatial region R, where h_{ab} is the metric induced by a space–time metric on a spatial slice $t = \text{const}$ with respect to some time coordinate. Volumes are surely sufficient to probe isotropic quantum geometries, in which case the classical phase space is small: as seen before, there is a single canonical pair (a, p_a). But if we allow all possible regions, spatial observables even in this simple geometry provide infinitely many numbers $V_R = \mathcal{V}(R)a^3$ for any given a, where $\mathcal{V}(R)$ is the coordinate-dependent, co-moving volume of the region measured just with the non-dynamical unit line element $d\sigma_k$ of constant curvature, (2.2). With a single dynamical degree of freedom, however, quantization can give wave functions only in one variable, such as $\psi(a)$ as used in a Wheeler–DeWitt quantization. It must then be ensured that wave functions have the correct scaling behavior under changing coordinates or $\mathcal{V}$ so that observables are invariant; otherwise one's quantization would not capture pure quantum geometry.

For a compact space (for instance the closed model), the total space would be a convenient choice to define $\mathcal{V}_R$. But one may still capture all isotropic degrees of freedom completely by any subspace, with a smaller value of $\mathcal{V}$. Moreover, in the closed model in its most general formulation one is not required to use the unit sphere as the total space multiplied by the scale factor, although it is certainly convenient. If a sphere of non-unit radius is used for the spatial coordinates, the coordinate volume changes to $\mathcal{V} = \lambda^3 \mathcal{V}_{\text{unit}}$. One can obtain the new coordinate system by the transformation $r \mapsto \lambda r$ while the angular coordinates do not change. The parameter $k \mapsto \lambda^{-2}k$, obeying the scaling law of curvature, remains positive but now differs from one. The line element is invariant with the usual transformation $a \mapsto \lambda^{-1}a$ of the scale factor under rescaling the coordinates, and so the curvature term k/a^2 in the Friedmann equation is invariant. Notice that this rescaling of coordinates and the sphere is not the same as changing a smaller integration region within the unit sphere even if $\lambda < 1$, if one chooses a smaller integration region within the sphere, k and a do not change. Just as in the flat model, also in the closed model rescaling the coordinates and choosing different integration regions is allowed by independent choices. The parameter $\mathcal{V}$ retains a free value and is not fixed. Sometimes, $\mathcal{V}$ is called a "regulator" and then argued to require the limit $\mathcal{V} \to \infty$ (or the maximum value in a compact space) rather than full $\mathcal{V}$-independence of wave functions. However, $\mathcal{V}$ is not a regulator because the classical predictions do not depend on the chosen value. It is simply a parameter that labels different but equivalent formulations of the symmetry reduction within one and the same model. While one may have reasons to restrict all attention to a specific simple value, in doing so one loses access to an important consistency check. Especially in isotropic and homogeneous models, in which the anomaly problem, to be discussed later, trivializes, making sure that the proper $\mathcal{V}$-behavior is realized is the only remaining test of consistency.

The scaling issue turns out to be a manifestation of a more general problem: How do we form a complete set of coordinate-independent measures of spatial geometry (intrinsic as well as extrinsic for the whole phase space) while retaining an algebra simple enough for further constructions? This problem, as of now, has not been solved in Wheeler–DeWitt-type quantizations, which when applied beyond homogeneous models remain formal. But at the level of spatial quantum geometry it can be solved after a change of phase-space variables.

3.2.2 Canonical Gravity

For a Hamiltonian formulation of general relativity (see [5]), one first brings the space–time metric in a form

$$ds^2 = -N^2 dt^2 + h_{ab}(dx^a + N^a dt)(dx^b + N^b dt) \qquad (3.14)$$

where h_{ab}, the only part contributing on a spatial slice $t = $ const, is the spatial metric. The lapse function N and shift vector N^a provide the remaining components of a space–time metric, and can be seen to contain information about the spatial foliation in space–time: The unit normal vector n^a to a spatial slice $t = $ const satisfies $N n^a = (\partial/\partial t)^a - N^a$, and the inverse space–time metric is $g^{ab} = h^{ab} + n^a n^b$.

The spatial metric h_{ab} can serve as a set of configuration variables, and a suitable geometric notion of its "velocities" is the extrinsic curvature

$$K_{ab} = \frac{1}{2N}(\dot{h}_{ab} - D_a N_b - D_b N_a). \qquad (3.15)$$

It does indeed have a time derivative of the spatial metric, denoted by the dot, and extra contributions if the shift vector is not constant and thus the spatial slice is deformed as seen from the time coordinate t: the normal vector is not proportional to $(\partial/\partial t)^a$. The symbol D_a denotes the spatial covariant derivative operator compatible with the metric h_{ab}.

3.2.2.1 Action and Constraints

In these variables, the Einstein–Hilbert action of general relativity (ignoring boundary terms) can be expressed as

$$
\begin{aligned}
L_{\text{grav}} &= \frac{1}{16\pi G} \int d^3 x \sqrt{-\det g}\, R \\
&= \frac{1}{16\pi G} \int d^3 x\, N \sqrt{\det h}\, \left({}^{(3)}R + K_{ab}K^{ab} - (K_a{}^a)^2 \right)
\end{aligned}
\qquad (3.16)
$$

with the three dimensional Ricci scalar ${}^{(3)}R$ computed from h_{ab}. In this way, the action looks like a complicated version of the general form known from classical field theories, with a kinetic term quadratic in the velocities K_{ab} and a potential depending only on configuration variables as well as their spatial derivatives, here given by the spatial Ricci scalar. From the kinetic term, we first compute the momentum

$$p^{ab}(x) = \frac{\delta L_{\text{grav}}}{\delta \dot{h}_{ab}(x)} = \frac{1}{2N}\frac{\delta L_{\text{grav}}}{\delta K_{ab}} = \frac{\sqrt{\det h}}{16\pi G}(K^{ab} - K_c^c q^{ab}) \qquad (3.17)$$

conjugate to h_{ab} : In this field-theoretical context, we have Poisson brackets

$$\{h_{ab}(x), p^{cd}(y)\} = \delta^c_{(a}\delta^d_{b)}\delta(x, y). \tag{3.18}$$

From the action we obtain the Hamiltonian

$$
\begin{aligned}
H_{grav} &= \int d^3x \left(\dot{h}_{ab}p^{ab} - L_{grav} \right) \\
&= \int d^3x \left(\frac{16\pi GN}{\sqrt{\det h}} \left(p_{ab}p^{ab} - \frac{1}{2}(p^c_c)^2 \right) + 2p^{ab}D_a N_b - \frac{N\sqrt{\det q}}{16\pi G}{}^{(3)}R \right) \\
&=: \int d^3x (NC^{grav} + N^a C^{grav}_a).
\end{aligned}
\tag{3.19}
$$

All terms in the Hamiltonian are linear in the lapse function N and the shift vector N^a, but time derivatives of these fields do not appear. As components of the space–time metric, the action is to be extremized with respect to them, too, not just with respect to the spatial metric h_{ab}. While variation with respect to h_{ab} and p^{ab} provides equations of motion due to the canonical piece $\dot{h}_{ab}p^{ab}$, variation by lapse and shift leads to constraints on the phase-space variables: the diffeomorphism constraint

$$C^{grav}_a = -2D_b p^b_a \tag{3.20}$$

and the Hamiltonian constraint

$$C^{grav} = \frac{16\pi G}{\sqrt{\det h}} \left(p_{ab}p^{ab} - \frac{1}{2}(p^a_a)^2 \right) - \frac{\sqrt{\det h}}{16\pi G}{}^{(3)}R. \tag{3.21}$$

Matter terms will contribute extra pieces, which is why we denote the pure gravitational terms with the superscript "grav". In vacuum, $C^{grav}_a = 0$ and $C^{grav} = 0$.

If these constraints are solved and objects invariant under the Hamiltonian flow they generate are considered, we are dealing with space–time observables independent of any coordinate choices. As already mentioned, completing such a program is extremely difficult; we thus postpone a discussion of the constraints at the classical level, trying to represent the tensorial objects h_{ab} and p^{ab} as operators. For this, we first need extra structures to get rid of the indices and integrate to scalars, for only those can directly be operators. (Otherwise, an appropriate behavior under complicated tensor transformations in a quantum algebra of non-scalar objects must be ensured, which, as experience shows, is prone to becoming anomalous.) One possibility to remove indices is by contraction with other geometrical objects, not containing the dynamical fields so as to retain linear structures. For instance, one may associate the length $\ell_e[h_{ab}] = \int_e dt \sqrt{\dot{e}^a \dot{e}^b h_{ab}}$ with any differentiable curve e in space. For any given curve, this is a scalar object not changing under coordinate transformations. But it is not linear in phase-space variables due to the square root.

3.2.2.2 Ashtekar–Barbero Variables

No linear scalar representation of the full phase space of general relativity is known in terms of metric variables. But such a formulation does exist in terms of a new set, formed by the densitized triad E_i^a together with the Ashtekar–Barbero connection A_a^i [6, 7]. The densitized triad replaces the spatial metric and is defined in several steps: Instead of using the metric we first introduce the co-triad e_a^i via $h_{ab} = e_a^i e_b^i$. (The index i does not refer to the tangent space but simply labels the three co-triad co-vectors. Its position, unlike the one of $a, b, \ldots$, is thus not relevant and we will freely move it up or down as convenient. No spatial metric is required to do so. For repeated indices, as in the defining relation, we will understand the summation convention unless stated otherwise.) A given metric does not uniquely define a co-triad, which can be redefined by any orthogonal transformation $e_a^i \mapsto R_j^i e_a^j$, $R_j^i R_i^k = \delta_j^k$, without changing h_{ab}. There is thus more freedom than in metric formulations, which later on will be removed via additional constraints. From the co-triad, one then obtains the triad e_i^a as its matrix inverse: $e_i^a e_a^j = \delta_i^j$. Equivalently, the triad, as suggested by the notation, is obtained by raising the index of e_a^i using the inverse metric h^{ab}. (The co-triad and the triad form dual orthonormal bases of the co-tangent and tangent space, respectively.) Finally, we densitize the triad by multiplying it with the scalar density $\sqrt{\det h} = |\det(e_a^i)|$:

$$E_i^a := |\det(e_b^j)| e_i^a. \tag{3.22}$$

Notice the absolute value at this stage. Even after factoring out the rotational freedom of a triad, it does have more information than a metric. Unless it is degenerate, the triplet of triad vectors can be left- or right-handed, meaning that the determinant of e_a^i seen as a 3×3-matrix can take both signs. Changing the sign corresponds to a large gauge transformation in O(3)/SO(3) not connected to the unit. It is thus not removed by factoring out the flow generated by a constraint, and remains relevant for geometry. Its meaning is that of the orientation of space, which will become important later in quantum cosmology.

Similarly, we manipulate extrinsic curvature K_{ab} by first contracting with the triad on one index, defining $K_a^i := e_i^b K_{ab}$. This expression turns out to be canonically conjugate to the densitized triad,

$$\{K_a^i(x), E_j^b(y)\} = 8\pi G \delta_a^b \delta_j^i \delta(x, y). \tag{3.23}$$

To define scalar objects to be quantized, it is useful to do one final step and combine extrinsic curvature with the spin connection Γ_a^i that is compatible with the triad: $D_a e_i^b = \partial_a e_i^b + \Gamma_{ac}^b e_i^c - \varepsilon^{ijk} \Gamma_a^j e_k^b = 0$, solved by

$$\Gamma_a^i = -\varepsilon^{ijk} e_j^b \left(\partial_{[a} e_{b]}^k + \frac{1}{2} e_k^c e_a^l \partial_{[c} e_{b]}^l \right). \tag{3.24}$$

We then have the Asthekar–Barbero connection

$$A_a^i = \Gamma_a^i + \gamma K_a^i \tag{3.25}$$

with the Barbero–Immirzi parameter $\gamma > 0$ [8]. This provides the final canonical pair with

$$\{A_a^i(x), E_j^b(y)\} = 8\pi\gamma G\delta_a^b\delta_j^i\delta(x, y),\tag{3.26}$$

a relation which directly follows from (3.23) since Γ_a^i is a functional of the triad and thus Poisson-commutes with it. These canonical variables have a structure analogous to what is known in gauge theories: a connection and a densitized vector field. They are subject to a Gauss constraint

$$D_a^{(A)}E_i^a := \partial_a E_i^a + \varepsilon^{ijk}A_a^j E_k^a = D_a E_i^a + \gamma\varepsilon^{ijk}K_a^j E_k^a = 0\tag{3.27}$$

of the usual form. (In the last reformulation we used the fact that the covariant divergence of a densitized vector field equals the coordinate divergence. Since the spin connection is compatible with the triad, the Gauss constraint is equivalent to $\varepsilon^{ijk}K_a^j E_k^a = 0$. This relation implies that $K_{ab} = K_a^i e_b^i$ is a symmetric tensor.) Compared to those gauge theories that occur in particle physics, the gravitational diffeomorphism constraint, which now reads

$$C_a^{\mathrm{grav}} = F_{ab}^i E_i^b\tag{3.28}$$

with the Yang–Mills curvature

$$F_{ab}^i = \partial_a A_b^i - \partial_b A_a^i - \varepsilon^{ijk}A_a^j A_b^k\tag{3.29}$$

is new, and the Hamiltonian is quite different from the Yang–Mills form and now a constraint contribution

$$C^{\mathrm{grav}} = \left(\varepsilon^{ijk}F_{ab}^i - 2(1+\gamma^{-2})(A_a^i - \Gamma_a^i)(A_b^j - \Gamma_b^j)\right)\frac{E_j^{[a}E_k^{b]}}{\sqrt{|\det E|}}.\tag{3.30}$$

3.2.2.3 Holonomy-Flux Algebra

Since we are ignoring the constraints for now, their specific form does not play a role; we can thus define and use objects which are well-known from general gauge theories: holonomies

$$h_e[A_a^i] = \mathscr{P}\exp\int_e \dot{e}^a A_a^i \tau_i \mathrm{d}\lambda\tag{3.31}$$

for curves e in space, and fluxes

$$F_S^{(f)}[E_i^a] = \int_S n_a E_i^a f^i \mathrm{d}^2 y\tag{3.32}$$

through surfaces S in space with smearing functions f^i taking values in the internal space. (The co-normal $n_a = \frac{1}{2}\varepsilon_{abc}\varepsilon^{uv}\frac{\partial x^b}{\partial y^u}\frac{\partial x^c}{\partial y^v}$ of S: $y \mapsto x(y)$ is independent of any metric.)

For our purpose of finding a set of linear scalar quantities, these variables turn out to be immensely useful. Fluxes are already spatial scalars linear in one of the canonical variables, E_i^a. And while holonomies are not linear in A_a^i, they do form a linear algebra together with the fluxes:

$$\{h_e[A], F_S^{(f)}[E]\} = 8\pi\gamma G\eta(e, S)\mathcal{O}_{e,S}^{(f)}(\tau_i h_e[A]) \tag{3.33}$$

with a topological number $\eta(e, S)$ which determines how and how often the curve e and the surface S intersect, and $\mathcal{O}$ denoting a suitable ordering of $\tau_i \in su(2)$ and $h_e[A] \in SU(2)$ depending on where on the curve e its intersections with S lie. For instance, if the intersection is at the endpoint of e, we have $\mathcal{O}_{e,S}^{(f)}(\tau_i h_e[A]) = h_e[A]\tau_i f^i(e(1))$, and $\mathcal{O}_{e,S}^{(f)}(\tau_i h_e[A]) = f^i(e(0))\tau_i h_e[A]$ if the intersection is at the starting point. (Unless stated otherwise, we will always assume curves to be defined on the interval $[0, 1]$ of their parameter.) In general, we have

$$\mathcal{O}_{e,S}^{(f)} = \sum_{p\in e\cap S} f^i(p)h_{e\to p}[A]\tau_i h_{p\leftarrow e}[A] \tag{3.34}$$

with $h_{e\to p}$ the holonomy along the starting piece of e up to p, and $h_{p\leftarrow e}$ along the ending piece from p onwards. See also [9] for detailed calculations of the holonomy-flux algebra.

The algebra (3.33) can explicitly be represented as operators on a Hilbert space [10]. All quantities are coordinate independent, and they do not make use of any extra structures except their labels e, S and f. (They certainly make use of standard structures such as topological or differentiable ones of the underlying manifold. But no extra metric, for instance, is introduced which would make the construction background dependent.) The algebra of $F_S^{(f)}[E]$ and $h_e[A]$ replaces the algebra of annihilation and creation operators $\hat{a}_k$ and $\hat{a}_k^\dagger$ for quantum field theories on a background.

The specific form of the algebra suggests to view holonomies as creation operators, raising the excitation level of fluxes: a state annihilated by some $F_S[E]$ would, after being acted on by a holonomy along a curve intersecting S, have a non-vanishing flux through S due to $\hat{F}_S(\hat{h}_e|\psi\rangle) = \hat{h}_e\hat{F}_S|\psi\rangle + [\hat{F}_S, \hat{h}_e]|\psi\rangle = [\hat{F}_S, \hat{h}_e]|\psi\rangle \neq 0$ with a non-vanishing commutator. What is needed for a construction of all possible excited states in this way is also a state to start with, from which holonomies can then generate new excitations. In common quantum field theories, this state would be the vacuum devoid of particles; here it would be a state where not even fluxes, and thus the densitized triad or spatial metric, would be present. It is a state in which geometry itself is highly quantum and only lowly excited, unlike any classical geometry. Such a state is extremely difficult to imagine physically, but it has a very simple mathematical expression: if we choose the connection representation of states $\psi[A_a^i]$, it is a mere constant. Then indeed, fluxes which would be derivative operators in such a representation, all vanish.

Let us thus define this state as the quantum geometrical "vacuum", $\psi_0(A_a^i) = 1$. Since holonomies only depend on the connection, they will become multiplication operators, directly showing their action on the state. More precisely, as basic operators we should allow all matrix elements of holonomies, which are in SU(2). Multiplying several ones of them can be tedious due to the group structure, but what is relevant now is already illustrated by the simpler case of holonomies taking values in the Abelian group U(1). (This group would be obtained in a loop quantization of electromagnetism [11].) Then, each holomony is a simple connection-dependent phase factor $h_e[A_a] = \exp(i \int_e d\lambda \dot{e}^a A_a)$, and excited states can be written as

$$\psi_{e_1, n_{e_1}; \ldots; e_i, n_{e_i}} = \hat{h}_{e_1}^{n_{e_1}} \cdots \hat{h}_{e_i}^{n_{e_i}} \psi_0. \tag{3.35}$$

As functionals, these states look like

$$\psi_{g,n}(A_a) = \prod_{e \in g} h_e(A_a)^{n_e} = \prod_{e \in g} \exp(in_e \int_e d\lambda \dot{e}^a A_a) \tag{3.36}$$

In this notation, each occurrence of a curve e_i in space signals that geometry is excited along that curve: fluxes through surfaces intersected by the curve will be non-zero. Moreover, each curve e_i can be excited several times, as indicated by the integer n_{e_i}. Thus, curves and the integers technically play roles analogous to particle wave numbers and occupation numbers in quantum field theories on a background space–time.

These constructions are used in a more general setting in the diverse models of loop quantum gravity. We now assume that we have a d-dimensional spatial manifold Σ (which in the concrete applications of symmetry-reduced models will be $d < 3$, but formally the dimension could be larger than three). Furthermore, we assume a compact structure group G, and as fields (i) a G-connection A_a^i and a densitized $\mathscr{L}G$-valued vector field E_i^a forming the gauge part of the theory (with E_i^a dual to an $\mathscr{L}G$-valued $(d-1)$-form $\Sigma_{a_1 \ldots a_{d-1}}^i = \varepsilon_{a_1 \ldots a_d} E_i^{a_d}$), and (ii) scalars $\phi_I : \Sigma \to \mathbb{R}$ with densitized momenta $p^I : \Sigma \to \mathbb{R}$ forming the "matter" part of the theory. In the actual models, the scalars may arise as some of the components of the full gravitational connection, rather than playing the role of physical matter. The fields form canonical variables

$$\Theta_{\text{gauge}} = \frac{1}{\kappa} \int_\Sigma d^3x \, \dot{A}_a^i E_i^a \longrightarrow \{A_a^i(x), E_j^b(y)\} = \kappa \delta_j^i \delta_a^b \delta(x, y)$$

$$\Theta_{\text{scalar}} = \int_\Sigma d^3x \, \dot{\phi}_I p^I \longrightarrow \{\phi_I(x), p^J(y)\} = \delta_I^J \delta(x, y)$$

with a coupling constant κ, and are subject to certain constraints $C_\alpha[A_a^i, E_j^b, \phi_I, p^J] = 0$. Examples for this general setting of fields are Yang–Mills theory, where (A_a^i, E_j^b) are subject to the Gauss law $\mathscr{G}_i := \partial_a E_i^a + \varepsilon_{ij}{}^k A_a^j E_k^a = 0$ (for $G = SU(2)$), or general relativity in Ashtekar–Barbero variables. Later chapters will provide a large set of further examples in which scalars arise from symmetry reduction. In this context, we start with a gauge theory (A_a^i, E_i^a) in $3+1$ dimensions and impose invariance under some symmetry group S acting on the principal fiber bundle $P \to \Sigma$ that underlies the gauge theory. An example which will be discussed in more detail later is spherically symmetric gravity; see also Sect. 9.1. It turns out that spherically symmetric SU(2)-connections and densitized triads can always be written as

$$^{3D}A_a^i \tau_i dx^a = A_x(x)\tau_3 dx + A_\varphi \bar{\Lambda}_\varphi^A d\vartheta + A_\varphi(x)\Lambda_\varphi^A \sin\vartheta \, d\varphi + \tau_3 \cos\vartheta \, d\varphi$$

$$^{3D}E_i^a \tau^i \frac{\partial}{\partial x^a} = E^x(x)\tau_3 \sin\vartheta \frac{\partial}{\partial x} + E^\varphi \bar{\Lambda}_E^\varphi \sin\vartheta \frac{\partial}{\partial \vartheta} + E^\varphi(x)\Lambda_E^\varphi \frac{\partial}{\partial \varphi}$$

with a U(1)-connection A_x, a densitized triad E^x, real-valued scalars A_φ, E^φ, and the angles α and β in $\Lambda^A_\varphi = \cos\beta(x)\tau_1 + \sin\beta(x)\tau_2$, $\Lambda^\varphi_E = \cos(\alpha(x) + \beta(x))\tau_1 + \sin(\alpha(x) + \beta(x))\tau_2$. The remaining fields are fixed by the conditions $\text{tr}(\bar{\Lambda}^A_\varphi \Lambda^A_\varphi) = 0 = \text{tr}(\bar{\Lambda}^\varphi_E \Lambda^\varphi_E)$. All free fields depend only on the radial coordinate x, and are thus defined on a 1-dimensional manifold.

Loop quantization then presents a specific way of canonical quantization, turning the Poisson algebra of basic fields into an operator algebra. Any such quantization requires smearing for field theories to remove delta-functions in the elementary Poisson brackets, usually done using a background metric, as in $\int_\Sigma d^d x \sqrt{\det h} \phi(x)$ for a scalar field. But such a procedure is not suitable if the metric itself (or a densitized triad) is to be quantized: no linear algebra of basic smeared objects would result. The advantage of connection variables is that they have a natural smearing without having to make use of a fixed metric: holonomies (3.31) along curves e in space, fluxes (3.32), in general through surfaces of codim$(S, \Sigma) = 1$, scalar values $\phi_I(x)$, and integrated momenta $\int_R p^J(y) d^d y$ can all be defined without an extra metric, and the integrations they contain remove all delta-functions from their Poisson brackets.

Constructing a Hilbert-space representation leads to states in a space of square-integrable functions $L^2(\bar{\mathscr{A}} \times \bar{\Phi}, d\mu_{AL})$ with a compact space $\bar{\mathscr{A}} \times \bar{\Phi}$ of generalized connections and scalars [12]. For (finite analytical) graphs $g \subset \Sigma$ with edge set $E(g)$ and vertex set $V(g)$, we define spaces of g-connections

$$\mathscr{A}_g = \{A_g \colon E(g) \to G\} \qquad \text{(holonomies along the graph g)}$$

and g-scalars

$$\Phi_g = \{\phi_g \colon V(g) \to \bar{\mathbb{R}}_{\text{Bohr}}\} \qquad \text{(vertex values on g)}$$

taking values in the Bohr compactification of the real line (see below). For $g \subset g'$, projections $\pi^{\mathscr{A}}_g \colon \mathscr{A}_{g'} \to \mathscr{A}_g$ and $\pi^\Phi_g \colon \Phi_{g'} \to \Phi_g$ are defined by restriction, and they allow the definition of the full space of generalized connections and scalars by projective limits to arrive at $\bar{\mathscr{A}}$ and $\bar{\Phi}$ as the spaces of fields "on an arbitrarily fine graph". To define generalized scalars we use a certain compactification of the real line, the so-called Bohr compactification $R \subset \bar{\mathbb{R}}_{\text{Bohr}}$. In this way, generalized scalars, just like generalized connections, take values in a compact set. This feature will allow us to provide a consistent definition of the inner product on $\bar{\mathscr{A}} \times \bar{\Phi}$. The Bohr compactification is a topological space such that all continuous functions are the almost-periodic ones:

$$f(\phi) = \sum_{\mu \in \mathscr{I} \subset \mathbb{R} \text{ countable}} f_\mu \exp(i\mu\phi)$$

The set of almost-periodic functions forms an Abelian C^*-algebra, and as a consequence the space $\bar{\mathbb{R}}_{\text{Bohr}}$ on which these functions are defined (the Gel'fand spectrum) is compact. The Bohr compactification also inherits an Abelian group structure from $\mathbb{R}$, allowing us to introduce the Haar measure

$$\int_{\bar{\mathbb{R}}_{\text{Bohr}}} d\mu_{\text{Haar}}(\phi) f(\phi) = \lim_{T \to \infty} \frac{1}{2T} \int_{-T}^{T} d\phi f(\phi).$$

An orthonormal basis with respect to this measure is given by $\{\phi \mapsto \exp(i\mu\phi) : \mu \in \mathbb{R}\}$; the Hilbert space $L^2(\bar{\mathbb{R}}_{\text{Bohr}}, d\mu_{\text{Haar}})$ is non-separable. For more information on $\bar{\mathbb{R}}_{\text{Bohr}}$, see Sect. 3.2.3.4. With these constructions we proceed to defining the inner product on our states. We focus on the dense set of cylindrical states: the projection $\pi_g \colon \bar{\mathscr{A}} \times \bar{\Phi} \to \mathscr{A}_g \times \Phi_g$, obtained by combining $\pi^{\mathscr{A}}_g$ and π^Φ_g, lifts any $f_g \colon \mathscr{A}_g \times \Phi_g \to \mathbb{C}$ to the cylindrical state $\psi = f_g \circ \pi_g$ such that

$$\psi(A, \phi) = f_g(A(e_1), \ldots, A(e_n), \phi(v_1), \ldots, \phi(v_m)).$$

On these states, the inner product is obtained from the Haar measures on G and $\bar{\mathbb{R}}_{\text{Bohr}}$ [13]: If $\psi^{(1)}$ and $\psi^{(2)}$ are cylindrical with respect to the same graph g,

$$\langle \psi^{(1)}, \psi^{(2)} \rangle = \int_{G^n \times \bar{\mathbb{R}}_{\text{Bohr}}^m} \prod_{i=1}^{|E(g)|} d\mu_{\text{Haar}}(h_i) \prod_{j=1}^{|V(g)|} d\mu_{\text{Haar}}(\phi_j)$$

$$\times f_g^{(1)}(h_1, \ldots h_n, \phi_1, \ldots, \phi_m)^* f_g^{(2)}(h_1, \ldots h_n, \phi_1, \ldots, \phi_m).$$

If states are not based on the same graph, one can embed both graphs in a larger one by subdivision, or by an extension of the graphs by "dummy" edges without connection dependence. From properties of the Haar measure one quickly sees that states are orthogonal if they are cylindrical with respect to graphs such that there is an edge e for which $\psi^{(1)}$ depends non-trivially on $A(e)$ while $\psi^{(2)}$ does not. On the resulting Hilbert space holonomies act by multiplication, fluxes as derivative operators measuring the excitation level of geometry.

How exactly flux values are increased by excitations of geometry is derived from the action of the flux operator. We already know the states, and a flux, which is linear in the densitized triad, becomes a simple functional derivative operator by the connection. Again in the U(1)-example (with $\{A_a(x), E^b(y)\} = 8\pi\gamma G \delta_a^b \delta(x, y)$),

$$\hat{F}_S \psi_{g,n} = \frac{8\pi\gamma G\hbar}{i} \int_S d^2y\, n_a \frac{\delta\psi_{g,n}}{\delta A_a(y)} = \frac{8\pi\gamma\ell_P^2}{i} \sum_{e \in g} \int_S d^2y\, n_a \frac{\delta h_e}{\delta A_a(y)} \frac{\partial \psi_{g,n}}{\partial h_e}$$

$$= 8\pi\gamma\ell_P^2 \sum_{e \in g} n_e \int_S d^2y \int_e dt\, n_a \dot{e}^a \delta(y, e(t)) h_e \frac{\partial \psi_{g,n}}{\partial h_e} = 8\pi\gamma\ell_P^2 \sum_{e \in g} n_e \eta(e, S) \psi_{g,n}$$

$$(3.37)$$

with the intersection number $\eta(e, S)$. Since such a state is reproduced after acting with a flux operator, we can directly read off the flux eigenvalues, which are proportional to sums over integers. Thus, the flux spectrum is discrete, providing a detailed realization of discrete spatial geometry [10].

Returning to SU(2)-valued variables, as required for general relativity, we have slightly more complicated expressions for states and operators. Instead of multiplying phase factors, as elements of irreducible U(1)-representations which all are 1-dimensional, we now multiply all possible matrix elements of SU(2)-holonomies along a set of edges. Such states can conveniently be expressed in terms of spin network states [14]

$$\psi_{g,j,C}(A_a^i) = \prod_{v \in g} C_v \prod_{e \in g} \rho_{j_e}(h_e[A]) \tag{3.38}$$

where g is a graph in space, labelled with spins j_e on its edges for irreducible SU(2)-representations ρ_j and with projection matrices C_v in the vertices of the graph which tell us how to pick and combine matrix elements of the holonomies used.

If we consider the example of an n-valent vertex v in which n edges $e_1, \ldots, e_n$ meet and, to be specific, all have the vertex as their endpoint, a suitable projection matrix C_v has n indices

Fig. 3.1 Splitting an
n-valent vertex into unique
trivalent vertex contractions

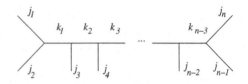

such that the incoming holonomies $(h_{e_i})^{A_i}{}_{B_i}$ are multiplied to $C_{v,A_1,...,A_n}\rho_{j_1}(h_{e_1})^{A_1}{}_{B_1}\cdots$
$\rho_{j_n}(h_{e_n})^{A_n}{}_{B_n}$. (The remaining indices B_i will be contracted with projection matrices in the
vertices corresponding to starting points of the edges incoming at v.) For a gauge-invariant
state, the projection matrices have to satisfy certain conditions. A gauge transformation
maps internal vectors v^A at a point to $g^A{}_B v^B$ with $g \in SU(2)$. Holonomies along a curve
$e\colon [0, 1] \to \Sigma$ transform as $h_e \mapsto g(e(1))h_e g(e(0))^{-1}$ such that $(h_e v)^A = (h_e)^A{}_B v^B$
transforms as an internal vector at $e(1)$ if v^A is an internal vector at $e(0)$. The spin-
network vertex v considered here, with $v = e_i(1)$ for all edges, thus receives a gauge
transformation $C_{v,A_1,...,A_n}\rho_{j_1}(g(v))^{A_1}{}_{C_1}\cdots\rho_{j_n}(g(v))^{A_n}{}_{C_n}$ by moving gauge factors from
the incoming holonomies to the projection matrix. For the spin network to be gauge invariant,
$C_{v,A_1,...,A_n}\rho_{j_1}(g(v))^{A_1}{}_{C_1}\cdots\rho_{j_n}(g(v))^{A_n}{}_{C_n} = C_{v,C_1,...,C_n}$ must hold, which can be realized
only if the trivial representation is contained in the tensor product of the ρ_{j_n}. For a trivalent
vertex, for instance, a gauge-invariant contraction exists if there is an integer $0 \le k \le 2j_1$
such that $j_3 = j_2 - j_1 + k$, where we assume $j_1 \le j_2$. If this condition is satisfied, there
is a unique gauge-invariant contraction. Higher-valent vertices do not have unique contractions. One can parameterize spaces of contraction matrices by integer spins, splitting the
n-valent vertex into subsequent trivalent contractions as illustrated in Fig. 3.1. All intermediate spins k_i can take values only in finite ranges, and spaces of contraction matrices are
finite dimensional.

Holonomies then act by contributions of new factors, changing some of the labels
j_e in an original state by tensor-product decomposition. Fluxes become intersection
sums of derivative operators on SU(2), of the well-known angular-momentum form:
By analogy with (3.37), functional derivatives by the connection can be written in
terms of

$$\hat{J}^i_e = \mathrm{tr}\left((h_e\tau^i)^T \partial/\partial h_e\right) \tag{3.39}$$

or using invariant derivative operators on SU(2). Since angular-momentum operators
have discrete spectra and we now sum finitely many such contributions over intersections of the graph and a surface, SU(2)-fluxes have discrete spectra, too. (If an edge
lies entirely on the surface, thus having infinitely many intersections, the flux vanishes thanks to a product $n_a\dot{e}^a = 0$ in (3.37).) Also the action of holonomies shows
a key feature: While holonomy operators are well-defined, one cannot extract a connection operator from them. Trying to do so, for instance by applying the classical
identity

$$\dot{e}^a(p)A^i_a(p)\tau_i = \lim_{t\to 0}\frac{dh_e|_{[0,t]}}{dt}$$

in the limit where e approaches $e(0) = p$, fails because $h_e|_{[0,t_1]}\psi$ and $h_e|_{[0,t_2]}\psi$ are
orthogonal for $t_1 \ne t_2$: they are cylindrical with respect to different graphs.

From the elementary fluxes one can construct more familiar geometrical objects [15–17], such as the area $A_S[E] = \int_S d^2y\sqrt{n_a E_i^a n_b E_i^b}$ of a surface S or the volume $V_R[E] = \int_R d^3x\sqrt{|\det E|}$ of a region R. Area eigenvalues, just like fluxes, depend on spin labels on curves in the graph intersecting the surface; volume eigenvalues depend on the contraction in vertices within the region. Also these operators have discrete spectra, which in the case of area follow easily from the square of derivative operators. For volume, the spectrum is more difficult to compute since the determinant of E_i^a involves products of three factors of triad components, resulting in couplings of different SU(2)-representations. Nevertheless, recoupling theory allows one to derive matrix elements [18, 19], and powerful computer codes now exist to analyze the eigenvalues [20, 21]. This is expected to be of particular importance for quantum cosmology since the volume spectrum can show how a discrete growing universe must refine its structure as it expands. We will come back to refinement in more detail once we have introduced the dynamics.

3.2.3 Isotropic Models

Many constructions characteristic of canonical quantum gravity can conveniently be illustrated and explicitly be evaluated in symmetric models. The simplest case is that of isotropy, where the spatial geometry is determined by a single number: the scale factor a. Quantum cosmology has traditionally been formulated in this context, and also much work in loop quantum cosmology has been done in an isotropic setting.

3.2.3.1 Symmetry Reduction

Classical symmetry reduction is performed by restricting fields to those left invariant by a set of symmetries (possibly up to gauge transformations only). For invariant connections, for instance, one is looking for the general form of 1-forms ω on a principal fiber bundle $P = (\Sigma, G, \pi)$ with structure group G and base manifold Σ such that $s^*\omega = \omega$ for any element $s \in S$ of a symmetry group S acting on P.

This general definition has two important consequences:

1. An action on the principal fiber bundle P is required, while one usually starts with a desired symmetry on the base manifold, such as isotropy on the space Σ. Lifts of the symmetry action to the whole bundle must thus be found, which are often non-unique. (They are classified by conjugacy classes of homomorphisms $\lambda\colon F \to G$, where F is the isotropy subgroup of S and G the structure group.) This lifting procedure gives rise to different inequivalent classes of invariant connections, which in physical terms can be classified by topological charges. (An example is magnetic charge as the quantity characterizing topologically inequivalent spherically symmetric U(1)-connections; see the following example.)

2. In terms of local connection 1-forms, invariance implies that a connection may change under a symmetry transformation, but only by a gauge transformation: $s^*A = g(s)^{-1}Ag(s) + g(s)^{-1}dg(s)$ where $g\colon S \to G$ is a mapping between the symmetry and structure groups. This mapping is not a group homomorphism, but must satisfy certain other conditions. Solving these conditions is equivalent to determining the possible lifts of symmetry actions to the bundle. In the bundle language, invariant connections are given by (i) a connection $A_{S/F}$ on a reduced bundle Q over Σ/S whose structure group $Z_G(\lambda(F))$ is the centralizer of $\lambda(F)$ in G, and (ii) scalar fields $\phi\colon Q \times \mathcal{L}F_\perp \to \mathcal{L}G$ subject to $\phi(\mathrm{Ad}_f X) = \mathrm{Ad}_{\lambda(f)}(\phi(X))$ for $f \in F$, $X \in \mathcal{L}S$. For more details of the bundle formulation, see [5].

Example 3.1 (Magnetic charge) Magnetic monopoles are spherically symmetric configurations of the magnetic field, which can be described by a U(1)-connection A_a, the vector potential. We are thus interested in the general form of spherically symmetric connections on U(1)-principal fiber bundles, on which the symmetry group SU(2) is acting. In general, inequivalent lifts of a symmetry group S from the base manifold, where its action is given and may have an isotropy subgroup F, to a principal fiber bundle with structure group G over the same base manifold, are classified by conjugacy classes of group homomorphisms $\lambda\colon F \to G$. They describe the twisting along fibers when the symmetry action is lifted from the base manifold to the bundle. In this example, we have $F \cong \mathrm{U}(1) \cong G$, all conjugacy classes are labelled by an integer $k \in \mathbb{Z}$ and can be represented as $\lambda_k(\exp(t\tau_3)) = \exp(ikt)$. For every k, invariant U(1)-connections must have the form of an arbitrary radial U(1)-connection $A_r dr$ plus a contribution of $\Lambda\colon \mathcal{L}F \to \mathcal{L}G$ with $\Lambda|_{\mathcal{L}F_\perp} = 0$ for U(1), $\Lambda|_{\mathcal{L}F} = d\lambda_k\colon \tau_3 \mapsto ik$ applied to the pull-back of the Maurer–Cartan form on S under an embedding of S/F in S. For spherical symmetry with $S = \mathrm{SU}(2)$ and $F = \mathrm{U}(1)$, this pulled-back form can be expressed as

$$A_{S/F} = (\tau_2 \sin \vartheta + \tau_3 \cos \vartheta)d\varphi + \tau_1 d\vartheta. \tag{3.40}$$

With these ingredients, we obtain generic spherically symmetric U(1)-connections of the form

$$A = A_r dr + k \cos \vartheta\, d\varphi.$$

We have a radial (densitized) magnetic field with $B^r = -k \sin \vartheta$, which implies a magnetic charge

$$Q = \frac{1}{4\pi} \int_{S^2} B^a n_a d\vartheta\, d\varphi = -k.$$

For a reduction of the full phase space, one must determine invariant forms of densitized vector fields as well. Once the form of invariant connections is known, for which a richer basis of mathematical results exists, the general form of invariant fields can uniquely be read off from the symplectic structure they must imply. For every free field A_I in the symmetric form of connections A_a^i, there is a conjugate field E^I in the invariant form of densitized triads E_i^a such that $\int_\Sigma \mathrm{d}^3x\, \dot{A}_a^i E_i^a = \int_{\Sigma/S} \dot{A}_I E^I$ when evaluated on invariant fields. For gravitational variables one must take into account an extra condition to ensure that momenta of A_a^i can be non-degenerate. This condition in most of the standard cases leads to a unique sector with no topological charge. Examples will be provided in Sect. 9.1 and the next section.

3.2.3.2 Isotropic Configurations

For isotropic connections invariant under arbitrary translations $\mathbb{R}^3$ and rotations, combining to the Euclidean group, one can always choose a gauge where they take the form

$$A_a^i = \tilde{c}\delta_a^i. \tag{3.41}$$

The single component $\tilde{c}$ is spatially constant but for general solutions to the equations of motion depends on time. A densitized triad of the same symmetry type is of the form

$$E_i^a = \tilde{p}\delta_i^a. \tag{3.42}$$

The reduced symplectic potential

$$\frac{1}{8\pi\gamma G}\int_R \mathrm{d}^3x\, E_i^a\delta A_a^i = \frac{3\mathscr{V}}{8\pi\gamma G}\tilde{p}\delta\tilde{c}, \tag{3.43}$$

where δ denotes a derivative on phase space as opposed to space Σ, then shows that $\tilde{c}$ and $\tilde{p}$ form a canonical pair,

$$\{\tilde{c}, \tilde{p}\} = \frac{8\pi\gamma G}{3\mathscr{V}}. \tag{3.44}$$

In this derivation, as before, we have selected a bounded region $R \subset \Sigma$ to make the spatial integration of homogeneous fields well-defined. If we are considering a cosmological model with compact spatial manifolds, we could choose $R = \Sigma$, but this is not possible for unbounded spatial manifolds. And even for compact spaces, we may as well choose a smaller region as long as it is non-empty; no information about homogeneous configurations is lost provided we just know them in an arbitrarily small neighborhood. The definition of our variables then depends on the choice of the region, and its coordinate size $\mathscr{V} = \int_R \mathrm{d}^3x$. Physical results, of course, must be independent of the choice.

In metric variables, isotropic models are formulated on the phase space with coordinates (a, p_a); triad variables differ from this description by a canonical transformation as well as an extension of the configuration space. From the general relationships between (A_a^i, E_j^b) and metric variables, one can directly derive the relation between $(\tilde{c}, \tilde{p})$ and the scale factor. For the triad component, we obtain $|\tilde{p}| = \frac{1}{4}\tilde{a}^2$, where the factor of $1/4$ can be seen to arise from matching variables of a closed model, not subject to arbitrary rescaling freedom of coordinates, at a fixed curvature scale [5, 22]. Computing the spin connection and extrinsic curvature for an isotropic metric, we combine them to obtain $\tilde{c} = \frac{1}{2}(k + \gamma\dot{a})$. For flat models, one may rescale the scale factor $\tilde{a}$ so as to eliminate the factor of $1/4$ in $\tilde{p}$, as often done. We will denote the rescaled parameter as $a = \frac{1}{2}\tilde{a}$, such that $|\tilde{p}| = a^2$, $\tilde{c} = \frac{1}{2}k + \gamma\dot{a}$. When a is rescaled, also coordinates and thus $\mathcal{V}$ are rescaled such that $a^3\mathcal{V}$ remains unchanged. Taking this into account, the Poisson-bracket relation (3.44) remains unchanged under any rescaling.

Before we describe possible quantizations of these variables, turning the Poisson bracket (3.44) into a commutator relationship, we should properly deal with the factor of $\mathcal{V}$. It merely multiplies the constant result for the Poisson bracket, but it is coordinate dependent. No such factors can be represented on a Hilbert space, which is defined independently of any coordinates chosen on space. We thus redefine our basic variables to absorb $\mathcal{V}$:

$$c := \mathcal{V}^{1/3}\tilde{c}, \qquad p := \mathcal{V}^{2/3}\tilde{p}. \tag{3.45}$$

The particular powers of $\mathcal{V}$ will turn out to be suitable later on in the context of a loop quantization. Moreover, they make the basic variables coordinate independent since $\tilde{p}$ and $\mathcal{V}^{2/3}$ change exactly in opposite ways when coordinates are rescaled, leaving the product invariant. Our new basic variables (c, p), being coordinate independent, should thus be representable on a Hilbert space. They do, however, depend on the size of the region R chosen which affects $\mathcal{V}$ but not $\tilde{p}$ or $\tilde{c}$. Care is then still needed in interpretations of our quantizations once they are formulated. In particular, although there is no explicit $\mathcal{V}$-dependence in the symplectic form

$$\Omega = \frac{3}{8\pi\gamma G}\mathrm{d}c \wedge \mathrm{d}p, \tag{3.46}$$

it must be rescaled proportionally by λ if the region R is enlarged to change $\mathcal{V}$ to $\lambda\mathcal{V}$. This rescaling will require a corresponding transformation on the resulting Hilbert-space representation.

3.2.3.3 Quantum Representation

Given just a pair of canonical variables allowed to take all real values, one possible quantum representation is a standard Schrödinger one as in quantum mechanics. Following this procedure will essentially be a Wheeler–DeWitt quantization, where we may choose either the triad representation with wave functions $\psi(p)$ or the connection representation with wave functions $\psi(c)$, required to be square integrable to make up a kinematical Hilbert space. Compared to the earlier choice of wave functions $\psi(a)$ in Chap. 2, this formulation has two minor advantages: (i) we use one of the coordinate-independent (but integration-region dependent) variables, and (ii) any of the two basic variables takes the full real line as its range, such that no boundary conditions are required, in contrast to $a > 0$ which can make adjointness conditions of operators difficult.

But this representation cannot be the final one since we know that a full quantization in inhomogeneous situations does not allow quantum representations of connection components directly, but only of their holonomies. If an isotropic model is to grasp any of these characteristic features, it should be based on variables analogous to holonomies. For an isotropic connection, it suffices to consider segments of straight lines (along generators of the homogeneity group). Only the length of the segment matters, but not its position or direction. A single parameter ℓ_0 can then be used to label all such straight-edged holonomies. For curves along integral vector fields with tangent $\dot{e}^a = X^a$, normalized with respect to a metric δ_{ab} on the homogeneous space, we have holonomies

$$\mathscr{P}\exp\int_e \tilde{c}\delta_a^i X^a \tau_i = \exp(\ell_0\tau_i X^i\tilde{c}) = \cos\left(\tfrac{1}{2}\ell_0\tilde{c}\right) + 2\tau_i X^i \sin\left(\tfrac{1}{2}\ell_0\tilde{c}\right). \quad (3.47)$$

Since $\tilde{c}$ (in contrast to X^i) is the only dynamical variable, we can express all relevant functions by the U(1)-holonomies $h_{\ell_0}(c) = \exp(\tfrac{1}{2}i\ell_0\tilde{c}) = \exp(\tfrac{1}{2}i\ell_0 c/\mathscr{V}^{1/3})$, where the length parameter first multiplies the original connection component $\tilde{c}$, which is then expressed in terms of the new c. Similarly, fluxes are integrated densitized triads, which for an isotropic configuration and a square surface of edge length given by the same parameter ℓ_0 is of the form $F_{\ell_0}(p) = \ell_0^2\tilde{p} = \ell_0^2 p/\mathscr{V}^{2/3}$. As is clear from the definitions, all these quantities are independent of coordinates, and they are independent of the region of size $\mathscr{V}$ chosen. In addition to the classical geometry given by a phase-space point $(\tilde{c}, \tilde{p})$, they only depend on the label ℓ_0 which alone now plays the role of all the curves and surfaces used in the full representation.

We could have chosen different parameters as labels for holonomies and fluxes at different places, instead of a single ℓ_0. The reason for not doing so is the understanding that an isotropic quantum configuration should require a rather regular graph, made of straight edges roughly of the same length ℓ_0. Such a graph also provides natural choices for similar-sized square surfaces filling a whole plane, such that each of them is transversal to one edge and intersects other surfaces at most at their boundaries. Regularity then requires all these surfaces to have the size ℓ_0^2, which we have used above.

In this picture, we have the added benefit of bringing in the number of discrete sites $\mathscr{N}$ in a natural way, allowing us to incorporate the Second Principle of Sect. 3.1: If there are $\mathscr{N}$ sites of linear dimension ℓ_0 in a region of size $\mathscr{V}$, then $\mathscr{N} = \mathscr{V}/\ell_0^3$ (Fig. 3.2). The geometrical size of each discrete site as measured by the metric to be quantized, moreover, is $\ell_0^3\tilde{p}^{3/2} = \ell_0^3 a^3$, which we can identify with the elementary size v in the refinement picture used before. Indeed, with the relations written here we identically satisfy (2.16). As we will see in the next chapter about dynamics, the

Fig. 3.2 A region of size $\mathscr{V}$, built from patches of linear size ℓ_0

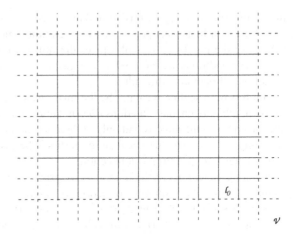

parameter ℓ_0, once it is allowed to become phase-space dependent, is in fact directly related to refinement.

We thus choose the 1-parameter family $\{\exp(\frac{1}{2}i\mu c),\, p\}|_{\mu\in\mathbb{R}}$ as our basic set of objects to construct the quantum representation. Here, we have defined $\mu = \ell_0/\mathscr{V}^{1/3}$, and have dropped the μ^2-factor of p since it would just be a multiplicative constant. The specific form of μ and its relationship with ℓ_0 and $\mathscr{V}$ are not relevant for basic operators, for which we can treat μ simply as a real-valued parameter. But the relationship to underlying discrete structures will become important for composite operators such as inverse triads or the Hamiltonian constraint.

As in the full theory, we construct the state space by starting from a basic state ψ_0, given in the connection representation by a mere constant, and generate all other states by multiplying with "holonomies" as creation operators. The result is a space of states all having the general form

$$\psi(c) = \sum_{\mu\in\mathscr{I}} \psi_\mu \exp\left(\tfrac{1}{2}i\mu c\right) \tag{3.48}$$

where $\mathscr{I} \subset \mathbb{R}$ is a countable index set. As already encountered in Sect. 3.2.2.3, all these functions are called almost periodic, forming a subset of all continuous functions on the real line. Since the space of these functions forms a C^*-algebra, there is a compact space such that almost-periodic functions give the set of *all* continuous functions on that space. This space is compact because its set of continuous functions is a unital C^*-algebra. It contains the real line because almost-periodic functions are functions of a real variable. The real line allows many more continuous functions, and so the space on which almost-periodic functions are all the continuous ones must be larger, with additional points making continuity conditions more restrictive. This larger space contains the real line as a dense subset; it is called the Bohr compactification $\overline{\mathbb{R}}_{\text{Bohr}}$ of $\mathbb{R}$.

Having based isotropic loop quantization on the space of almost-periodic functions, the quantum configuration space will be the Bohr compactification $\overline{\mathbb{R}}_{\text{Bohr}}$ rather than the real line itself. The usual integration on $\mathbb{R}$ also extends to $\overline{\mathbb{R}}_{\text{Bohr}}$, which is an Abelian group and thus has a unique Haar measure up to a constant factor. It can be written explicitly as

$$\int\limits_{\overline{\mathbb{R}}_{\text{Bohr}}} d\mu \, \bar{f}(c) = \lim_{T \to \infty} \frac{1}{2T} \int\limits_{-T}^{T} dc f(c) \tag{3.49}$$

where for any continuous function $\bar{f}$ on $\overline{\mathbb{R}}_{\text{Bohr}}$, f is its restriction to the dense subset $\mathbb{R}$ on which the ordinary integration measure is used.

As usual, holonomies then act on states simply by multiplication. We pick a basis given by the uncountable set $\{|\mu\rangle\}_{\mu \in \mathbb{R}}$ where

$$\langle c|\mu\rangle = \exp(\tfrac{1}{2}i\mu c) \tag{3.50}$$

in the connection representation. It is clear that these states span all states (3.48), and with the inner product based on (3.49) they are orthonormal. On the basis, holonomies act by shifting the labels:

$$\widehat{\exp(\tfrac{1}{2}i\delta c)}|\mu\rangle = |\mu + \delta\rangle. \tag{3.51}$$

To compare with basic holonomy operators in the full theory, one can think of this action as being analogous to changing the spin of an SU(2)-representation by coupling the edge spin of a spin network state with the spin of the holonomy used for the operator; or one can think of it as a holonomy operator extending an already existing edge, thus making the length parameter larger. In an isotropic context, these two interpretations (or any mixtures thereof) cannot be separated—a degeneracy which has to be taken into account for proper interpretations of operator actions.

The flux operator $\hat{p}$ can be expressed directly as a derivative operator

$$\hat{p} = \frac{8\pi \gamma \ell_{\text{P}}^2}{3i} \frac{d}{dc} \tag{3.52}$$

taking into account the factor in the symplectic structure (3.46) and introducing the Planck length $\ell_{\text{P}} = \sqrt{G\hbar}$. Its action on the basis states follows directly as

$$\hat{p}|\mu\rangle = \frac{4\pi \gamma \ell_{\text{P}}^2}{3} \mu |\mu\rangle \tag{3.53}$$

which shows that we picked the flux eigenbasis with (3.50). Just as in the full case, (3.37), the flux operator has a discrete spectrum: all its eigenstates are normalizable. Unlike with the full spectrum, however, every real number is an eigenvalue. These two properties are consistent with each other in the present case of a non-separable Hilbert space. As we will see later, the mathematical definition of a discrete spectrum via the normalizability of eigenstates turns out to be the appropriate one here, too, because it shows that crucial features of the full theory are realized in isotropic models as well. Also our isotropic quantum geometry is thus atomic in the sense of discrete flux spectra.

Now having the basic representation of the isotropic holonomy-flux algebra at our disposal, we can analyze its rescaling behavior. We had absorbed factors of $\mathscr{V}$ in the original canonical variables $\tilde{c}$ and $\tilde{p}$ in order to deal with coordinate-invariant quantities. But then the variables c and p, as well as their resulting quantum theory, become dependent on the volume $\mathscr{V}$ of a region chosen arbitrarily. Changing the region must result in a well-defined transformation between the quantum representations obtained for different values of $\mathscr{V}$; otherwise there would be no way of eventually testing whether observables are indendent of the rescaling.

To analyze this, let us rescale $\mathscr{V}$ to $\lambda\mathscr{V}$ with a positive real number λ. Then, c and p change to $\lambda^{1/3}c$ and $\lambda^{2/3}p$, respectively. Also the symplectic structure changes to $\lambda\Omega$ which indeed follows from the classical rescaling. Thus, the rescaling transformation is not canonical. Since the symplectic structure of basic variables is rescaled, also the quantum representation must change: commutators of basic operators, following from the basic Poisson brackets, are to be divided by λ. Instead of modifying commutators, which are universally defined, we can formally implement this rescaling of the representation by changing Planck's constant to $\lambda\hbar$ in all equations where it enters. (Of course, the physical value of $\hbar$ is fixed. The scaling can be normalized to the correct value only if the underlying discreteness scale of quantum gravity is taken into account. Such a scale is not explicitly available in minisuperspace models, but will be included at a later stage once we discuss inhomogeneity; Sect. 10.1.) This indeed provides the correct rescaling relationship between c and p. For instance, in (3.52) we have $\mathrm{d}/\mathrm{d}c \mapsto \lambda^{-1/3}\mathrm{d}/\mathrm{d}c$, while $\ell_{\mathrm{P}}^2 \mapsto \lambda\ell_{\mathrm{P}}^2$, combining to the correct $\hat{p} \mapsto \lambda^{2/3}\hat{p}$. From (3.53) we then read off that the state parameter μ must rescale to $\lambda^{-1/3}\mu$. (In particular, μ rescales like the classical c^{-1}, not like the flux p whose eigenvalues it provides after multiplying it with $4\pi\gamma\ell_{\mathrm{P}}^2/3$.) Also this is consistent, for (3.51) tells us that μ and the holonomy coefficient $\delta \propto \ell_0/\mathscr{V}^{1/3}$ must rescale in the same way for $|\mu + \delta\rangle$ to have a meaningful rescaling.

In quantum cosmology we are thus dealing with a family of models $(\hat{c}, \hat{p}, [\cdot, \cdot])_{\mathscr{V}}$. Classically, changing $\mathscr{V}$ is not a canonical transformation; in quantum theory there is no unitary relationship between models of different $\mathscr{V}$ with the same Hilbert-space representation. Still, there are simple means to check the $\mathscr{V}$-independence of results, as we will use them in what follows.

As a final remark, we notice that we could have decided to fix ℓ_0 once and for all, as it was indeed done in the first formulations of loop quantum cosmology [23, 24]. There is a disadvantage of implying, essentially, that one makes the configuration space of connections strictly periodic. From holonomies of this type, we could not reconstruct $\tilde{c} \in \mathbb{R}$ completely, but only up to integer multiples of $4\pi/\ell_0$. Almost-periodic functions, on the other hand, do not require a periodification of the configuration space but rather compactify it by an enlargement. At the kinematical level, no freedom is lost by using the Bohr compactification. Still, it remains to be seen at this stage what the dynamics actually requires: it will also rely on holonomies as basic building blocks of the Hamiltonian; and if there is a choice of ℓ_0 to be made there, as it will turn out to be indeed the case, one could have made that restriction already at the kinematical level. Dynamics, in this case, would see only one sector of periodic connections, and nothing would be lost by fixing attention to one periodic sector from the outset. To allow the full freedom of how this might turn out, we do not fix ℓ_0 for now, permitting all $\exp(\frac{1}{2}i\ell_0\tilde{c})$ as operators. (In fact, lattice refinement especially for anisotropic models will require this general attitude, as we will see later.) Quantum dynamics will then be based on the Bohr compactification of the real line as the configuration space. Since we already indicated the relationship between ℓ_0 and lattice refinement, this issue will naturally be revisited in the chapter on dynamics.

3.2.3.4 More on Bohr

The Bohr compactification of the real line plays an important role in loop quantum cosmology to include all kinematical sectors of the models in one non-separable Hilbert space. The dynamics sometimes picks a separable sector, but since the one that is realized depends on the specific form of the dynamics and is subject to quantization ambiguities, as we will see, it is useful to consider all quantum configurations in one setting even though this information is only kinematical. In this context, several properties and characterizations of the Bohr compactification are of interest. For additional mathematical discussions, see [25, 26].

- The Bohr compactification of the real line contains $\mathbb{R}$ densely. It is thus not a periodification which would identify different points in $\mathbb{R}$ and not contain all the original ones. It is also different from the one-point compactification, which, too, contains the real line densely but adds just one point at infinity. The difference can be seen by the set of continuous functions on these two compact spaces. For the Bohr compactification, as used in the definition, this is the set of almost-periodic functions; for the one-point compactifications, this is the set of functions $f(c)$ for which $\lim_{c\to-\infty} f(c)$ and $\lim_{c\to\infty} f(c)$ exist and agree. The only functions continuous on the Bohr compactification as well as the one-point compactification are the constant ones.
- One can visualize the Bohr compactification by subsets in a torus $[0, 1]^2$. The whole real line can be embedded in the torus as a straight line $x = \omega y$ if ω is irrational. In the trace topology, the resulting subset of the compact torus can be completed to a compact space containing the real line (the original embedding) densely.
- The Fourier space of the Bohr compactification is the discrete real line whose open neighborhoods are arbitrary unions of single points. This is indeed the space of momenta as it arises from (3.48) with (3.53).
- If we contrast a periodification with the Bohr compactification, the configuration space in one case is a circle with Fourier space given by the integers $\mathbb{Z}$. The enlargement of the Bohr compactification as compared to the periodification allows more momenta filling all of $\mathbb{R}$. As a trace of the compactness of the configuration space, the momentum space is the real line with discrete topology.
- A Wigner function for states supported on $\bar{\mathbb{R}}_{\text{Bohr}}$ can be defined as [25]

$$W(c, p) = \frac{1}{2\pi \ell_P^2} \int\limits_{\mathbb{R}_{\text{Bohr}}} \psi^* \left(p - \tfrac{1}{2}\alpha\right) \psi \left(p + \tfrac{1}{2}\alpha\right) h^{(2\alpha/\ell_P^2)}(c) \mathrm{d}\alpha \qquad (3.54)$$

with holonomies $h^{(\delta)}(c) = \exp(\mathrm{i}\delta c/2)$. The Wigner function of a triad eigenstate, for instance, is then a delta-function peaked at the triad eigenvalue, and independent of c. Further properties of the Bohr-Wigner function are discussed in [25].

3.2.3.5 Inverse-Triad Operators

For a first glimpse on the singularity issue we now have a look at suitable quantizations of a^{-1} or any other inverse power, which would diverge at the classical singularity of isotropic cosmology. In a Wheeler–DeWitt quantization, a^{-1} can easily be quantized as a multiplication operator acting on wave functions $\psi(a)$. It is densely defined and thus suitable. It certainly fails to be a bounded operator, but so does $\hat{a}$ itself. Kinematically, the classical singularity does not appear to be different in Wheeler–DeWitt quantum cosmology.

At first, the problem seems worse in loop quantum cosmology: we would now have to find an inverse of the triad operator $\hat{p}$ which has a discrete spectrum containing zero. No such operator has a densely defined inverse. One could define an inverse using multiplication with μ^{-1} on all states except $|0\rangle$, but since $|0\rangle$ is orthogonal to all states $|\mu\rangle$ with $\mu \neq 0$,[1] this procedure would not make the inverse densely defined. It thus does not correspond to an operator on the Hilbert space. This issue would not just be a problem indicating a singularity, it would even prevent us from quantizing Hamiltonians, including the gravitational one (3.30), which all contain some form of inverse-triad components.

Nevertheless, operators whose classical limit is a^{-1} do exist. To see this, we follow a construction which is also available in the full theory [27, 28] and which in the next chapter will allow us to quantize Hamiltonian constraints and matter Hamiltonians. Applied in an isotropic setting [29], we have the identity

$$\frac{2\mathrm{i}}{\delta} e^{\mathrm{i}\delta c/2} \{e^{-\mathrm{i}\delta c/2}, |p|^{r/2}\} = \{c, |p|^{r/2}\} = \frac{4\pi \gamma G r}{3} |p|^{r/2-1} \operatorname{sgn} p \qquad (3.55)$$

If we choose the real parameter r to be in the range $0 < r < 2$, we have an inverse power of the triad component p on the right-hand side, while no inverse is required on the left. The left-hand side of this equation can easily be quantized by using the volume operator $\hat{V} = \hat{p}^{3/2}$, holonomy operators for $\exp(\mathrm{i}\delta c/2)$ and turning the Poisson bracket into a commutator divided by $\mathrm{i}\hbar$. A densely defined operator with an inverse of p as the classical limit results.

To be closer to operators of the full form we first replace the exponentials by SU(2)-holonomies evaluated in isotropic connections:

---

[1] Assuming a sequence of states in $|\psi_n\rangle \in \mathscr{H}/(\mathbb{C}|0\rangle)$ such that $\lim_{n\to\infty} |\psi_n\rangle = |0\rangle$, we obtain the contradiction $0 = \lim_{n\to\infty}\langle 0|\psi_n\rangle = \langle 0| \lim_{n\to\infty} \psi_n\rangle = 1$.

Fig. 3.3 Correction function (3.59) from inverse-triad corrections

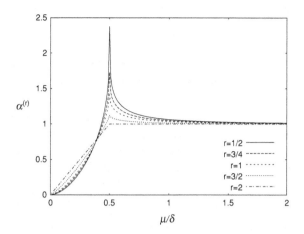

$$\mathrm{tr}(\tau_3\exp(\delta c\tau_3)[\exp(-\delta c\tau_3), \hat{V}^r]) = \sin(\delta c/2)\hat{V}^r\cos(\delta c/2)-\cos(\delta c/2)\hat{V}^r\sin(\delta c/2).$$
(3.56)

(The trace can be evaluated explicitly after inserting $\exp(A\tau_3) = \cos\left(\frac{1}{2}A\right) + 2\tau_3\sin\left(\frac{1}{2}A\right)$.) Using the basic operators, it is easy to see that the resulting

$$\widehat{|p|^{r/2-1}\mathrm{sgn}(p)} = \frac{3}{4\pi\gamma\ell_{\mathrm{P}}^2\delta r}\left(\widehat{e^{\mathrm{i}\delta c/2}}[\widehat{e^{-\mathrm{i}\delta c/2}}, \widehat{|\hat{p}|^{r/2}}] - \widehat{e^{-\mathrm{i}\delta c/2}}[\widehat{e^{\mathrm{i}\delta c/2}}, \widehat{|\hat{p}|^{r/2}}]\right)$$
(3.57)

has the same eigenbasis $|\mu\rangle$ as the triad operator, with eigenvalues

$$\left(\widehat{|p|^{r/2-1}\mathrm{sgn}(p)}\right)_\mu = \frac{1}{\delta r}\left(\frac{4\pi\gamma\ell_{\mathrm{P}}^2}{3}\right)^{r/2-1}(|\mu + \delta/2|^{r/2} - |\mu - \delta/2|^{r/2}) \quad (3.58)$$

clearly well-defined even at $\mu = 0$. At $\mu = 0$, in fact, the eigenvalue vanishes instead of being divergent like the classical value. For large $|\mu| \gg \delta$, on the other hand, the classical expression $\mathrm{sgn}(\mu)|4\pi\gamma\ell_{\mathrm{P}}^2\mu/3|^{r/2-1}$ is approached. The difference gives rise to correction functions

$$\alpha^{(r)}(\mu) = \frac{\left(\widehat{p^{r/2-1}\mathrm{sgn}p}\right)_\mu}{p_\mu^{r/2-1}\mathrm{sgn}p_\mu} = \left(\widehat{|p|^{r/2-1}\mathrm{sgn}(p)}\right)_\mu\left|\frac{4\pi\gamma\ell_{\mathrm{P}}^2}{3}\mu\right|^{1-r/2}\mathrm{sgn}(\mu)$$

$$= \frac{1}{\delta r}|\mu|^{1-r/2}\left(|\mu + \delta/2|^{r/2} - |\mu - \delta/2|^{r/2}\right)\mathrm{sgn}(\mu) \neq 1$$
(3.59)

in quantizations of expressions that classically contain inverses of densitized-triad components; see Fig. 3.3. (The δ-dependence is not explicitly noted as an ambiguity in $\alpha^{(r)}$ because it simply rescales μ.)

This calculation demonstrates that densely defined operators with the classical limit of an inverse power of p do exist; we will later use these constructions for

Hamiltonians. The classical divergence at the singularity implies that these operators cannot be inverse operators of $\hat{p}$:

$$\widehat{p^{r/2-1}}\hat{p}^{1-r/2} \neq 1;$$

they only approach the inverse in the classical limit. In this way, the initial problem of $\hat{p}$ having zero in its discrete spectrum is overcome.

Instead of being singular, the small-μ behavior is bounded and approaches zero at $\mu = 0$ as already seen. Around $\mu \sim \delta/2$, a peak is reached demarkating the strong quantum-geometrical behavior for small μ and the nearly classical behavior for large values. The position of the peak is not unique, but depends on quantization ambiguities. For instance, one can use different values of δ and also different r in the specified range without changing the crucial properties. As we will see later, the precise form of the function enters cosmological and other equations, such that ambiguities can in principle be fixed by phenomenology. The freedom is also reduced by considering the anomaly problem of quantum constraints, where inverse-triad operators enter, too; see Sect. 10.3.

For phenomenology, it will be important to consider the typical size of deviations of inverse-triad operators from the classical expectation. In an effective formulation, we would refer not to eigenvalues μ as in (3.59) but to an effective geometry reconstructed from μ via $p_\mu = 4\pi \gamma \ell_{\mathrm{P}}^2 \mu/3$. This relationship leads to correction functions

$$\alpha^{(r)}(p_\mu) = \frac{1}{\delta r}\left(\frac{4\pi\gamma\ell_{\mathrm{P}}^2}{3}\right)^{-1} |p_\mu|^{1-r/2}\mathrm{sgn}(p_\mu)$$
$$\times \left(|p_\mu + 2\pi\gamma\delta\ell_{\mathrm{P}}^2/3|^{r/2} - |p_\mu - 2\pi\gamma\delta\ell_{\mathrm{P}}^2/3|^{r/2}\right) \qquad (3.60)$$

with strong corrections setting in at $p_\mu \sim 2\pi\gamma\delta\ell_{\mathrm{P}}^2/3$. At this stage lattice refinement again becomes relevant. If corrections are to arise from a general quantum state, we should use in expressions such as (3.56) not the total volume V of our region of coordinate size $\mathcal{V}$, but the elementary volume of coordinate size ℓ_0^3. (Otherwise, the expressions we obtain cannot be considered local. Further justification comes from the analogous operators in the full theory, where only local vertex terms touched by the holonomies contribute. See [30] for a calculation using kinematical coherent states in the full theory.) The conversion from p_μ to the scale factor then does not come from $|p_\mu| = \mathcal{V}^{2/3}a^2$, but from $|p_\mu| = |F_{\ell_0}(p)| = \ell_0^2 a^2$; see also Sect. 10.1. Once the replacement of the cell volume by the patch volume is made, we refer to almost-local phase-space variables; Planck's constant or the Planck length in commutators or other quantum formulas no longer rescale when $\mathcal{V}$ changes. In terms of a, we have a correction function

$$\alpha^{(r)}(a) = \frac{\ell_0^2}{\delta r}\left(\frac{4\pi\gamma\ell_{\mathrm{P}}^2}{3}\right)^{-1} a^{2-r} \left(|a^2 + 2\pi\gamma\delta\ell_{\mathrm{P}}^2/3\ell_0^2|^{r/2} - |a^2 - 2\pi\gamma\delta\ell_{\mathrm{P}}^2/3\ell_0^2|^{r/2}\right)$$
$$(3.61)$$

with strong corrections setting in at $a \sim a_* := \sqrt{2\pi\gamma\delta/3}\ell_{\mathrm{P}}/\ell_0$.

This refined procedure has two consequences: (i) While using the region of size $\mathcal{V}$ would make expressions dependent on $\mathcal{V}$, which is not allowed for observables, the elementary sizes F_{ℓ_0} are independent of $\mathcal{V}$ and instead refer to an underlying discrete state via the quantity ℓ_0; and (ii) with the region ℓ_0^3 being smaller than $\mathcal{V}$ we get deeper into the small-scale regime and inverse-triad corrections will become comparatively larger (a_* being proportional to ℓ_0^{-1}). When elementary sizes are used in expressions for correction functions, the latter peak for values of the discrete increment δ of about the elementary plaquette size $a^2\ell_0^2/\ell_{\mathrm{P}}^2$ relative to the Planck scale, not for $\delta \sim a^2\mathcal{V}^{2/3}/\ell_{\mathrm{P}}^2$ which could be huge, and is even subject to coordinate and other choices. These features must be taken into account for consistent formulations of models as well as reliable phenomenology, but also for a meaningful realization of inverse-triad corrections. We will provide an inhomogeneous calculation in Sect. 10.1.4.2, exhibiting these properties explicitly. As another consequence of the fuller treatment, the range of values for the ambiguity parameter δ will be restricted.

The explicit formulas provided here rely on the Abelianization of the full theory when it is reduced to isotropy. Several new features arise if one tries to construct inverse-triad operators in an SU(2)-setting and to evaluate the characteristic commutators [31]. First, the commutators quantizing $\mathrm{tr}(h\{h^{-1}, V\})$ no longer commute with the volume operator and it becomes less clear how to compare spectra when they refer to different eigenbases. Secondly, inverse-triad operators, though still densely defined, are no longer bounded [32]. The latter is a feature which is shared by some related operators in anisotropic models discussed later, and is thus not only a consequence of non-Abelian behavior. The non-commutativity of inverse-triad operators with the volume operator, on the other hand, is directly related to the full non-Abelian nature. It is probably the most serious issue that suggests some caution toward results obtained only in isotropic models.

References

1. Giulini, D., Kiefer, C., Joos, E., Kupsch, J., Stamatescu, I.O., Zeh, H.D.: Decoherence and the Appearance of a Classical World in Quantum Theory. Springer, Berlin (1996)
2. Bojowald, M., Brizuela, D., Hernandez, H.H., Koop, M.J., Morales-Técotl, H.A.: arXiv:1011.3022
3. Gazeau, J.P., Klauder, J.: J. Phys. A: Math. Gen. **32**, 123 (1999)
4. Novello, M., Bergliaffa, S.E.P.: Phys. Rep. **463**, 127 (2008)
5. Bojowald, M.: Canonical Gravity and Applications: Cosmology, Black Holes, and Quantum Gravity. Cambridge University Press, Cambridge (2010)
6. Ashtekar, A.: Phys. Rev. D **36**(6), 1587 (1987)
7. Barbero G.J.F., (1995) Phys. Rev. D 51(10): 5507
8. Immirzi, G.: Class. Quantum Grav. **14**, L177 (1997)
9. Lewandowski, J., Newman, E.T., Rovelli, C.: J. Math. Phys. **34**, 4646 (1993)
10. Rovelli, C., Smolin, L.: Nucl. Phys. B **331**, 80 (1990)
11. Corichi, A., Krasnov, K.: Mod. Phys. Lett. A **13**, 1339 (1998). hep-th/9703177
12. Ashtekar, A., Lewandowski, J., Marolf, D., Mour ao, J., Thiemann, T.: J. Math. Phys. **36**(11), 6456 (1995). gr-qc/9504018
13. Ashtekar, A., Lewandowski, J.: J. Math. Phys. **36**(5), 2170 (1995)
14. Rovelli, C., Smolin, L.: Phys. Rev. D **52**(10), 5743 (1995)

15. Rovelli, C., Smolin, L.: Nucl. Phys. B **442**, 593 (1995). gr-qc/9411005
16. Ashtekar, A., Lewandowski, J.: Class Quantum Grav. **14**, A 55 (1997). gr-qc/9602046
17. Ashtekar, A., Lewandowski, J.: Adv. Theor. Math. Phys. **1**, 388 (1998). gr-qc/9711031
18. De Pietri, R., Rovelli, C.: Phys. Rev. D **54**(4), 2664 (1996)
19. Brunnemann, J., Thiemann, T.: Class Quantum Grav. **23**, 1289 (2006). gr-qc/0405060
20. Brunnemann, J.., Rideout, D.: Class. Quant. Grav. **25**, 065001 (2008). arXiv:0706.0469
21. Brunnemann, J., Rideout, D.: Class. Quant. Grav. **25**, 065002 (2008). arXiv:0706.0382
22. Bojowald, M.: Living Rev. Relativity **11**, 4 (2008). gr-qc/0601085
23. Bojowald, M.: Class. Quantum Grav. **17**, 1489 (2000). gr-qc/9910103
24. Bojowald, M.: Class. Quantum Grav. **19**, 2717 (2002). gr-qc/0202077
25. Fewster, C., Sahlmann, H.: Class Quantum Grav **25**, 225015 (2008). arXiv:0804.2541
26. Velhinho, J.M.: Class. Quantum Grav. **24**, 3745 (2007). arXiv:0704.2397.
27. Thiemann, T.: Class Quantum Grav **15**, 839 (1998). gr-qc/9606089
28. Thiemann, T.: Class. Quantum Grav. **15**, 1281 (1998). gr-qc/9705019
29. Bojowald, M.: Phys. Rev. D **64**, 084018 (2001). gr-qc/0105067
30. Brunnemann, J., Thiemann, T.: Class. Quantum Grav. **23**, 1395 (2006). gr-qc/0505032
31. Bojowald, M.: Class. Quantum Grav. **23**, 987 (2006). gr-qc/0508118
32. Brunnemann, J., Thiemann, T.: Class. Quantum Grav. **23**, 1429 (2006). gr-qc/0505033

Chapter 4
Dynamics: Changing Atoms of Space–Time

In the previous chapter, we have seen some aspects of spatial quantum geometry, with its characteristic discrete spectra, emerge even in isotropic models. Now, these spatial structures have to fit into a consistent quantum space–time which in a certain sense reduces to a solution of Einstein's equation in a semiclassical or low-curvature limit. Only completing this most challenging step will make the theory one of quantum gravity, rather than of spatial quantum geometry.

4.1 Refinement and Internal Time

On general grounds, we expect discrete growth of the expanding universe in any theory with a microstructure of space–time. This has been anticipated several times [1, 2, 3] but never fully formulated. Loop quantum gravity makes progress at least in describing discrete spatial structures, and in providing means to analyze their evolution. A handle on these inhomogeneous features is necessary even for the consistency of the quantum cosmological theory: any kind of discreteness is in danger of being enlarged by the expansion of the universe and of becoming macroscopic unless its structure is being dynamically refined. It is not guaranteed that refinement happens in a suitable way so as to be consistent with the possibility of a macroscopic, very nearly continuous universe. Whether suitable refinement is realized can be checked in model systems, thereby providing feedback on the correct low-curvature limit of the full theory.

The picture we will take to develop dynamics is that of an evolving quantum geometry. Mathematically, it is implemented by constructing states in the Hilbert spaces seen before, but subject to the condition that they be annihilated by a quantization of the classical constraints: $\hat{C}|\psi\rangle = 0$. As the classical constraints make sure that initial data indeed are as they can be derived by restricting the geometry of a space–time solution to one of its spatial slices, quantum constraints ensure that there is no spurious gauge-dependent information in the quantum states we use. Discussing quantum constraints can be quite involved not just because the expressions

M. Bojowald, *Quantum Cosmology*, Lecture Notes in Physics 835,
DOI: 10.1007/978-1-4419-8276-6_4, © Springer Science+Business Media, LLC 2011

of constraints are complicated, but also because they are not always guaranteed to
have zero as an eigenvalue in their discrete spectrum. In fact, especially for the
Hamiltonian constraint of quantum gravity one expects interesting physical states to
correspond to zeros in the continuous part of the spectrum. (There may be states in
the discrete part as well.) Solutions then will not be normalizable in the spaces we
constructed so far, and new inner products would have to be introduced to extract
observable information from the states. Technical issues of constructing physical
Hilbert spaces are discussed in Chap. 12.

Also the issue of observables strikes again: A general operator on the kinematical
Hilbert space will not map a solution of $\hat{C}|\psi\rangle = 0$ to another such solution; it is
guaranteed to do so only if it commutes with $\hat{C}$: $[\hat{O}, \hat{C}] = 0$. Such operators are
called Dirac observables, and thanks to the commutation property they correspond
to classical observables as phase-space functions invariant under the flow generated
by the constraints: their classical correspondents satisfy $\{O, C\} = 0$. Unfortunately,
complications in constructing general observables at the classical level become even
more severe at the quantum level. For instance, one would have to be very careful
with factor orderings to make a classical observable commute exactly with the constraints after quantization. (And exact commutation is required when it comes to
removing gauge.)

We will discuss issues of physical Hilbert spaces and quantum observables in
more detail in a later chapter; here we are primarily interested in the fact that the
non-commutativity of the Hamiltonian constraint and operators for spatial geometry
in general implies that physical states must be superpositions of different eigenstates
of any operator of spatial geometry. Explicitly, this follows from the fact that the
Hamiltonian constraint contains connection components, quantized to holonomies,
which as creation operators excite the spatial geometry. No state with a sharp non-
degenerate spatial geometry can be an eigenstate of such creation operators, and
general superpositions are required.

Choosing a particular non-observable operator $\hat{\phi}$, for instance the volume or a
matter field, and its eigenbasis, we expand a physical state as $|\psi\rangle = \sum_\phi |\psi\rangle_\phi$
where ϕ is a label for the eigenstates of $\hat{\phi}$. (Terms $|\psi\rangle_\phi$ in such a superposition
are not physical states unless $\hat{\phi}$ is an observable.) We may then view the family
$|\psi\rangle_\phi$ as an evolving quantum geometry, or a "state-time" [4]. It represents physical
information, for it is just another way of writing the original physical state $|\psi\rangle$,
supposed to solve the quantum constraint equation. Evolution is described not in
coordinate time, as one usually does it in classical solutions, but with respect to an
internal time ϕ. One could use a similar picture in classical gravity: The classical
variable ϕ corresponding to the operator $\hat{\phi}$ we picked for the decomposition fulfills
an equation of motion $d\phi/dt = \{\phi, C\} = f$ with some phase-space function f. At
any phase-space point where $f \neq 0$, the usual t-derivatives in classical equations of
motion can be traded for ϕ-derivatives by dividing all equations of motion by f. While
such a procedure would usually make classical calculations more complicated than
necessary, not even being too useful since one is mostly interested in the dependence
of quantities on variables such as proper time, it is a more natural viewpoint to take in

quantum gravity. The quantum representation of geometry, after all, does not contain any space–time coordinates; it can refer only to degrees of freedom as found in the classical phase space, and so the internal-time picture is the only one available for evolution. Once physical states are parameterized with respect to an internal time, one can compute time-dependent expectation values of operators other than time $\hat{\phi}$ and their fluctuations or other moments. Even if such an operator $\hat{O}$ would not be an observable, the expectation value $_\phi\langle\psi|\hat{O}|\psi\rangle_\phi$ is meaningful. In general, however, care must be taken as to the inner product used here. For instance, one would like a sense of unitarity of ϕ-evolution to be realized, which then implies that $_\phi\langle\psi|\psi\rangle_\phi$ is constant in ϕ. (This is, of course, not guaranteed if the family $|\psi\rangle_\phi$ comes from an arbitrary decomposition of some state in any eigenbasis.)

Effective techniques allow one to formulate the concept of local internal time systematically, and to address the possible failure of unitarity of the evolution meaningfully. These methods and consequences such as the complexity of time will be discussed in Chap. 13; for now it suffices to know that consistent formulations can be provided, and we turn to the construction of the constraint operators themselves.

4.2 Constructions in the Full Theory

We will make use of the concept of internal time mainly in cosmological models where it is rather tractable. But we will first have to see what the Hamiltonian constraint and the dynamics it provides look like in the full theory, since this will be our guideline in reduced constructions just as it was for states and basic operators. Also for the dynamics, the key will be the use and form of holonomies, their appearance in the Hamiltonian constraint, and the way they influence superpositions in the averaging of kinematical to physical states. Remaining non-holonomy terms in the constraint are also important and provide clues to the small-scale behavior of quantum gravity, but they merely influence the concrete coefficients rather than the general form of such superpositions.

4.2.1 Gravitational Constraint

The classical expression of the Hamiltonian constraint in Ashtekar–Barbero variables is

$$C_{\text{grav}}[N] = \frac{1}{16\pi\gamma G} \int_\Sigma d^3x N \left(\varepsilon_{ijk} F^i_{ab} \frac{E^a_j E^b_k}{\sqrt{|\det E|}} \right.$$

$$\left. - 2(1+\gamma^{-2})(A^i_a - \Gamma^i_a)(A^j_b - \Gamma^j_b) \frac{E^{[a}_i E^{b]}_j}{\sqrt{|\det E|}} \right) \qquad (4.1)$$

Fig. 4.1 A loop used to quantize curvature components

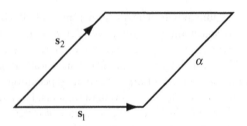

smeared with the lapse function N. As the main term responsible for the creation of geometrical excitations, we first turn to the curvature components F_{ab}^i. They are functionals of the connection components A_a^i, which in a loop representation cannot be quantized directly. Instead we have to use holonomies, providing curvature components via the well-known expression

$$s_1^a s_2^b F_{ab}^i \tau_i = \Delta^{-1}(h_\alpha - 1) + O(\Delta) \tag{4.2}$$

for the leading-order term of a holonomy along a closed loop α of square shape, of coordinate area Δ and along the tangent vectors s_1^a and s_2^a; see Fig. 4.1.

The loop will give rise to holonomies and thus, when acting with the constraint, excitations of geometry [5]. There are different specific constructions, with different general behaviors. For instance, when acting on a given state the loop α may lie entirely on the graph of edges already excited for the state. In this case, the graph itself would not be changed, and no new vertices are being created (see e.g. [6]); only the excitation levels of existing edges are raised or lowered. Otherwise, if α does not lie entirely on the state's graph, new vertices and new edges will emerge [7]. In the picture of an evolving geometry, the number $\mathcal{N}$ of discrete sites may or may not change in the internal evolution parameter ϕ. But in all possible constructions put forward so far, individual sizes of elementary sites, earlier called v, must change: since the volume operator does not commute with all holonomies, $[\hat{V}, h] \neq 0$, vertex volumes of states must change in ϕ. While we do not yet have a systematic way of averaging these inhomogeneous results to isotropic geometries, the behavior seen here provides strong motivation for a lattice-refinement picture, and it does put limits on some parameters.

As the next important term in the classical constraint we have the triad-dependent function $\varepsilon^{ijk} E_j^b E_k^c / \sqrt{|\det E|}$. Fluxes cannot directly be used to quantize this expression since they have discrete spectra containing zero, making the determinant potentially degenerate. Instead, as already seen in the isotropic setting, one applies the classical identity [7]

$$\left\{ A_a^i, \int \sqrt{|\det E|} \mathrm{d}^3 x \right\} = 2\pi \gamma G \varepsilon^{ijk} \varepsilon_{abc} \frac{E_j^b E_k^c}{\sqrt{|\det E|}} \mathrm{sgn} \det(E_l^d) \tag{4.3}$$

(or another form since it is not unique) and quantizes the left-hand side regularly by using holonomies for A_a^i, the volume operator, and turning the Poisson bracket into a commutator divided by $i\hbar$. A densely defined operator results. This key result of loop quantum gravity shows that Hamiltonians in quantum gravity, with similar constructions for matter Hamiltonians [8], can be made well-defined by particular properties of quantum geometry [7]. We do not have to normal-order Hamiltonian operators, which would be difficult in a background-independent way that lacks the ordinary creation and annihilation operators.[1] Coefficients in any superposition of physical states are automatically well-defined. In particular, the specific form of triad operators appearing in the constraint, containing commutators such as $h[h^{-1}, \hat{V}]$, changes vertices and their intertwiners for states they act on; thus another reason to expect a changing $v(\phi)$ in a physical state.

The remaining terms in the constraint are more difficult since they contain the spin connection (3.24), a complicated functional of E_i^a. Fortunately, they can be reduced to what has already been quantized by employing another Poisson identity [7]:

$$K_a^i = \gamma^{-1}(A_a^i - \Gamma_a^i) \propto \left\{ A_a^i, \left\{ \int d^3x F_{ab}^i \frac{\varepsilon^{ijk} E_j^a E_k^b}{\sqrt{|\det E|}}, \int \sqrt{|\det E|} d^3x \right\} \right\}. \quad (4.4)$$

Putting all these ingredients together, one constructs a well-defined Hamiltonian operator, schematically of the form

$$\hat{C}_{\text{grav}}[N]\psi_g \propto \sum_{v \in V(g)} N(v) \sum_{e_I, e_J, e_K \in E(g)} \varepsilon^{IJK} \text{tr} \left(h_{IJ} h_{e_K} \left[h_{e_K}^{-1}, \hat{V} \right] \right) \psi_g \quad (4.5)$$

summing over vertices v and triples of edges $e_I \ni v$ of a graph g, with h_{IJ} the holonomy around a closed loop tangent to e_I and e_J in a vertex v. The specific form is subject to ambiguity, but the general form is characteristic and, as used later, easily extended to model systems.

There are several characteristic features implied by the construction steps, which give rise to corrections to the classical dynamics explored in quite some detail in this book. Also several fundamental properties arise in a specific form. For instance, for the resulting operator to be well-defined it is crucial that one orders the commutators quantizing inverse-triad components to the right of holonomies representing the curvature. When the commutator acts first on a cylindrical state, only vertex contributions appear. If holonomies act first, on the other hand, they create new vertices on arbitrary points of edges existing in the initial state, and on those new vertices (which are trivalent but not immediately gauge-invariant) the commutator has non-trivial action. Such an ordering would not give rise to a cylindrically consistent operator [7]. Once a consistent operator is defined, one may order it symmetrically by averaging with its adjoint. But this procedure is different from ordering the holonomies for curvature to the right [9]: for an operator creating new edges, as the Hamiltonian constraint does

---

[1] Notice that this sense of regularity does not by itself imply UV-finiteness in the usual meaning of quantum field theory. To test finiteness, one would have to compute scattering amplitudes of particle excitations on a quantum geometry state, which is difficult. It thus remains open how exactly a fundamentally finite version of loop quantum gravity could resolve non-renormalizability issues of perturbative quantum gravity.

by multiplying with holonomies, the adjoint removes edges. A symmetric ordering of such operators is different from simply reordering factors in its regularization.

The behavior of vertices under the action of the Hamiltonian constraint also plays a role for an argument of anomaly-freedom which one may make [10]. Commutators of Hamiltonian constraint operators should mimic the classical constraint algebra for the dynamics and gauge aspects to be consistent. (This theme will be discussed in more detail in the part on inhomogeneities, Chap. 10.) From the form of full Hamiltonian constraint operators one can argue that this is realized at least on diffeomorphism-invariant states, which are annihilated by the commutator. However, several difficulties with this very general statement arise: first, it has been pointed out that an extension to states that are not fully diffeomorphism invariant, and on which the off-shell constraint algebra should nevertheless be represented faithfully for anomaly-freedom, is difficult [11, 12]. Secondly, anomaly-freedom in the sense proposed in [10], even if it would work at a fundamental level, does not at all guarantee that effective geometries can be formulated with a consistent dynamics, a question which we will address later by other means. Thirdly, the same argument of anomaly-freedom does not apply in midisuperspace models such as spherically symmetric ones, see Chap. 9, whose vertex structure is different from the full one. While this failure of the argument may be due to a possible inadequacy of midisuperspace models regarding questions of fundamental anomaly-freedom, it should also be taken as a warning sign.

4.2.2 Matter Hamiltonian

Matter fields are quantized by similar means in a loop quantization, using graph states, and then coupled dynamically to the geometry by adding the matter Hamiltonian to the constraint. For a scalar field φ, the density-weighted momentum $\pi = \sqrt{|\det E|}\dot{\varphi}/N$ is a density of weight one. In the φ-representation, states, as described in Sect. 3.2.2.3 will simply be of the form already used for the gravitational field, except that each vertex now also carries a label $\nu_v \in \mathbb{R}$ describing the dependence on the scalar field $\varphi(v)$ through $\exp(i\nu_v\varphi(v))$ [13]. (The scalars take values in the Bohr compactification of the real line.) Well-defined lattice operators are given by $\widehat{\exp(i\nu_0\varphi(v))}$, for any $\nu_0 \in \mathbb{R}$, which shifts the label ν_v by ν_0. The momentum, with its density weight, has to be integrated before it can meaningfully be quantized. We introduce

$$P_R := \int_R \mathrm{d}^3x\, \pi$$

where R is a spatial region. Since we have $\{\varphi(v), P_R\} = \chi_R(v)$ in terms of the characteristic function $\chi_R(v) = 1$ if $v \in R$ and zero otherwise, a momentum operator P_R has eigenvalue $\sum_{v \in R} \hbar \nu_v$ in a state introduced above.

For the matter Hamiltonian of a scalar field φ with momentum π and potential $W(\varphi)$ we have the classical expression

$$H_\varphi[N] = \int d^3x N(x) \left(\frac{1}{2\sqrt{|\det E|}} \pi(x)^2 + \frac{E_i^a E_i^b}{2\sqrt{|\det E|}} \partial_a \varphi(x) \partial_b \varphi(x) + \sqrt{|\det E|} W(\varphi) \right)$$

containing inverse powers of the metric. It can be quantized by loop techniques [8, 14] making use of identities similar to (4.3). One first generalizes the identity to arbitrary positive powers of the volume in a Poisson bracket,

$$\{A_a^i, V_v^r\} = 4\pi \gamma G r V_v^{r-1} e_a^i \tag{4.6}$$

with V_v the volume of a small region including only the vertex v, and then combines such factors with suitable exponents r to produce a given product of triad and co-triad components. Since such identities would be used only when inverse components of densitized triads are involved and a positive power of volume must result in the Poisson bracket, the allowed range for r is $0 < r < 1$. Any such Poisson bracket will be quantized to

$$\dot{e}^a \{A_a^i, V_v^r\} \mapsto \frac{-2}{i\hbar\delta} \mathrm{tr} \left(\tau^i h_{v,e} \left[h_{v,e}^{-1}, \hat{V}_v^r \right] \right) \tag{4.7}$$

using holonomies $h_{v,e}$ in direction $\dot{e}^a$, starting at v and of parameter length δ. These general parameterized expressions are also useful for alternative quantizations of terms in the gravitational part of the constraint, where different choices of r would represent quantization ambiguities.

The exponent used for the gravitational part is $r = 1$, while the scalar Hamiltonians introduced in [8, 14] use $r = 1/2$ for the kinetic term and $r = 3/4$ for the gradient term. With

$$\varepsilon^{abc} \varepsilon_{ijk} \{A_a^i, V_v^{1/2}\} \{A_b^j, V_v^{1/2}\} \{A_c^k, V_v^{1/2}\} = (2\pi \gamma G)^3 \varepsilon^{abc} \varepsilon_{ijk} \frac{e_a^i e_b^j e_c^k}{V_v^{3/2}}$$

$$= 6(2\pi \gamma G)^3 \frac{\det(e_a^i)}{V_v^{3/2}}$$

and

$$\varepsilon^{abc} \varepsilon_{ijk} \{A_b^j, V_v^{3/4}\} \{A_c^k, V_v^{3/4}\} = (3\pi \gamma G)^2 \varepsilon^{abc} \varepsilon_{ijk} \frac{e_b^j e_c^k}{V_v^{1/2}}$$

$$= 6(3\pi \gamma G)^2 \frac{E_i^a}{V_v^{1/2}} \mathrm{sgn} \det(e_b^j)$$

one can replace all inverse powers in the scalar Hamiltonian:

$$\hat{H}_\varphi[N]\psi_g = \sum_{v\in g} N(v) \left(\frac{1}{2}\hat{P}_v^2 \left(\frac{1}{48} \sum_{e_I,e_J,e_K} \varepsilon^{IJK} \hat{B}_{v,e_I}^{(1/2)} \hat{B}_{v,e_J}^{(1/2)} \hat{B}_{v,e_K}^{(1/2)} \right)^2 \right.$$

$$\left. + \frac{1}{2} \left(\frac{1}{48} \sum_{e_I,e_J,e_K} \varepsilon^{IJK} (\Delta_{e_I}\varphi)(v) \hat{B}_{v,e_J}^{(3/4)} \hat{B}_{v,e_K}^{(3/4)} \right)^2 + \hat{V}_v W(\varphi(v)) \right) \psi_g \quad (4.8)$$

where $B_{v,e}^{(r)}$ denotes the vertex contribution at vertex v of a quantization of (4.7); $\Delta_e\varphi$ denotes the difference operator along the edge e, and $\hat{P}_v$ is the vertex contribution to the momentum operator of the scalar.

In this way, inverse-triad corrections arise in matter Hamiltonians as well as in the gravitational part of the constraint, affecting the dynamics. What kind of effective action this might correspond to and whether these corrections can be consistent (not spoiling general covariance) will be discussed later in Chap. 10 about inhomogeneous perturbations. For now, we point out the relationship of full inverse-triad expressions based on (4.7) with the corresponding behavior in isotropic models. In isotropic models, we have seen that it is important to use the local patch volume V_v (as a power of the elementary fluxes F_{ℓ_0}) in order to ensure the correct scaling dependence. In (4.7), or the Hamiltonians (4.5) and (4.8), we could use V_v as well as the total volume V or that of any other region: volume contributions from vertices not lying on the edge used for holonomies in the commutator drop out, anyway. In homogeneous models, on the other hand, all vertices are equivalent and no such difference arises. It is then important to ensure that only local vertex contributions feature in the homogeneous expressions, which as seen in [15] has several other important implications regarding consistency.

4.2.3 Problem of Dynamics

Handling the dynamics is the key problem of any approach to quantum gravity. It splits into quite different but closely related subissues, the consistent construction of the dynamics on one hand, and the evaluation on the other. In isotropic quantum cosmology, the construction simplifies considerably and some of the key problems, such as that of anomalies, trivialize. One can thus focus on developing methods for the evaluation of quantum gravitational dynamics, which in the full setting has gained only preliminary insights. Important issues of the evaluation, which we will see in detail, are the role of lattice refinement, quantum back-reaction, and the implications of changing states. Later constructions in models including inhomogeneity will allow us to discuss also quantum space–time structure.

4.3 Isotropic Universes

A reduction of the Hamiltonian constraint from the full theory to symmetric models at the quantum level is difficult, but there is information about the reduction of states and spatial quantum geometry, as discussed in detail in Sects. 8.2.5, 9.1.3, and 10.1.2. This formalism allows us to derive the basic quantum geometrical principles for isotropic loop quantum cosmology and then construct a Hamiltonian constraint operator along the lines used in the full theory. Since general spectral properties of basic operators, holonomies and fluxes, are preserved in this reduction, one can expect to capture important features of full loop quantum gravity at least at a qualitative level. Specific results seen in models can then be used to focus further derivations in the full theory.

4.3.1 Symmetry

Different procedures exist and are still being elaborated to reduce states from the full stage to a reduced setting [16–22]. A natural definition of a symmetric state is one that, as a functional on the space of connections, is supported only on the subset of invariant connections [16]. By this procedure, one can directly reduce any given state by evaluating it in the general expression of invariant connections for a symmetry type of interest, such as isotropy. The result will not be a spin network state, but rather a distribution on the full state space; it can be interpreted as being obtained by averaging an inhomogeneous state over the symmetry group. Such states are not embedded in the full Hilbert space but rather constitute truly reduced states. They comprise a minisuperspace quantization, but one obtained with input from the full theory. In contrast to Wheeler–DeWitt quantizations, crucial quantum space–time structures are still realized in the models derived in this way.

Suitable combinations of basic operators can then be found which map those states to others of this form, defining the basic representation of the model by reduction from the full holonomy-flux algebra. Difficulties arise when one tries to implement the full Hamiltonian constraint in such a way since it will not preserve the space of symmetric states, and projections to ensure this cannot easily be introduced at the distributional level. Constructions which may seem unnatural or contrived within a minisuperspace model but are required to incorporate full features thus remain required to implement the dynamics in a way mimicking the full one. No unique dynamics can result in this way, which even with a complete reduction would not be possible since the full dynamics is not unique in the first place. But with sufficiently general parameterizations one can explore the range of possibilities and extract general phenomena. As always with coarse-grained descriptions of microscopic physics, one has to suitably parameterize one's ignorance. If this is not done at a sufficiently general level, results will be spurious. No guidelines are provided in Wheeler–DeWitt quantizations, but much information is available from loop quantum gravity.

In particular, we will have to account for the crucial properties seen in the full theory. Every proposal for the full Hamiltonian constraint so far changes the individual sizes of discrete building blocks of geometry. In an isotropic setting this means that the function $v(\phi)$ in (2.16), now written in an internal time ϕ, must indeed be a function and cannot be constant in general. Although the precise form of $v(\phi)$ and the associated $\mathcal{N}(\phi)$ cannot easily be derived as mean fields of the refinement, a quantization of isotropic dynamics must take the non-constant behavior of $v(\phi)$ into account in a sufficiently general way.

4.3.2 Models

Since holonomies are the basic operators exciting and refining the geometry, they will be crucial also in isotropic models. As already seen in Sect. 3.2.3.3, holonomies refer to the number $\mathcal{N} = \mathcal{V}/\ell_0^3$ of lattice states; now in a dynamical situation, $\mathcal{N}(\phi)$ will be a function of internal time. As in the full case, holonomies initially enter the Hamiltonian constraint via curvature components of the Ashtekar–Barbero connection, which in an isotropic setting reduces essentially to the Hubble parameter. Classically, we have a Hamiltonian constraint

$$C_{\text{grav}} = -\frac{3}{8\pi G \gamma^2} c^2 \sqrt{|p|} + H_{\text{matter}}(p, \varphi, p_\varphi) = 0 \qquad (4.9)$$

written for simplicity only in the spatially flat case (and without curvature coupling in the matter Hamiltonian). Once the variables c and p are inserted in terms of the scale factor a, this constraint indeed gives rise to the Friedmann equation with energy density

$$\rho = \frac{H_{\text{matter}}}{a^3 \mathcal{V}} \qquad (4.10)$$

in terms of a matter Hamiltonian for the region of size $\mathcal{V}$.

The constraint also generates Hamiltonian equations of motion $\dot{f}(c, p) = \{f(c, p), C_{\text{grav}}[N]\}$ for a general phase-space function f. To specify the choice of time coordinate, referred to by the dot, we have introduced the lapse function N, $C_{\text{grav}}[N] = NC_{\text{grav}}$. Its main choices usually are $N = 1$ for proper time and $N = a = \sqrt{|\bar{p}|}$ for conformal time. Using the appearance of the lapse function in the canonical form of the metric, this indeed provides the correct isotropic line elements $\mathrm{d}s^2 = -\mathrm{d}t^2 + a^2 (\mathrm{d}x^2 + \mathrm{d}y^2 + \mathrm{d}z^2)$ in terms of proper time t, and $\mathrm{d}s^2 = a^2 (-\mathrm{d}\eta^2 + \mathrm{d}x^2 + \mathrm{d}y^2 + \mathrm{d}z^2)$ in terms of conformal time η. Equations of motion for the basic variables read

$$\dot{p} = 2N\sqrt{|p|}c/\gamma, \qquad (4.11)$$

implying the relationship $c = \mathcal{V}^{1/3}\gamma\dot{a}/N$ already seen, and

$$\dot{c} = \{c, C_{\text{grav}}[N]\} = -\frac{c^2}{2\gamma\sqrt{|p|}} (N + 2p\mathrm{d}N/\mathrm{d}p) + \frac{8\pi\gamma G}{3} \frac{\partial H_{\text{matter}}}{\partial p}. \qquad (4.12)$$

This equation is the Raychaudhuri equation of the isotropic model, which can be brought to the standard form after using the thermodynamical relation

$$P = -\frac{\partial E}{\partial V} = -\frac{1}{3a^2 \mathcal{V} N} \frac{\partial H_{\text{matter}}}{\partial a} \tag{4.13}$$

for pressure. At this time, one should note that both equations, (4.11) and (4.12), are derived from the Hamiltonian constraint. When quantized, the Hamiltonian most likely receives quantum corrections, such that also these two equations, even the innocent-looking (4.11), will be corrected.

The classical Hamiltonian constraint is not directly representable in terms of holonomies due to the appearance of the non-almost periodic c^2. However, almost periodicity can be checked only when the full real range of the curvature variable c is taken, which includes arbitrarily large values of curvature where one would no longer trust the classical dynamics. At this time, the fundamental requirement that operators in an isotropic loop quantization must be representable through holonomies which are almost periodic in c becomes a guiding principle for deciding how the classical dynamics must change in quantum gravity. Restricting the function $c^2/p \propto (\dot{a}/a)^2$ to just a finite range, for instance up to Planckian curvatures, still allows one to extend it to an almost-periodic function in c over the full range. Such an extension is not unique, $\sin c$ and $\frac{1}{2}\sin(2c)$ being just two examples to extend c almost-periodically, but the resulting quantization ambiguity, as always, can be parameterized and then tested.

If we choose a class of almost-periodic extensions of c^2 by $\delta^{-2}\sin^2(\delta c)$ with a parameter δ, we can directly use the basic actions of $\exp(i\delta c)$ and $\hat{p}$ to write the action of a Hamiltonian constraint operator. As expected from the presence of holonomies, it is a shift operator with terms raising and lowering the isotropic triad levels μ. Expanding a general state in the triad eigenbasis, $|\psi\rangle = \sum_\mu \psi_\mu |\mu\rangle$, we obtain the state's triad representation as the set of coefficients ψ_μ, which may also depend on other fields if different kinds of matter are present.

The construction proceeds along the following steps, which are analogous to those from the full theory but can now be performed very explicitly. We represent $\sqrt{|p|}$ via a commutator $e^{i\delta c}\left[e^{-i\delta c}, \hat{V}\right]$ to mimic the full treatment in the presence of inverse powers of the densitized triad. (In isotropic models the inverse powers cancel completely in the gravitational part of (4.9), and the treatment of inverse triads is not required at this stage. Crucial properties remain unchanged if one uses a more straightforward quantization of $\sqrt{|p|}$.) The action of the commutator on triad eigenstates follows as before for inverse-triad operators. Also the action of the holonomy term can be computed easily: $\sin^2(\delta c)$ maps $|\mu\rangle$ to $-\frac{1}{4}(|\mu + 4\delta\rangle - 2|\mu\rangle + |\mu - 4\delta\rangle)$. In the full theory, a consistent Hamiltonian constraint operator requires an ordering in which holonomies quantizing the curvature components F^i_{ab} appear to the left. Taking the same ordering in reduced models, we arrive at an operator with action

$$\hat{C}_{\text{grav}}|\mu\rangle \propto \delta^{-3}\text{sgn}(\mu)\left(V_{\mu+\delta} - V_{\mu-\delta}\right)(|\mu + 4\delta\rangle - 2|\mu\rangle + |\mu - 4\delta\rangle). \tag{4.14}$$

The constraint equation $\hat{C}_{\text{grav}}|\psi\rangle = \hat{C}_{\text{grav}}\sum_\mu \psi_\mu|\mu\rangle = \sum_\mu(\hat{C}_{\text{grav}}\psi)_\mu|\mu\rangle = 0$ then implies a difference equation $(\hat{C}_{\text{grav}}\psi)_\mu = 0$, of the form

$$\text{sgn} (\mu + 4\delta) \left(V_{\mu+5\delta} - V_{\mu+3\delta} \right) \psi_{\mu+4\delta}(\varphi) - 2\text{sgn}(\mu) \left(V_{\mu+\delta} - V_{\mu-\delta} \right) \psi_\mu(\varphi)$$

$$+ \text{sgn}(\mu - 4\delta) \left(V_{\mu-3\delta} - V_{\mu-5\delta} \right) \psi_{\mu-4\delta}(\varphi) = -\frac{4}{3}\pi G\gamma^3\delta^2\ell_{\mathrm{P}}^2 \hat{H}_{\mathrm{matter}}(\mu)\psi_\mu(\varphi)$$

$$(4.15)$$

for the wave function $\psi_\mu(\varphi)$ in the triad representation. Its coefficients, written here in terms of volume eigenvalues $V_\mu = \left(4\pi\gamma\ell_{\mathrm{P}}^2|\mu|/3\right)^{3/2}$, follow from the representation of $\sqrt{|p|}$ as commutators.

The difference equation (4.15) follows from a constraint operator in which the commutator $h\left[h^{-1}, \hat{V}\right]$ quantizing triad components is ordered entirely to the right of holonomies quantizing curvature components. This ordering is in fact suggested by the full theory as explained in Sect. 4.2.1. But once the operator has been defined, one can always order it symmetrically by replacing $\hat{C}_{\mathrm{grav}}$ with $\frac{1}{2}\left(\hat{C}_{\mathrm{grav}} + \hat{C}_{\mathrm{grav}}^\dagger\right)$. In this case coefficients of the difference equation change but the structure remains intact:

$$\frac{1}{2}\left(\text{sgn}(\mu + 4\delta)\left(V_{\mu+5\delta} - V_{\mu+3\delta}\right) + \text{sgn}(\mu)\left(V_{\mu+\delta} - V_{\mu-\delta}\right)\right)\psi_{\mu+4\delta}(\varphi)$$

$$- 2\text{sgn}(\mu)\left(V_{\mu+\delta} - V_{\mu-\delta}\right)\psi_\mu(\varphi) + \frac{1}{2}(\text{sgn}(\mu)\left(V_{\mu+\delta} - V_{\mu-\delta}\right)$$

$$+ \text{sgn}(\mu - 4\delta)\left(V_{\mu-3\delta} - V_{\mu-5\delta}\right))\psi_{\mu-4\delta}(\varphi)$$

$$= -\frac{4}{3}\pi G\gamma^3\delta^2\ell_{\mathrm{P}}^2 \hat{H}_{\mathrm{matter}}(\mu)\psi_\mu(\varphi).$$

$$(4.16)$$

There are other ways to order an operator symmetrically, such as $\sqrt{\hat{C}_{\mathrm{grav}}^\dagger \hat{C}_{\mathrm{grav}}}$, but this would be more complicated to compute. One might also be tempted to define an operator symmetrically from the outset by splitting the $\sin^2(\delta c)$ in two factors to be positioned to the two sides of the commutator: $\sin(\delta c)h\left[h^{-1}, \hat{V}\right]\sin(\delta c)$. (The commutator $h\left[h^{-1}, \hat{V}\right] = \hat{V} - h\hat{V}h^{-1}$ automatically gives rise to symmetric operators when an su(2)-trace is taken.) However, this procedure does not properly mimic constructions in the full theory where h_α for a whole loop as in (4.2) could not be split into two equal factors.

Depending on the treatment of extrinsic-curvature contributions to the Lorentzian constraint (4.1), using (4.4), a difference equation of higher order than shown here may result. Such equations have been derived in [23, 24, 25].

4.3.3 Quantum-Geometry Corrections

In addition to obvious quantum features, two classes of quantum-geometry corrections are present: holonomy and inverse-triad corrections. Both of them are based directly on properties of the holonomy-flux algebra and the accompanying discreteness of space, but there is a difference in their realization. Inverse-triad corrections result from a quantization of the classical reformulation (4.3) or (4.7) without regularization. Holonomy corrections, on the other hand, are obtained after replacing c^2 in the isotropic Hamiltonian constraint by some almost-periodic function approximated by c^2 when δc is small. After this modification of connection terms, the dynamics is no longer identical to the classical one. One may understand the higher-order terms as some of the contributions expected from higher-curvature effective actions,[2] but the modification is done at whim. It disappears in the classical limit only if $\delta \rightarrow 0$ in the classical limit, which requires one to relate the curve parameter to the Planck length. This relation, as seen in the context of lattice refinement, can be done only by reference to an underlying inhomogeneous state, not within the symmetric model; see Sect. 4.4.1.

The parameter δ is a regulator because it modifies the classical theory before it can be quantized by loop methods. Since the limit $\delta \rightarrow 0$ cannot be taken at the operator level—otherwise an operator for $\hat{c}$, not just for holonomies would exist—one must give the extra terms arising from the regularization some physical meaning. In loop quantum cosmology, this is done by interpreting them as higher powers of curvature which are not relevant for the small-curvature regime of the classical theory but are required for a well-defined quantum representation. At this stage, as always in loop quantizations, one takes a considerable risk: one assumes that the classical constraints can be modified in this way and still produce a reasonable theory. In particular covariance is at stake here, for one is adding only higher powers of the Ashtekar–Barbero connection at the Hamiltonian level, not higher powers of covariant curvature contractions at the level of an action. We will come back to the important related problem of anomalies later in this book (Chap. 10). For now, we notice that the key and still outstanding test of holonomy corrections is not to evaluate their dynamical implications in homogeneous models where they can be implemented freely, thanks to a trivialization of the anomaly problem, but rather a consistent extension to inhomogeneities.

In some models, notably isotropic ones with a free, massless scalar, it is possible to quantize the Hamiltonian constraint in exponentiated form, rather than exponentiating just the connection components [26]. One obtains an unregularized loop quantization without higher-order corrections. Such a quantization would be preferred compared to regulated ones because it would allow the anomaly-free inclusion of inhomogeneities in much simpler terms. It would also eliminate all holonomy effects. At the present stage of developments, however, the models in which such an unregulated loop quantization can be applied appear too special to rule out holonomy corrections altogether.

One can remove the regulator δ at the level of the difference equation provided one requires solutions to be smooth enough [27]. The difference equation then becomes a differential equation, essentially reproducing the Wheeler–DeWitt equation as the continuum limit. However, with the limit $\delta \rightarrow 0$ taken only at the level of

---

[2] These cannot be all contributions because higher-derivative terms of the metric do not arise by the holonomy replacement. See Chap. 13 for a general treatment of effective canonical dynamics which introduces new quantum degrees of freedom analogously to higher-derivative actions.

equations of motion for states, not in a full quantization, the resulting theory cannot be considered as fundamental.

It is sometimes suggested that the coordinate volume $\mathcal{V}$, which appears in δ but also in inverse-triad corrections, is a regulator and should be taken to infinity or the total size of a compact space. This is not correct because the classical theory simply does not depend on the value of $\mathcal{V}$. Sometimes in this context it is claimed that inverse-triad corrections disappear, just as holonomy corrections would disappear if the limit $\delta \to 0$ could be taken. Also here, the limit could only be taken at the level of equations, not at the level of operators where no inverse of the flux operator exists.

4.4 The Role of an Underlying State

So far, we have treated the parameter δ for different almost-periodic extensions as a constant. In a derivation of holonomies directly from isotropic connections in the general expression, we have $\delta = \ell_0/\mathcal{V}^{1/3}$ as seen earlier. In terms of the mean fields of lattice refinement, this can be written as $\delta = \mathcal{N}^{-1/3}$, which shows that a lattice-refinement model must be based on a parameter $\delta(\phi)$ depending on the internal time ϕ used to realize the refinement. In other words, while the size $\mathcal{V}$ is that of a constant region chosen once and for all to set up the quantization, the coordinate length ℓ_0 of curves used for holonomies can in general not be considered as a constant. Physically, the elementary discrete geometry must be refined during expansion to avoid that the discreteness scale is blown up to macroscopic sizes by the expanding universe. Such refinement indeed happens by a fundamental Hamiltonian constraint operator, which may generate new vertices and always changes the elementary sizes $v(\phi)$ of discrete building blocks.

In isotropic models, it is often convenient to use the scale factor a as internal time, such that we will have a function $\delta(a)$. In the mean-field picture, this will make the step-size δ of the difference equation dependent on the label μ: the equation is no longer one of constant step-size. We will later discuss how such equations can be dealt with; for now we are only interested in their general form.

4.4.1 Refinement Models

For specific difference equations based on refinement models, we would need $\mathcal{N}(a)$ or $v(a)$, related by $\mathcal{N}v = \mathcal{V}a^3$, as it arises from the genuine full dynamics in a suitable state. Lacking a derivation, we parameterize $\mathcal{N}(a) = \mathcal{N}_0 a^{-6x}$ such that $\delta(a) \propto a^{2x}$ and $v(a) = \mathcal{V}a^{3(1+2x)}/\mathcal{N}_0$. This is only an ansatz to probe different behaviors; in general $\mathcal{N}(a)$ need not be of power-law form for all a. But power-laws provide insights into the possible behaviors and can, at least for small ranges for a to vary, be used as approximations of general functions. Looking at different values of x, this procedure will show the behavior in different phases of refinement. For

$x > 0$, $\mathcal{N}$ is decreasing with expansion; for $x = 0$ it is constant and the discrete volume $v \propto a^3$ is proportional to the total volume. For $0 > x > -1/2$, both v and $\mathcal{N}$ are increasing, which makes it the regular range expected from the full behavior of the discrete dynamics.

> Another, independent argument for $x < 0$ is that $\delta(a)$ then depends on a via a negative power, whose dimension can be compensated for only by a positive power of a parameter with dimension length. If the Planck length is used as this parameter, $\delta \to 0$ in the classical limit, removing holonomy corrections.

If $x = -1/2$, the local sizes v remain constant while the number $\mathcal{N}$ of sites increases proportionally to volume. As already seen, this behavior is unlikely from the point of view of the full theory because local vertex contributions to volume are always changed by the Hamiltonian constraint. It could at best be a coarse-grained, averaged description in a special case of the full dynamics. If $x < -1/2$, finally, the discrete sizes v must decrease when the universe expands. Since both $\mathcal{N}$ and v are bounded from below, the generic range where power-law behavior can be realized for long times is $-1/2 < x < 0$. Near the upper bound, however, the discrete volume contributions v are in danger of increasing too fast, or $\mathcal{N}$ is not increasing fast enough and refinement is too weak. In this regime, phenomenological restrictions on x can easily be found [28–32].

More generally, $\mathcal{N}$ may be a generic function, which one can think of as being composed of different power-law phases each parameterized by its own x. At this place we can see the real strength of using the Bohr compactification of the real line as kinematical quantum configuration space: for a single power-law form, the dynamical equation would always be periodizable by using $|p|^{1-x}$ instead of p as the state label, and $|p|^x c$ as the canonically conjugate curvature parameter in holonomies. Since the dynamics splits the range of all values into distinct sectors connecting only countably many values by the difference equation, we could from the outset have worked with states periodic in $\ell_0 \tilde{c} = c/\mathcal{N}^{1/3} \propto a^{2x} c$. Irrespective of whether we assume periodicity when choosing the dynamics or when formulating the kinematical states, such an assumption will always seem ad-hoc. If different power-laws or a non-power law function are involved, however, no periodicity occurs at all. This shows the real strength of the Bohr compactification by providing a repository for all possible refinement cases. (In a later chapter we will see that anisotropic models also make use of the full Bohr compactification without implicitly using periodic sectors, even if they are based on power-law behaviors of $\mathcal{N}$, crucially leading to non-equidistant difference equations.)

> At this stage, it is worth commenting on differences between the mean-field picture and a pure minisuperspace quantization. We have first derived the constraint operator for a constant δ, and are then putting in the μ-dependence to ensure that full properties are reliably captured. One can also take the point of view that $\delta(p)$ is used from the outset, having to quantize a more complicated phase-space function $\exp(i\delta(p)c)$ instead of ordinary holonomies. For simple choices of $\delta(p)$ of power-law form, such quantizations can be performed: one employs a canonical transformation such that the product $\delta(p)c$ which appears in the exponential is now one of the basic variables, and considers "holonomies" to be written in this variable. If

$\delta(p) \propto |p|^x$ is of power-law form, such a transformation can easily be done with $|p|^{1-x}$ being the new momentum conjugate to $|p|^x c$. For $x = -1/2$, for instance, this momentum would be the volume $V = |p|^{3/2}$. Pretending that $c/\sqrt{|p|}$ is to be used in almost-periodic holonomies would thus lead to a difference equation which is equidistant in volume values rather than densitized-triad values. For large μ, this procedure is equivalent to inserting $\delta(\mu) \propto |\mu|^{-1/2}$ directly in the basic difference equation (4.16), as one can see by substitution; for smaller μ, one can think of the equation obtained after a canonical transformation as providing one specific factor ordering of $\exp(i\delta(p)c)$ which unlike the basic $\exp(i\delta c)$ with constant δ is not defined uniquely. As one can see, however, there are difficulties in particular around $\mu = 0$ due to the absolute value used in fractional powers and the appearance of inverses of μ. Physically, this is not surprising since the μ-dependence of δ arises from lattice refinement, and around $\mu = 0$ not many lattice sites are excited. The lattice is thus very irregular, and one cannot expect to describe the behavior well by using simple power-laws for $\mathcal{N}(p)$. Keeping a general function $\delta(\mu)$ for basic properties of the dynamics, and specializing to simple forms only to analyze concrete cases at larger μ, is then the best way to shed light on the discrete dynamics.

Refinement is also relevant for inverse-triad operators and the corrections they imply, which contain the basic holonomies and fluxes (or the volume operator). In the commutator (4.7) used crucially in their definition, the local vertex volume V_v features, corresponding to the plaquette sizes in a refinement model.[3] As already discussed in Sect. 3.2.3.5, it is then not $p = \mathcal{V}^{2/3} a^2$ (the total box size) which features in commutators or correction functions but the plaquette size $\mathcal{N}^{-2/3} p = \ell_0^2 a^2$. This change makes the flux contribution, appearing as the argument of correction functions, smaller, and the correction correspondingly larger. For instance for $x = -1/2$ in a power-law form $\mathcal{N}(p)^{-1/3} = \delta(p) \propto |p|^x$, the plaquette size $\ell_0^2 a^2 = \mathcal{N}^{-2/3} |p|$ is constant. Large p do not make inverse-triad corrections shrink in this case, as it would happen without lattice refinement. Note that for a correct treatment of inverse-triad corrections with refinement, implemented by a phase-space dependent $\mathcal{N}(a)$, the refinement function is not to be inserted in the commutator before its computation, which would give the wrong result since the a-dependence of the factor would change the classical Poisson bracket. Refinement and the a-dependence is a mean-field effect, and thus to be inserted in the final expressions of an effective or coarse-grained theory. Since minisuperspace models by definition constitute coarse-grained descriptions of the full theory, mean-field treatments cannot be avoided altogether.

4.4.2 Interpretations

Different viewpoints have emerged in the development of loop quantum cosmology, all rooted in the original constructions of [23, 33–36]:

---

[3] In such commutators in the full theory, a single V_v gives the same contribution as the volume operator of all of space: contributions from vertices not lying on the edges used for the holonomy in the commutator cancel. But this observation does not change the fact that inverse-triad operators receive contributions only from local vertex contributions of the volume operator. In reduced models, homogeneity implies that all vertex contributions must contribute equally; one can properly capture the full behavior only by using single vertex or plaquette contributions from the outset.

Pure minisuperspace quantization[4]: One may argue that the curvature relation (4.2), $\delta^2 s_1^a s_2^b F_{ab}^i \tau_i \sim h_\alpha - 1$ with $h_\alpha \sim \sin^2(\delta c)$ in isotropic models, used crucially in quantizing the Hamiltonian constraint, should be evaluated by fixing the geometric area $A = a^2 \delta^2$ as an ambiguity parameter rather than the coordinate area δ^2. If A indeed takes a fixed value, one obtains $\delta \propto 1/\sqrt{|p|}$, corresponding to the refinement scheme $x = -1/2$. In particular, by fixing A one trivially ensures that local patches of the underlying lattice state are constant, and thus $\mathcal{N} \propto \mathcal{V}$. (Most such realizations make use of the further ad-hoc assumption that this fixed value of A should be the minimal or some close-to-minimal non-zero eigenvalue of the full area spectrum. This condition is ad-hoc because it brings in the full area spectrum—the area operator in a reduced model does not have a non-zero minimal eigenvalue—and in that it crucially refers to the area operator at a place where it is not made use of in full constructions. Moreover, the resulting size of holonomy corrections is incompatible with inverse-triad corrections [15], implying an ℓ_0 that gives rise to large inverse-triad corrections even at large volume.)

The main (and perhaps only) justification of such a viewpoint toward the dynamics of loop quantum cosmology, compared to a constant δ, is a posteriori: A behavior of $\delta \propto 1/\sqrt{|p|}$ in holonomies appearing in the dynamics provides an additional suppression of higher powers of curvature at large volume, thus making it easier to comply with semiclassical and near-continuum behavior (provided inverse-triad corrections are ignored, which is often done in this context). Holonomy corrections depending on c^2/p rather than just c^2 are better behaved at large volume especially if there is a positive cosmological constant (in which case $c \propto \sqrt{|p|}$ at large volume is growing) or in the presence of intrinsic curvature. Moreover, holonomy corrections automatically depend on the scale-independent Hubble parameter $\dot{a}/a$. The pure minisuperspace viewpoint, without additional mean-field effects, thus presents a valid refinement scheme regarding curvature, but there are no strict arguments why it should be the preferred one. In fact, this viewpoint fails in Kantowski–Sachs models for the Schwarzschild interior, where a refinement with $\mathcal{N} \propto \mathcal{V}$ is not viable near the horizon; see Sect. 8.3. The scheme is also inconsistent with slow-roll inflation in the presence of inhomogeneities [39].

The starting point itself of the pure minisuperspace view is not fully convincing: curvature components F_{ab}^i are coordinate dependent, so why should one not refer to coordinate areas in their regularization? A reference to coordinates is indeed what happens in the full regularization, where using geometrical areas such as A would not be possible. Finally, the improved view in holonomies goes only half-way toward a consistent regularization. It uses refinement ideas in holonomies, where large-volume effects are potentially most problematic, but not for fluxes. Accordingly, minisuperspace quantizations typically produce wrong inverse-triad corrections by

---

[4] These models were initially introduced under the name "improved" quantization [37, 38], indicating advantages in certain regimes of low curvature and large volume. However, the modifications introduced in these models turned out to be rather ad-hoc. (To appreciate this realization, the models are sometimes called "improvised.") By now, what goes by the name "improved dynamics" is under strong pressure from different types of inconsistencies. The improved dynamics is itself to be improved, giving the name a rather misleading connotation.

overlooking the refinement for the volume operator appearing in commutators (see the Second Principle). In the pure minisuperspace context, it is then often stated erroneously that inverse-triad corrections play no role. By insisting that all quantum effects be realized in reduced operators, rather than partially in an underlying state, this view remains stuck in a mere minisuperspace picture, ignoring lessons from the full theory.

One undeniable advantage of the improved dynamics is that it removes ambiguities, if only by ad-hoc choices, and thus tends to provide very specific schemes and equations. For this reason, this special kind of the dynamics in loop quantum cosmology is often explored.

Refined loop quantum cosmology: A general and consistent viewpoint realizes that refinement is mainly implemented by properties of an underlying state: it is a mean-field effect that appears in minisuperspace models but crucially rests on behavior in the full theory. The incompleteness of the full theory and the complicated nature of symmetry reduction at the quantum level make it difficult to derive refinement models, but characteristic properties can be implemented by means of sufficiently general parameterizations; see Sect. 9.1.6.3 for a model that illustrates how refinement schemes could be derived. The state dependence will further be illustrated below. As the main justification for this refined view we state that it is required in general models such as black-hole interiors, and that only such a treatment can produce a consistent form of inverse-triad corrections.

4.4.3 Refinement from Reduction

Refinement arises in reduced models as a consequence of properties of underlying states used in the reduction. One may wonder why state-dependence should arise at all, and why effects cannot be captured completely in a reduction of operators. Properties of solutions to constraint equations certainly depend on the operators used, but the definition of reduced operators is normally not expected to depend on properties of *solutions* to the full operator. And refinement, ultimately a consequence of the generation of new vertices or lattice sites by full Hamiltonian constraint operators, is a property of solutions to the full constraint equation.

If quantum cosmological models were fundamental, they should in fact be defined fully in terms of their operator algebra, and no free function such as $\mathcal{N}(p)$ or extra parameters such as $\mathcal{N}_0$ or x could appear. But quantum cosmological models are not fundamental; they provide an average description of full quantum gravity. In the averaging, additional features not captured fully by the operator algebra arise as mean fields.

A further crucial property especially in loop quantum gravity is the fact that regularizations of the full Hamiltonian constraint depend on the state the operator is going to act on. We refer to the graph when defining the loops used to rewrite the curvature components of the constraint. In the end, cylindrical consistency allows us to extend the graph-dependent definition to that of a consistent operator on the full Hilbert space, but the state dependence of it remains. This state dependence can then reappear in reduced models in a way whose explicit details may be rather hidden by the reduction procedure but can still be parameterized. Such a reduction is complicated to do for the actual problem of interest, the Hamiltonian constraint operator of loop quantum gravity. But the resulting state dependence can easily be seen in an example:

Example 4.1 Consider a classical phase space with three canonical degrees of freedom $(q_1, p_1; q_2, p_2; q_3, p_3)$, subject to the first-class constraints $D = p_2$, $C = p_3$. Instead of performing a symmetry reduction, we are going to look at consequences of different implementations of gauge fixing the "diffeomorphism" constraint D, thus reducing the "Hamiltonian" constraint C. As the classical gauge-fixing condition we choose $F = q_2 - 1 = 0$.

Let us propose the state-dependent quantization (with regularization)

$$\hat{C}_{(n)} = \hat{p}_3 + 1 - \frac{\hat{q}_2^2 + \hat{p}_2^2}{2n + 1}$$

on the subspace of the Hilbert space formed by states $\psi_1 \otimes |n\rangle \otimes \psi_3$ with $(\hat{q}_2^2 + \hat{p}_2^2) |n\rangle = (2n + 1)|n\rangle$. By using any integer $n \geq 0$, a constraint operator with the correct classical limit is defined on the full Hilbert space. Note that the state-dependent regularization of $\hat{C}_{(n)}$ produces an anomalous quantization with $\left[\hat{D}, \hat{C}_{(n)}\right] = \hat{q}_2/(n + 1/2) \neq 0$, resembling what may happen with the diffeomorphism and Hamiltonian constraint in loop quantum gravity.

If we simply implement the gauge-fixing conditions for D in a reduced setting, n becomes inaccessible. Just as we insert symmetric forms for the phase-space variables in a minisuperspace quantization, we would require $p_2 = 0$ and $q_2 = 1$ to implement $D = 0$ and $F = 0$ in a constraint operator $\hat{C}_{\text{red}}$ for a reduced model, leaving us with the n-dependent

$$\hat{C}_{\text{red},(n)} = \hat{p}_3 + 1 - \frac{1}{2n + 1}.$$

We can recover n and thus uniquely fix the reduced Hamiltonian only if we know the state $|n\rangle$ that gives rise to the gauge fixing at the full quantum level, corresponding to the state of non-symmetric degrees of freedom in a minisuperspace quantization (the underlying state). At the state level, $p_2 = 0$ and $q_2 = 1$ is possible only for $n = 0$, which fixes n and brings the reduced Hamiltonian to the simple form $\hat{C} = \hat{p}_3$.

In this example, no trace of the state-dependent regularization is left if the reduction is done completely because no regularization of the operator was required in the first place; we just picked an unnecessarily contrived operator. For loop quantum gravity, the initial regularization and state dependence of the full Hamiltonian constraint is crucial, and so it should be expected to leave a trace in reduced models as well. It is also easy to see that the reduction parameters, as long as they are not eliminated by a complete reduction, may depend on the reduced or physical degrees of freedom, just as a general refinement function $\mathcal{N}$ does. In the example here, we may regularize C to

$$\hat{C}_{(n)} = \hat{p}_3 + \left(1 - \frac{\hat{q}_2^2 + \hat{p}_2^2}{2n + 1}\right)\hat{p}_1$$

and obtain a reduced Hamiltonian

$$\hat{C}_{\text{red},(n)} = \hat{p}_3 + \left(1 - \frac{1}{2n + 1}\right)\hat{p}_1$$

with a state-dependent correction depending on the physical observable $\hat{p}_1$.

In the preceding example, a full reduction, computing the exact state that implements the reduction condition, allowed us to arrive at a unique reduced Hamiltonian without any remnant of the state-dependent regularization. A further property that is realized in loop quantum gravity (the full diffeomorphism constraint relating symmetric and non-symmetric degrees of freedom) prevents this from happening, as a slight variation of the example shows:

Example 4.2 Now, take the constraints $D = p_2 + p_3$ with $C = p_3$ and the previous gauge-fixing condition $F = q_2 - 1 = 0$. Implementing $D = 0$ and $F = 0$ in a reduction with the same $\hat{C}_{(n)}$ as before then leads to the reduced Hamiltonian

$$\hat{C}_{\text{red},(n)} = \hat{p}_3 + 1 - \frac{1 + \hat{p}_3^2}{2n + 1}$$

and we are guaranteed a remnant of the state dependence in any reduced model: any value for n that may be obtained from a complete reduction can only depend on the (q_2, p_2)-state, and it cannot cancel the extra p_3 appearing in the reduced Hamiltonian.

As this set of examples shows, gauge-fixing the diffeomorphism constraint (for which symmetry reduction is one example), the regularization of the Hamiltonian constraint with its state dependence, and the anomaly issue all matter for a complete understanding of refinement. Currently, these issues are only poorly understood, and the best one can do is a sufficiently general parameterization of refinement options.

4.5 Basic Singularity Removal: Quantum Hyperbolicity

Given the difference equation in any form of lattice refinement, we use it to understand quantum evolution in the small-volume regime, near the singularity $\mu = 0$. First, in contrast to a Wheeler–DeWitt quantization based on metric variables and the scale factor a, the singular state is now in the interior of the configuration space rather than at a boundary: the freedom of having two possible orientations of the triad makes μ take values of both signs. This fact opens up a direct way of testing whether there is a singularity in the sense of breakdown of evolution, since the difference equation for a state in the triad representation provides a natural evolution scheme via its recurrence. Irrespective of what internal time one would choose to describe evolution, that is whether it is indeed the triad value μ or something else entirely, we can test whether physical wave functions satisfying the constraint equation remain well-defined in a neighborhood of the singularity [40, 41].

Going through the recurrence, starting with suitable initial values on one side of positive μ, say, it is indeed easy to see, using (4.15) that $\mu = 0$ does not pose any obstruction. The matter Hamiltonian remains regular at $\mu = 0$, just as it does in the full theory at degenerate triads. Nevertheless, it could happen that the recurrence stops at $\mu = 0$ if coefficients of the difference equation vanish, preventing one from determining the next values of the wave function. For the form written here, coefficients can indeed vanish just at the classical singularity: for backward evolution toward smaller μ, the relevant term is $V_{\mu-3\delta(\mu)} - V_{\mu-5\delta(\mu)}$, which vanishes if $\mu = 4\delta(\mu)$. It multiplies ψ_0, the value of the wave function at the classical singularity, which thus remains undetermined by initial data. This would be a disaster if that value would be required to continue the recurrence: the wave function at negative μ, "at the other side of the singularity," would not be determined by initial data at positive μ. At vanishing volume, evolution would still break down, and we would have a singularity as classically.

It turns out, however, that ψ_0 is not needed for the further recurrence: it completely decouples from the rest of the wave function. Whenever it would show up in the difference equation, it is multiplied with a coefficient that vanishes at this rung of the ladder. (For the symmetric ordering of the constraint giving rise to the difference equation (4.16), we move through $\mu = 0$ without decoupling. Other orderings lead to singularities since ψ_0 or other values of the wave function near $\mu = 0$ could not be determined but would be needed for further recurrence.) In this way, the recurrence can be undertaken through the place of the classical singularity; there is no singularity anymore and our evolving quantum space–time instead extends to a new region not seen classically. "Before" the big bang, in this internal-time picture and now back in forward evolution, we have a contracting universe since $V(\mu)$ decreases with increasing μ for $\mu < 0$. The change in sign of μ means that the orientation of the triad reverses in the big bang transition where the universe, as it were, turns its inside out.

Instead of determining the value of ψ_0, the difference equation taken at $\mu = 4\delta(\mu)$ provides a linear equation for two values of the wave function that would be used in the recurrence for ψ_0 if the ψ_0-coefficient in the difference equation would not vanish. By the preceding recurrence steps, the linear equation can be traced back to one for the initial values chosen at some large μ, providing a dynamical initial condition [42, 43].

All this happens deeply in the quantum regime and different effects are at play. The difference-equation nature, crucial for the recurrence, relies on the use of holonomies. Comparing this to the classical expression of the constraint, we are considering higher-order corrections to the classically quadratic term in c, as it is required for almost-periodic expressions. We also have crucially made use of the regular behavior of inverse-triad operators, especially in the matter Hamiltonian (which then annihilates $|0\rangle$). We have not specified a refinement scheme but just used the property that around $\mu = 0$, $\delta(\mu)$ is regular. Since singularity-traversal requires only a few μ-steps, possible changes in $\delta(\mu)$ can be ignored. The argument is independent of the refinement scheme. Finally, since we are considering the evolution of an entire wave function, there are implicit quantum back-reaction effects of the whole state on its expectation values, implying further deviations from classical evolution. All these effects are at play in the highly quantum phase around $\mu = 0$, which makes the development of an intuitive picture difficult. There are, however, simple special cases where one of the effects is dominant, or where the different effects can be separated from one another. Such models provide more intuitive pictures of at least some aspects of singularity resolution in loop quantum cosmology, and they allow one to develop effective descriptions, the topic of the next part.

References

1. Unruh, W.: Time, Gravity, and Quantum Mechanics pp. 23–94. Cambridge University Press, Cambridge (1997)
2. Weiss, N.: Phys. Rev. D **32**, 3228 (1985)
3. Jacobson, T.: (2000). hep-th/0001085
4. Bojowald, M., Singh, P., Skirzewski, A.: Phys. Rev. D **70**, 124022 (2004). gr-qc/0408094
5. Rovelli, C., Smolin, L.: Phys. Rev. Lett. **72**, 446 (1994). gr-qc/9308002
6. Giesel, K., Thiemann, T.: Class Quantum Grav. **24**, 2465 (2007). gr-qc/0607099
7. Thiemann, T.: Class Quantum Grav. **15**, 839 (1998). gr-qc/9606089
8. Thiemann, T.: Class Quantum Grav. **15**, 1281 (1998). gr-qc/9705019
9. Thiemann, T.: Class Quantum Grav. **15**, 875 (1998). gr-qc/9606090
10. Thiemann, T.: Phys. Lett. B **380**, 257 (1996). gr-qc/9606088
11. Lewandowski, J., Marolf, D.: Int. J Mod. Phys. D **7**, 299 (1998). gr-qc/9710016
12. Gambini, R., Lewandowski, J., Marolf, D., Pullin, J.: Int. J Mod. Phys. D **7**, 97 (1998). gr-qc/9710018
13. Ashtekar, A., Lewandowski, J., Sahlmann, H.: Class Quantum Grav. **20**, L11 (2003). gr-qc/0211012
14. Sahlmann, H., Thiemann, T.: Class Quantum Grav. **23**, 867 (2006). gr-qc/0207030
15. Bojowald, M.: Class Quantum Grav. **26**, 075020 (2009). arXiv:0811.4129
16. Bojowald, M., Kastrup, H.A.: Class Quantum Grav. **17**, 3009 (2000). hep-th/9907042
17. Engle, J.: Class Quantum Grav. **23**, 2861 (2006). gr-qc/0511107
18. Engle, J.: Class Quantum Grav. **24**, 5777 (2007). gr-qc/0701132
19. Koslowski, T.: (2006). gr-qc/0612138
20. Koslowski, T.: (2007). arXiv:0711.1098
21. Brunnemann, J., Fleischhack, C.: (2007). arXiv:0709.1621
22. Brunnemann, J., Koslowski, T.A. arXiv:1012.0053
23. Bojowald, M.: Class Quantum Grav. **19**, 2717 (2002). gr-qc/0202077
24. Hinterleitner, F., Major, S.: Phys. Rev. D **68**, 124023 (2003). gr-qc/0309035
25. Henriques, A.B.: Gen. Rel. Grav. **38**, 1645 (2006). gr-qc/0601134
26. Varadarajan, M.: Class Quantum Grav. **26**, 085006 (2009). arXiv:0812.0272
27. Bojowald, M.: Class Quantum Grav. **18**, L109 (2001). gr-qc/0105113
28. Nelson, W., Sakellariadou, M.: Phys. Rev. D **76**, 104003 (2007). arXiv:0707.0588

29. Nelson, W., Sakellariadou, M.: Phys. Rev. D **76**, 044015 (2007). arXiv:0706.0179
30. Grain, J., Barrau, A., Gorecki, A.: Phys. Rev. D **79**, 084015 (2009). arXiv:0902.3605
31. Grain, J., Cailleteau, T., Barrau, A., Gorecki, A.: Phys. Rev. D **81**, 024040 (2010). arXiv:0910.2892
32. Shimano, M., Harada, T.: Phys. Rev. D **80**, 063538 (2009). arXiv:0909.0334
33. Bojowald, M.: Class Quantum Grav. **17**, 1489 (2000). gr-qc/9910103
34. Bojowald, M.: Class Quantum Grav. **17**, 1509 (2000). gr-qc/9910104
35. Bojowald, M.: Class Quantum Grav. **18**, 1055 (2001). gr-qc/0008052
36. Bojowald, M.: Class Quantum Grav. **18**, 1071 (2001). gr-qc/0008053
37. Ashtekar, A., Pawlowski, T., Singh, P.: Phys. Rev. D **74**, 084003 (2006). gr-qc/0607039
38. Ashtekar, A., Wilson-Ewing, E.: Phys. Rev. D **79**, 083535 (2009). arXiv:0903.3397
39. Bojowald, M., Calcagni, G.: JCAP **1103**, 032 (2011). arXiv:1011.2779
40. Bojowald, M.: Phys. Rev. Lett. **86**, 5227 (2001). gr-qc/0102069
41. Bojowald, M.: Proceedings of the XIIth Brazilian School on Cosmology and Gravitation. In: AIP Conference Proceedings, vol. 910, pp. 294–333 (2007). gr-qc/0702144
42. Bojowald, M.: Phys. Rev. Lett. **87**, 121301 (2001). gr-qc/0104072
43. Bojowald, M.: Gen. Rel. Grav. **35**, 1877 (2003). gr-qc/0305069

Part II
Effective Descriptions

Effective descriptions provide powerful tools for analyzing quantum systems as well as for understanding them in intuitive terms. Well-known examples are effective potentials in condensed-matter physics, or the low-energy effective action in particle physics. The general setting and applicability of effective descriptions is, however, much wider. They can be used for conceptual questions, as a shortcut to derive the behavior of certain quantum states, and for numerical purposes. Effective descriptions are, moreover, much more amenable to certain approximation schemes, such as semiclassical ones, than direct quantum states. Even issues of constrained systems, including properties of physical states, anomalies or the problem of time, can be addressed at this level.

It is not always easy to derive effective formulations for a regime of interest in a given quantum system. But once models have been identified in which they can be computed, more general properties can be added on by perturbation theory. This formulation has been achieved for quantum cosmology, as described in this part. Chapter 13 will deal with general issues of effective descriptions such as the relation to the low-energy effective action, more general types of effective actions, and physical Hilbert-space issues for constrained systems.

Part II
Effective Descriptions

Chapter 5
Effective Equations

With (4.15), and perhaps a lattice-refining $\delta(\mu)$, we have derived a fundamental difference equation that governs the dynamical behavior of a wave function for an isotropic evolving quantum geometry. The equation's structure shows that, at a basic level, there is no big bang singularity in this model since dynamics does not stop where the classical singularity would be. At this level, however, it is difficult to isolate specific mechanisms of singularity prevention active in general, or to describe the kind of (quantum) geometry that replaces the classical singular space-time: Conceptually, one would have to face the thorny issues of interpreting the wave function appropriately or, more precisely, of determining observables of interest and computing their expectation values by a suitable inner product to normalize the wave function. (See Sect. 7.3 and Chap. 12 for more on these topics.) Technically, explicit solutions in all but the simplest models are hard to come by; and even if some could be found (for instance by numerical methods), all the quantum geometry and quantum dynamics effects we have seen—holonomies, inverse-triad corrections as well as ubiquitous quantum back-reaction—usually occur at once, blurring the overall picture. While one can say that the big bang, according to the difference equation, is a transition from collapse to expansion, with a role played by reversal of orientation, specific statements about the universe before the big bang, such as the sizes it reached, its semiclassical properties or even the notion of time, require more details of the solutions to be known.

5.1 Quantum Effects in Separation

As we have seen in the constructions of loop quantum gravity, the main classes of quantum corrections in this setting are those due to holonomies, due to inverse-triad operators, and due to quantum back-reaction. None of them is, a priori, to be underestimated in its direct influence on the wave function and in changes to the classical physics it implies. But quantum physics, anyway, is different from classical physics already in its basic mathematical formulation. Thus, it is not immediately

M. Bojowald, *Quantum Cosmology*, Lecture Notes in Physics 835,
DOI: 10.1007/978-1-4419-8276-6_5, © Springer Science+Business Media, LLC 2011

clear what implications exactly a certain change in a Hamiltonian operator may entail. Quantum features are brought out and analyzed much more clearly in effective equations: equations of motion of the classical type, which nevertheless incorporate quantum corrections such as effective potentials or new interactions and degrees of freedom. Such equations directly arise if one can manage to formulate quantum dynamics for expectation values of operators of interest in certain states, rather than for the states or wave functions. As such, they deal with the full quantum dynamics, but organized in a way different from the standard one. A derivation, however, is complicated since expectation values are not the only degrees of freedom in quantum physics; they usually couple to and interact with all kinds of moments of a state such as its fluctuations or correlations.

An expectation value $\langle \hat{O}^n \rangle$ of some integer power ($n > 1$) of an operator $\hat{O}$ cannot, in general, be reduced to expectation values of lower powers. There are infinitely many independent moments measuring the differences $\langle (\hat{O} - \langle \hat{O} \rangle)^n \rangle$, whose dynamics in general can be formulated only by a partial differential equation such as the Schrödinger or Wheeler–DeWitt equation rather than a finite set of ordinary differential equations as in classical mechanics or homogeneous cosmology. Expectation-value dynamics, subject to equations of motion on an infinite-dimensional space, then cannot always be useful, but there are solvable models in which expectation values decouple from the moments, and there are semiclassical or other regimes in which the magnitude of higher-order moments is systematically suppressed by factors of $\hbar$ (or rather, as we will see, $\sqrt{\hbar}$). In those cases, much information can be gained by analyzing equations for expectation values, or possibly those equations in combination with additional ones for a finite number of the moments. This method provides means to study at least the approximate behavior in certain classes of states, incorporating all quantum corrections.

The strategy for effective descriptions based on expectation-value dynamics will thus begin with an attempt to find the simplest model system of decoupled dynamics, which might not be realistic but opens up a route to more general situations. If there is such a highly controlled system, one can proceed by perturbation theory to extend it to cases closer to what one is interested in. Such a strategy is analogous to what is done in much of particle physics: start with free quantum field theories which can be quantized completely and non-perturbatively, and add to them the effects of interactions by perturbation theory. An analogous procedure for systems with finitely many classical degrees of freedom is applied in this chapter to quantum cosmology, followed later in Chap. 13 by a more general discussion of its parallel nature to what is done in quantum field theories.

For a tractable model, we should first reduce the number of relevant corrections to be taken into account. While we cannot arbitrarily change the quantum dynamics by removing some quantum corrections, there are often parameter choices (such as matter contents, quantization ambiguities or the class of states considered) which render some effects subdominant. In particular, we have different ranges of the geometrical variables—in isotropic settings chiefly the curvature or the Hubble parameter $\dot{a}/a$ together with the scale factor itself—for holonomy corrections on one hand and inverse-triad corrections on the other to enter the game:

- Holonomy corrections are weak if the argument of the holonomy's exponential function is small compared to one, $\ell_0 \dot{a} \ll 1$. Formulated for the Hubble parameter and using $\mathcal{N} = \mathcal{V}/\ell_0^3$ with (2.16), we obtain

$$\frac{\dot{a}}{a} \ll \mathcal{H}_* := \frac{1}{\ell_0 a} = \left(\frac{\mathcal{N}(a)}{\mathcal{V}a^3}\right)^{1/3} = \frac{1}{v(a)^{1/3}}. \tag{5.1}$$

If this inequality holds for $\mathcal{H} = \dot{a}/a$, holonomy corrections are very small. In other words, and rather intuitively, the Hubble distance should be much larger than the discreteness scale $v(a)^{1/3}$.

- Inverse-triad corrections are weak if elementary fluxes are large compared to Planckian sizes, $\ell_0^2|\tilde{p}| \gg \ell_P^2$; see (3.59) and Fig. 3.3. Thus,

$$a \gg a_* := \frac{\ell_P}{\ell_0} = \frac{\ell_P \mathcal{N}(a)^{1/3}}{\mathcal{V}^{1/3}} = \frac{a}{v(a)^{1/3}}\ell_P \tag{5.2}$$

or simply $v(a) \gg \ell_P^3$ ensures small inverse-triad corrections.

Quantum-geometry corrections of either type are small if $\ell_P \ll v(a)^{1/3} \ll 1/\mathcal{H}$. In both cases, it is thus the elementary size v of discrete building blocks that, in comparison with classical phase-space values or fundamental constants, determines the relevance of quantum-geometry corrections. As one can directly see, the equations are invariant under the rescaling options which we always have: changing coordinates or changing the region of size $\mathcal{V}$ used to formulate the homogeneous phase space and its quantization. As written, no side of either equation depends on $\mathcal{V}$; and while (5.2) has factors depending on the coordinates, both sides of that equation behave in the same way when coordinates are changed. These equations, just like the quantum corrections they describe, are scaling invariant. In fact, they can be fully formulated in terms of the scaling-invariant $v(a)$ and $\mathcal{H}$ (and the constant ℓ_P).

If we assume an underlying discrete state in which $v(a)$ is sufficiently large, $v(a) \gg \ell_P^3$, for all values of a considered, inverse-triad corrections can be ignored. In the simplest case of a constant v, which with $x = -1/2$ is possible as a limiting case of the generic region of lattice refinement, the size of inverse-triad corrections is preserved during evolution. If they are suppressed once, for instance at large volume, they will not become significant even if one evolves toward the big bang. For other cases of x, or even non-power law forms of $\mathcal{N}(a)$, the significance of inverse-triad corrections will, however, change during evolution, and must be analyzed carefully.

Holonomy corrections become important only if the energy density $G\rho$, which by the Friedmann equation is proportional to $\dot{a}^2/a^2$, comes close to the characteristic (a-dependent) density $\mathcal{H}_*^2 = v(a)^{-2/3}$ of discrete patches. Physically one clearly expects quantum effects due to the discreteness to set in at this stage, which is also borne out directly by (5.1). Again, we may initially choose $x = -1/2$ for the simplest case of a constant $v(a)$, but now the matter density itself, and thus $\dot{a}/a$ via the Friedmann equation, is changing dynamically. For a general scalar field source φ we would have to know the behavior of $\varphi(a)$ and its momentum $p_\varphi(a)$ to see when its energy density becomes large. We may pick the simplest case of a free, massless scalar whose potential vanishes, thus implying a constant momentum $p_\varphi = \mathcal{V}a^3\dot\varphi$ and a purely kinetic energy density behaving as $\rho_{\text{free}} = \frac{1}{2}p_\varphi^2/a^6\mathcal{V}^2$. The a-dependence of ρ_{free} is evident, which can easily be compared with one's choice of $v(a)^{-2/3}$ to see when (5.1) gets violated. (The choices of this example, $x = -1/2$ with a scalar free and massless, were also made in [1], but for other reasons).

5.2 Wave-Function Dynamics

The Wheeler–DeWitt equation for a wave function of the universe can be solved explicitly in several cases. In most of them, quantum back-reaction results: expectation values of operators of interest in evolving states do not follow the classical trajectories but show deviations depending on the amount of quantum fluctuations and other quantum parameters. Especially near the singularity at vanishing scale factor, a boundary in the phase space underlying Wheeler–DeWitt quantizations, the form of the wave function can matter significantly; explicit examples, in which the guiding equation of a Bohmian viewpoint was taken to express scale-factor dynamics, have for instance been analyzed in [2, 3]. Even more situations can be studied numerically, also including several of the corrections of loop quantum gravity.

Example 5.1 For a free, massless scalar as the matter source, we have the purely kinetic energy density $\rho_{\text{free}} = p_\varphi^2/2a^6\mathcal{V}^2$. In terms of triad variables (but not using holonomies, thus assuming a Wheeler–DeWitt representation) and a specific ordering to make it soluble, the Wheeler–DeWitt equation

$$\left(-\gamma^{-2}c^2\widehat{\sqrt{|p|}} + \frac{4\pi G}{3}\frac{\widehat{p_\varphi^2}}{|p|^{3/2}}\right)\psi = 0$$

becomes

$$-\frac{8\pi G}{3}p\frac{\partial}{\partial p}\left(p\frac{\partial}{\partial p}\psi\right) + \frac{1}{2}\frac{\partial^2}{\partial\varphi^2}\psi = 0.$$

With $\ell := \sqrt{3/16\pi G}\log|p|$, the equation reduces to the Klein–Gordon equation with general solution $\psi(\ell, \varphi) = \psi_+(\varphi + \ell) + \psi_-(\varphi - \ell)$ for arbitrary functions $\psi_\pm$. Normalizability restrictions on the admissible functions that arise from a physical inner product on the solution space will be discussed in Sect. 12.3.2. Left- and right-moving, or collapsing and expanding, solutions $\psi_\pm$ are independent in this Wheeler–DeWitt model. An interesting interpretation of holonomy corrections from loop quantum cosmology is that they introduce scattering between these solutions at high density, allowing the transition from collapse to expansion [4].

Once the form of the wave function becomes relevant, systematic investigations—analytical or numerical—are rather involved. The problem is not so much handling the equations, but rather gaining sufficient control over the large parameter space that is available for an initial state. Moreover, in a quantum cosmological situation it is not always straightforward to guess what form a good semiclassical state may have, or in which way it would evolve into a stronger quantum state. If a whole wave function, or even a sufficiently general class, must be provided, there are hardly any guidelines. At this stage, a systematic organization of all the parameters of a generic wave function becomes important. A general formulation of these properties, together with implications for quantum corrections at the level of equations of motion, is provided by effective descriptions.

5.3 Solvable Models for Cosmology

It turns out that a free, massless scalar in a spatially flat isotropic universe provides a model much simpler than one could have hoped for, in a sense very different from what is illustrated by the availability of explicit wave-function solutions in Example 5.1: it is a free quantum model in which even the unavoidable interactions that happen between matter and gravity do not cause quantum back-reaction: in suitable factor orderings, expectation values of basic operators satisfy equations of motion completely independent of the behavior of moments of a state: they can be solved for exactly without knowing much else about the state, which could even be highly quantum.

5.3.1 Wheeler–DeWitt Quantization

Formulating the flat Friedmann equation in canonical variables shows that the momentum of the scalar φ is a quadratic expression of the gravitational canonical variables. We will use the latter in a general form

$$V = \frac{3\mathcal{V}}{8\pi G f_0(1-x)}a^{2(1-x)}, \quad P = -f_0 a^{2x}\dot{a} \quad \text{such that} \quad \{V, P\} = 1 \quad (5.3)$$

which later on will allow us to encompass all x-power-law refinement schemes $\delta(a) = f_0 a^{2x}/\gamma \mathcal{V}^{1/3}$ at once (if holonomies e^{iP} are used). We can then write the Friedmann equation as

$$p_\varphi = \mp\sqrt{\frac{3}{4\pi G}}a^2|\dot{a}|\mathcal{V} = \mp\sqrt{\frac{16\pi G}{3}}(1-x)|VP| \quad (5.4)$$

which is indeed quadratic in the canonical variables (except for the absolute value, see below). The right-hand side corresponds to an upside-down harmonic-oscillator Hamiltonian after a linear canonical transformation of variables.

Here, the parameter f_0 which determines the discreteness scale of lattice refinement drops out, as appropriate for the classical theory. The parameter x enters only as a constant rescaling of p_φ, which is itself a constant. Both parameters, f_0 and x, will play a more quantitative role once we loop-quantize this model, even though the qualitative behavior of solutions in the exactly solvable model will be insensitive to their values.

As for dimensions, $f_0 \mathscr{V}^{-1/3} a^{2x} = \gamma \delta(a)$ is dimensionless and so is P, while V has the dimension of an action. When $\mathscr{V}$ is changed, V changes in the same way while P remains unchanged. Note that f_0 is independent of $\mathscr{V}$ because $f_0 a^{2x} = \gamma \ell_0$ must be independent of $\mathscr{V}$.

We interpret $p_\varphi(V, P)$ as the Hamiltonian of the system, generating the dynamical flow of V and P with respect to the internal time φ. In fact, writing (5.4) as a constraint

$$C := p_\varphi \pm H(V, P) = 0 \quad \text{with} \quad H(V, P) := \sqrt{\frac{16\pi G}{3}}(1 - x)|VP| \qquad (5.5)$$

as a positive Hamiltonian, equations of motion are

$$\frac{d\varphi}{d\varphi} = \{\varphi, C\} = 1, \quad \frac{dp_\varphi}{d\varphi} = \{p_\varphi, C\} = 0 \qquad (5.6)$$

$$\frac{dV}{d\varphi} = \{V, C\} = \pm\frac{\partial H}{\partial P}, \quad \frac{dP}{d\varphi} = \{P, C\} = \mp\frac{\partial H}{\partial V}. \qquad (5.7)$$

The first equation means that the gauge parameter used here as time is identical with the scalar φ; the following equations then provide evolution of V and P with respect to φ in the ordinary Hamiltonian way. To translate back to proper time, one would use the original Hamiltonian constraint (4.9) with lapse function $N = 1$. Multiplying the φ-equations of motion by $d\varphi/d\tau = \{\varphi, C_{\text{grav}}[1]\}$ then provides the equations of motion in proper time τ.

The Hamiltonian can easily be quantized, choosing the symmetric ordering

$$\hat{H} = \sqrt{\frac{4\pi G}{3}}(1 - x)|\hat{V}\hat{P} + \hat{P}\hat{V}|. \qquad (5.8)$$

Comparing with the quantization in Example 5.1, which corresponds to $\hat{p}_\varphi^2 \propto \hat{V}\hat{P}\hat{V}\hat{P}$ (for $x = 0$), we have

$$\hat{p}_\varphi^2 \propto \frac{1}{4}(\hat{V}\hat{P} + \hat{P}\hat{V})^2 = \hat{V}\hat{P}\hat{V}\hat{P} - i\hbar\hat{V}\hat{P} - \frac{1}{4}\hbar^2,$$

constituting a different ordering. Here, the operator is symmetric with respect to an inner product with integration measure dV; in Example 5.1 the choice resulted in an operator symmetric with the measure $d\log|p|$. Logarithmic scale factors are often used in Wheeler–DeWitt quantizations to map the singular boundary $a = 0$ to infinity (which, however, does not remove the singularity).

We are using a variable V which takes values only on the positive real line. At the effective level, we can take the required boundary conditions into account, as alluded to in Sect. 2.2 and Example 2.1, by using the "affine" phase-space variables V and $D := VP$ with the non-canonical algebra $\{V, D\} = V$. These variables generate a free, transitive and symplectomorphic group action on $\mathbb{R}^+ \times \mathbb{R}$ and can be used for a group-theoretical quantization of the phase space [5]. A unitary representation on a Hilbert space then provides self-adjoint operators $\hat{V}$ and $\hat{D}$, while a self-adjoint operator for P does not result. The model is thus an example for the more general affine quantization program [6, 7].

The Hamiltonian $H \propto |D|$ is now linear, but formulated in non-canonical variables. The solvability of the system is still realized because it is based on a linear algebra (V, D, H) of basic variables together with the Hamiltonian. Any quantization respecting the linearity will lead to the same solvability properties by which expectation values and moments of a state can be explicitly solved for. The affine variables are preferable from a kinematical perspective. However, dynamical equations of the model show that the boundary $V = 0$ is not crossed for any finite φ; thus we may keep using the simpler canonical variables. In particular, this will allow us access to quantum fluctuations of the curvature parameter P, not just of D.

In $\hat{H}$, after taking the square root of p_φ^2, the absolute value required for a positive Hamiltonian can make it difficult to progress further if our aim was to find complete wave functions as solutions or analyze the Hamiltonian operator directly. This difficulty is a very characteristic one in quantum gravity, where square-root Hamiltonians arise in many deparameterizations. (The theory is even more complicated if one does not or cannot deparameterize. In this context we will see a further, crucial advantage of effective descriptions in Sect. 13.2.3)

Thanks to the solvability, which will now be demonstrated, most of the cosmologically interesting information can be gained side-stepping wave functions. First turning to expectation values, we must derive and solve equations such as

$$\frac{\mathrm{d}}{\mathrm{d}\varphi} \langle \hat{V} \rangle = \frac{\langle [\hat{V}, \hat{H}] \rangle}{\mathrm{i}\hbar}. \tag{5.9}$$

Also here, the absolute value in $\hat{H}$ seems to make a derivation of the commutator with $\hat{V}$ difficult. But if we have a state that is supported only on the positive part of the spectrum of $\hat{V}\hat{P} + \hat{P}\hat{V}$, the expectation value $\langle [\hat{V}, \hat{H}] \rangle_+ \propto \langle \hat{V}(\hat{V}\hat{P} + \hat{P}\hat{V}) - (\hat{V}\hat{P} + \hat{P}\hat{V})\hat{V} \rangle_+$ in such a state, denoted by the subscript "+", can be computed without implementing the absolute value: $\hat{H}$ in both terms of the commutator acts directly on the state that obeys the positivity condition, where it agrees with the expression without the absolute value. After the absolute value can safely be dropped, the remaining expression for the Hamiltonian is quadratic and the commutator follows straightforwardly. Similarly, on a state supported only on the negative part of the spectrum of $\hat{V}\hat{P} + \hat{P}\hat{V}$, we have $\langle [\hat{V}, \hat{H}] \rangle_- \propto -\langle \hat{V}(\hat{V}\hat{P} + \hat{P}\hat{V}) - (\hat{V}\hat{P} + \hat{P}\hat{V})\hat{V} \rangle_-$.

Superpositions of these different kinds of states, supported on opposite signs of the spectrum of $\hat{V}\hat{P} + \hat{P}\hat{V}$, cannot be treated in the same way, but usually such information for expectation values would not be of much use, anyway: to see this, we can use the analogy of a free, relativistic particle with a wave function satisfying the Klein–Gordon equation

$$-\frac{\partial^2 \psi}{\partial t^2} + \frac{\partial^2 \psi}{\partial x^2} = 0. \tag{5.10}$$

Its classical formulation is presented by the constraint $E^2 - p^2 = 0$ relating energy and momentum. Writing energy as the momentum of time, we have $-p_t^2 + p^2 = 0$ which with p_t analogous to p_φ has a form similar to the Friedmann equation used here. We can solve for $p_t = \pm |p|$ where an absolute value appears, too.

After quantization, equations of motion for $\langle \hat{q} \rangle$ are computed as in quantum cosmology, making use of states supported entirely on the positive or negative, respectively, parts of the spectrum of $\hat{p}$. Since the sign of the momentum is definite in both cases, they correspond to purely right-moving or purely left-moving states, respectively. It now becomes clear why the evolution of an expectation value of $\hat{q}$ in a superposition of such states would be of low interest: as an expectation value in two wave packets moving in opposite directions, it would indicate a point somewhere between the wave packets, which in general can be very far away from the actual positions of the wave packets; it can even be at a place where the probability to observe the particle in a single measurement would be zero. Similarly, if we compute fluctuations as we will do it for cosmology, they can grow arbitrarily large just because the wave packets in the superposition are moving away from each other. Such cases of large fluctuations do not at all correspond to strong quantum behavior, and so their knowledge for a superposition would even be potentially misleading. If we restrict attention to purely right-moving or purely left-moving superpositions, on the other hand, no such issues arise. Thus, it is in fact much more suitable to consider only states of definite sign of the momentum, or of H in quantum cosmology, and compute expectation values or moments. Then, absolute values can be dropped, immensely simplifying calculations. If superpositions of solutions of opposite signs are desired, knowing the shape of individual wave packets still allows one to infer the behavior of superpositions.

For the absolute value to be dropped in the Hamiltonian, states used must always be supported on a definite sign of the spectrum of $\hat{V}\hat{P} + \hat{P}\hat{V}$. Since the φ-independent Hamiltonian is preserved during evolution, it is enough to ensure positivity (or negativity) for the initial state, thus posing conditions for initial values to be chosen. While expectation values and moments do evolve, the sign of $\hat{V}\hat{P} + \hat{P}\hat{V}$ will be preserved such that equations of motion valid at all times are obtained. There is a second sign choice involved in specifying the dynamics: the one in the relationship (5.5) between p_φ and H. This sign distinguishes negative-frequency and positive-frequency modes and plays a role quite different from the sign for left-moving and right-moving modes. It will be important in the discussion of physical inner products in Chap. 12; for now, we may simply fix it by choice, which we do as the positive sign. In particular, for expectation values of basic operators we have

$$\frac{d}{d\varphi}\langle \hat{V} \rangle = \frac{\langle [\hat{V}, \hat{H}] \rangle}{i\hbar} = \sqrt{\frac{4\pi G}{3}} \frac{1-x}{i\hbar} \langle [\hat{V}, \hat{V}\hat{P} + \hat{P}\hat{V}] \rangle = \sqrt{\frac{16\pi G}{3}}(1-x)\langle \hat{V} \rangle \tag{5.11}$$

$$\frac{d}{d\varphi}\langle \hat{P} \rangle = -\sqrt{\frac{16\pi G}{3}}(1-x)\langle \hat{P} \rangle \tag{5.12}$$

which can easily be solved:

$$\langle \hat{V} \rangle(\varphi) = V_{\mathrm{i}} \exp\left(\sqrt{16\pi G/3}(1-x)\varphi\right),$$
$$\langle \hat{P} \rangle(\varphi) = P_{\mathrm{i}} \exp\left(-\sqrt{16\pi G/3}(1-x)\varphi\right) \tag{5.13}$$

with initial values satisfying the constraint

$$\sqrt{\frac{16\pi G}{3}}(1-x)V_{\mathrm{i}}P_{\mathrm{i}} = -p_{\varphi}. \tag{5.14}$$

To obtain solutions in proper time τ rather than internal time φ, we solve the equation

$$
\begin{aligned}
p_{\varphi} = a^3 \dot{\varphi} \mathcal{V} &= \left(\frac{8\pi G f_0 (1-x)V}{3\mathcal{V}}\right)^{3/(2(1-x))} \frac{\mathrm{d}\varphi}{\mathrm{d}\tau}\mathcal{V} \\
&= -\left(\frac{8\pi G f_0 (1-x)V_{\mathrm{i}}}{3\mathcal{V}}\right)^{3/(2(1-x))} e^{\sqrt{12\pi G}\varphi} \frac{\mathrm{d}\varphi}{\mathrm{d}\tau}\mathcal{V} \tag{5.15}
\end{aligned}
$$

for $\varphi(\tau)$, using again that p_{φ} must be constant. With $\exp\left(\sqrt{12\pi G}\varphi(\tau)\right) \propto -\tau$ as the solution to (5.15) with constant p_{φ}, the proper-time solutions for $\langle \hat{V} \rangle(\tau)$ and $\langle \hat{P} \rangle(\tau)$ directly follow. In particular, $\langle \hat{V} \rangle(\tau) \propto (-\tau)^{2(1-x)/3}$ corresponds exactly to the classical behavior for a stiff fluid, for which $V \propto a^{2(1-x)} \propto (-\tau)^{2(1-x)/3}$.

Not only the solutions but also the equations themselves reveal the correct Friedmann dynamics of a free, massless scalar. This behavior follows from the construction together with the absence of quantum back-reaction in the present model, but for the later discussion of modified Friedmann equations in loop quantum cosmology it is instructive to derive the proper-time equations from those in internal time φ. Using the relationships between the proper-time derivative $\dot{\varphi}$ and the scalar momentum, and the one between a and V, we compute the Friedmann equation

$$
\begin{aligned}
\left(\frac{\dot{a}}{a}\right)^2 &= \left(\frac{\dot{\varphi}}{2(1-x)\langle \hat{V} \rangle}\frac{\mathrm{d}\langle \hat{V} \rangle}{\mathrm{d}\varphi}\right)^2 \\
&= \left(\frac{\mathcal{V}^{-1}a^{-3}p_{\varphi}}{2(1-x)\langle \hat{V} \rangle}\right)^2 \cdot \frac{16\pi G}{3}(1-x)^2 \langle \hat{V} \rangle^2 = \frac{4\pi G}{3}\frac{p_{\varphi}^2}{a^6 \mathcal{V}^2} \tag{5.16}
\end{aligned}
$$

and from

$$\frac{\mathrm{d}^2 \langle \hat{V} \rangle}{\mathrm{d}\varphi^2} = \frac{16\pi G}{3}(1-x)^2 \langle \hat{V} \rangle \tag{5.17}$$

(taking one more derivative of (5.11)) the equation

$$\frac{\ddot{a}}{a} + 2\left(\frac{\dot{a}}{a}\right)^2 + 2(1-x)\left(\left(\frac{\dot{a}}{a}\right)^2 - \frac{4\pi G}{3}\frac{p_{\varphi}^2}{a^6 \mathcal{V}^2}\right) = 0. \tag{5.18}$$

The second part, multiplying $(1 - x)$, vanishes by virtue of the Friedmann equation; the first one being zero is equivalent to the Raychaudhuri equation of a stiff fluid.

At this stage, we have provided a quantization of the model and determined the exact behavior of expectation values. The dynamics of states is based on the operator algebra that defines the quantum system and its dynamics via commutators with the Hamiltonian. By solving the equations of motion, we have computed expectation values ready to deliver physical properties of the model. Similarly, as seen later, we can compute fluctuations, correlations, and higher moments of an evolving state, providing information about the statistics of repeated measurements. What we do not provide in this way of describing the quantum system, not making use of a Hilbert-space representation, is a handle on eigenvalues of operators, or on the outcomes of single measurements. But for quantum cosmology, in which we never do a direct measurement on the whole system presented by the universe but rather averaged collections of small individual ones, the information we do obtain is clearly sufficient. We can then fully exploit the advantage of a representation-independent treatment, keeping the algebra of basic operators with the Hamiltonian fundamental.

The quantization analyzed so far corresponds to a Wheeler-DeWitt equation for the states used since a direct representation of the curvature variable P, rather than just holonomies, was assumed. This model thus cannot resolve the classical singularity, and indeed wave packets simply follow the classical trajectories as shown by the solutions for expectation values. (In interacting models quantum back-reaction might produce new quantum forces preventing the singularity; see Example 7.1. This possibility remains incompletely studied in Wheeler–DeWitt quantizations, but even if it could be realized, singularity avoidance would not be general: the free, massless scalar model used here remains singular in any case.)

5.3.2 Loop Quantization

Loop quantum cosmology shows more promise for resolving singularities, which we have in fact already seen at the level of the fundamental difference equation. Since we cannot directly quantize non-almost periodic expressions such as c^2 (or P^2), the previously realized solvable nature is no longer satisfied as in the Wheeler–DeWitt model. One could worry that the infinite number of higher-order terms when P^2 is extended to an almost-periodic function such as $\sin^2 P$ would make the model deviate strongly from the solvable one. Additional terms would become relevant just in the high-curvature regime of most interest, where $P \ll 1$ no longer holds. Quite surprisingly, it turns out that even the loop-quantized free, massless scalar model is of an exactly solvable nature. Its Hamiltonian is no longer quadratic, but in suitable variables it is realized as a linear model where basic operators together with the Hamiltonian still form a linear algebra. This realization of loop quantum cosmology [8] is called harmonic cosmology, whose cosmological implications we will explore further in the next chapter. Here, we derive its basic formulation and the relevant equations.

The solvable dynamics analyzed in the Wheeler–DeWitt setting is representation-independent and sensitive just to the fundamental algebra. With a loop quantization, we now have an inequivalent representation. Properties of the representation itself clearly cannot feature at a representation-independent level, but it comes along with a new algebra of basic operators, now including holonomies, and thus new physics. In particular, the Hamiltonian must be adapted to the basic algebra by including the quantum-geometry corrections of loop quantum gravity, thus changing the dynamics.

Solvable loop models can only be derived with suitable choices of variables that are not canonical; otherwise the non-quadratic Hamiltonian in V and P will hide any solvability features. In addition to V as defined before, let us introduce the variable $J := V \exp(iP)$ and promote these basic variables to a quantum $*$-algebra generated by operators $\hat{V}$ and $\hat{J}$, such that $\hat{V}$ is self-adjoint, $\hat{V}^\dagger = \hat{V}$, while $\hat{J}$ is not due to the presence of the imaginary unit in its classical analog. We thus have two independent (anti-)self-adjoint combinations $\hat{J}_\pm := \hat{J} \pm \hat{J}^\dagger$, whose freedom will later have to be restricted by reality conditions. While these variables are not canonical, they satisfy a linear algebra

$$[\hat{V}, \hat{J}_\pm] = \hbar \hat{J}_\mp, \quad [\hat{J}_+, \hat{J}_-] = 4\hbar \hat{V} \tag{5.19}$$

which correctly quantizes the classical Poisson brackets between them. Most importantly, the loop-quantized Hamiltonian, again using evolution in internal time φ, is linear in these variables, $\hat{H} = \sqrt{16\pi G/3}(1-x)|\frac{1}{2}i\hat{J}_-|$, except for the absolute value which can be dealt with as in the preceding Wheeler–DeWitt case. There is then a linear algebra between the basic operators and the Hamiltonian, implying solvability and the absence of quantum back-reaction as it did for systems with a Hamiltonian quadratic in canonical variables. Holonomy corrections are fully included: with $-P = f_0 a^{2x} \dot{a} = \delta c = \ell_0 \tilde{c}$ we have $\ell_0 = \gamma^{-1} f_0 a^{2x}$ suitable for a power-law form of lattice refinement.

In accordance with the classical identity $JJ^* = V^2$ we have to impose a reality condition at the quantum level: $\hat{J}\hat{J}^\dagger = \hat{V}^2$ in this ordering. Once implemented, this condition brings us to the correct number of real degrees of freedom. As a condition quadratic in operators, it will imply relations between fluctuations and expectation values. But as a relationship that does not arise dynamically, it does not correspond to quantum back-reaction: imposed once for an initial state, it will remain true at all times.

We notice that solvability is realized for a specific factor ordering between V and $\exp(iP)$ in the original quantum constraint, which does not agree with the ordering used in (4.15). Also here, we have chosen a very special ordering to realize solvability. In a Wheeler–DeWitt-type equation, the ordering corresponds to

$$\hat{p}_\varphi^2 \propto -\hat{J}_-^2 = -(\widehat{\hat{V}e^{iP}} - \widehat{e^{-iP}\hat{V}})^2 = -\widehat{\hat{V}e^{iP}}\,\widehat{\hat{V}e^{iP}} + \hat{V}^2 + \widehat{e^{-iP}\hat{V}}\,\widehat{e^{iP}\hat{V}} - \widehat{e^{-iP}\hat{V}}\,\widehat{e^{-iP}\hat{V}}$$

or to a difference operator

$$\hat{p}_\varphi^2|\omega\rangle \propto -(\omega+1)(\omega+2)|\omega+2\rangle + \omega(2\omega+1)|\omega\rangle - \omega(\omega-1)|\omega-2\rangle \tag{5.20}$$

in terms of $\hat{V}$-eigenstates $|\omega\rangle$. The resulting difference equation does not correspond to the non-singular one (4.15) arising from loop quantum cosmology, but away from $\omega = 0$ expectation-value solutions analyzed here capture its properties up to factor-ordering corrections. Moreover, we are ignoring inverse-triad corrections to realize exact solvability. (They will be dealt with in more detail in Chap. 10). All those additional effects can be included approximately by perturbations around the solvable model. Then, quantum back-reaction results once quantum effects become important. For a framework of such perturbations, we will first return to the Wheeler–DeWitt model and analyze effects it implies in weak-curvature regimes where no loop representation should be necessary.

5.4 Isotropic Perturbation Theory: Spatial Curvature, a Cosmological Constant, and Interacting Matter

For small-curvature regimes, holonomy corrections of loop quantum gravity are not important; and since we keep for now our assumption that inverse-triad corrections are weak as well, systems can be analyzed by a Wheeler-DeWitt model. In order to avoid interpretational issues of a wave function, we will work with the solvable quantum model obtained for a free, massless scalar in a spatially flat isotropic geometry. This will allow us to directly compute dynamical expectation values as well as properties of evolving states such as their fluctuations and correlations. However, only the spatially flat case with the specific matter content is exactly solvable; in other cases, which are certainly of high interest, too, quantum back-reaction results, implying that expectation values alone no longer obey a closed set of differential equations. In internal-time evolution, the Hamiltonian will no longer be quadratic in canonical variables, nor will it have the form of a linear system in new variables. Equations of motion of the form (5.9) will require expectation values of commutators non-linear in basic variables on the right-hand side. Such terms constitute independent quantum variables not reducible to expectation values of basic operators; they are subject to their own dynamics which must be known in order to be able to solve for the time dependence of expectation values.

5.4.1 Free Matter

If we have spatial curvature or a cosmological constant of either sign, the free, massless scalar still provides a global internal time. But its flow is now generated by a Hamiltonian $H(V, P)$ not quadratic in canonical variables: Solving the Friedmann equation for p_φ and introducing our earlier parameterization of variables, we have

$$p_\varphi = \mp\sqrt{\frac{3}{4\pi G}}a^2\sqrt{\dot{a}^2 + k - \Lambda a^2}\mathcal{V}$$

$$= \mp\sqrt{\frac{16\pi G}{3}}(1-x)V$$

$$\times\sqrt{P^2 + kf_0^2\left(\frac{8\pi G(1-x)f_0 V}{3\mathcal{V}}\right)^{\frac{2x}{1-x}} - \Lambda f_0^2\left(\frac{8\pi G(1-x)f_0 V}{3\mathcal{V}}\right)^{\frac{1+2x}{1-x}}}. \qquad (5.21)$$

As a function of V and P, this expression is not even polynomial, let alone quadratic, and its derivatives as they would enter Poisson brackets are certainly non-linear. The square root may in fact be difficult to quantize at an operator level since doing so via the spectrum would require one to determine the equivalent of the energy spectrum of a particle in a general power-law potential (depending on x). And even if one can find the spectrum in special cases and define the square root via the spectral decomposition, the computation of commutators with $\hat{V}$ and $\hat{P}$ as required for equations of motion would be even more involved. This is another place where an effective treatment via expectation values and moments will be of much use to shed light on evolving state properties.

Instead of quantizing the Hamiltonian and solving evolution equations for full wave functions, we will exploit the relation to the solvable model realized before, and approximately determine the evolution of state parameters. To do so concretely, we will make use of a "background-state method", analogous to a procedure well-known from quantum field theory. In our case, this method entails that the quantum Hamiltonian operator is expanded around expectation values $\langle\hat{V}\rangle$, $\langle\hat{P}\rangle$ in a background state that corresponds to a solution of the solvable model. Perturbation theory then allows one to add quantum corrections due to the interacting nature of the non-solvable model. The background-state expansion can formally be written for the operator $H(\hat{V}, \hat{P}) = H(\langle\hat{V}\rangle + (\hat{V} - \langle\hat{V}\rangle), \langle\hat{P}\rangle + (\hat{P} - \langle\hat{P}\rangle))$ itself:

$$H(\hat{V}, \hat{P}) \text{``} = \text{''} \sum_{n=0}^{\infty}\sum_{k=0}^{n}\frac{1}{n!(n-k)!}\frac{\partial^n H(\langle\hat{V}\rangle, \langle\hat{P}\rangle)}{\partial\langle\hat{V}\rangle^k\langle\hat{P}\rangle^{n-k}}\frac{1}{n!}\sum_{symm}(\hat{V}-\langle\hat{V}\rangle)^k(\hat{P}-\langle\hat{P}\rangle)^{n-k}$$

$$= H(\langle\hat{V}\rangle, \langle\hat{P}\rangle) + \frac{\partial H(\langle\hat{V}\rangle, \langle\hat{P}\rangle)}{\partial\langle\hat{V}\rangle}(\hat{V}-\langle\hat{V}\rangle) + \frac{\partial H(\langle\hat{V}\rangle, \langle\hat{P}\rangle)}{\partial\langle\hat{P}\rangle}(\hat{P}-\langle\hat{V}\rangle)$$

$$+ \frac{1}{2}\frac{\partial^2 H(\langle\hat{V}\rangle, \langle\hat{P}\rangle)}{\partial\langle\hat{V}\rangle^2}(\hat{V} - \langle\hat{V}\rangle)^2$$

$$+ \frac{1}{2}\frac{\partial^2 H(\langle\hat{V}\rangle, \langle\hat{P}\rangle)}{\partial\langle\hat{V}\rangle\langle\hat{P}\rangle}\left((\hat{V}-\langle\hat{V}\rangle)(\hat{P}-\langle\hat{P}\rangle)+(\hat{P}-\langle\hat{P}\rangle)(\hat{V}-\langle\hat{V}\rangle)\right)$$

$$+ \frac{1}{2}\frac{\partial^2 H(\langle\hat{V}\rangle, \langle\hat{P}\rangle)}{\partial\langle\hat{P}\rangle^2}(\hat{P} - \langle\hat{P}\rangle)^2 + \cdots \qquad (5.22)$$

with quotation marks indicating that this is not a strict equation for operators on the full Hilbert space of states. It is rather an identity valid in suitable regimes such as semiclassical ones, as we will justify in more detail in Chap. 13 on general effective descriptions.

For a totally symmetric Hamiltonian, we have inserted a summation over the permutation group, summing up all possible reorderings of the following operators. If the right-hand side of (5.22) is truncated to a certain polynomial order in basic operators, with coefficients $\partial^n H(\langle\hat{V}\rangle, \langle\hat{P}\rangle)/\partial^k\langle\hat{V}\rangle\partial^{n-k}\langle\hat{P}\rangle$ depending only on expectation values, commutators with $\hat{V}$ and $\hat{P}$ can easily be computed. Once expectation values are taken, quantum corrections depend on the fluctuations $(\Delta V)^2 = \langle(\hat{V} - \langle\hat{V}\rangle)^2\rangle$ and $(\Delta P)^2 = \langle(\hat{P} - \langle\hat{P}\rangle)^2\rangle$ as well as the covariance $C_{VP} = \frac{1}{2}\langle(\hat{V} - \langle\hat{V}\rangle)(\hat{P} - \langle\hat{P}\rangle) + (\hat{P} - \langle\hat{P}\rangle)(\hat{V} - \langle\hat{V}\rangle)\rangle$. Higher-order corrections appear with coefficients of the general moments

$$
\begin{aligned}
\Delta(O_1 O_2 \ldots O_n) := & \left\langle(\hat{O}_1 - \langle\hat{O}_1\rangle)(\hat{O}_2 - \langle\hat{O}_2\rangle)\cdots(\hat{O}_n - \langle\hat{O}_n\rangle)\right\rangle_{\text{symm}} \\
= & \frac{1}{n!}\sum_{\pi\in S_n}\left\langle(\hat{O}_{\pi(1)} - \langle\hat{O}_{\pi(1)}\rangle)(\hat{O}_{\pi(2)} - \langle\hat{O}_{\pi(2)}\rangle)\cdots(\hat{O}_{\pi(n)} - \langle\hat{O}_{\pi(n)}\rangle)\right\rangle
\end{aligned}
$$

$$(5.23)$$

ordered totally symmetrically. In Chap. 13 we will see that formal expansions of operators as in (5.22) can be avoided (and simplified) using Poisson geometry on the quantum phase space of expectation values together with the moments (5.23) of basic operators.

If we keep only the linear operator order in this expansion, we can directly verify that the Ehrenfest-type equations

$$
\frac{\mathrm{d}\langle\hat{V}\rangle}{\mathrm{d}\varphi} = \frac{\langle[\hat{V}, H(\hat{V}, \hat{P})]\rangle}{i\hbar} = \frac{\partial H(\langle\hat{V}\rangle, \langle\hat{P}\rangle)}{\partial\langle\hat{P}\rangle} + \cdots
$$

$$(5.24)$$

$$
\frac{\mathrm{d}\langle\hat{P}\rangle}{\mathrm{d}\varphi} = \frac{\langle[\hat{P}, H(\hat{V}, \hat{P})]\rangle}{i\hbar} = -\frac{\partial H(\langle\hat{V}\rangle, \langle\hat{P}\rangle)}{\partial\langle\hat{V}\rangle} + \cdots
$$

$$(5.25)$$

will simply reproduce the classical equations of motion for the expectation values. The second-order terms written above do not contribute to equations of motion of expectation values such as $\langle\hat{V}\rangle$ since, e.g., $\langle[\hat{V}, (\hat{P}-\langle\hat{P}\rangle)^2]\rangle = i\hbar\langle(2\hat{P}-2\langle\hat{P}\rangle)\rangle = 0$. These terms do, however, contribute to the evolution of quantum fluctuations and correlations. Quantum corrections to the evolution of expectation values then arise from cubic operator terms in the background-state expansion. This will most clearly be seen by the following examples, or by the general presentation of Chap. 13.

Example 5.2 (Harmonic cosmology) For the solvable model of Sect. 5.3.1, we have

$$
\begin{aligned}
\hat{H} \propto \frac{1}{2}(\hat{V}\hat{P} + \hat{P}\hat{V}) = & \langle\hat{V}\rangle\langle\hat{P}\rangle + \langle\hat{P}\rangle(\hat{V} - \langle\hat{V}\rangle) + \langle\hat{V}\rangle(\hat{P} - \langle\hat{P}\rangle) \\
& + \frac{1}{2}\left((\hat{V} - \langle\hat{V}\rangle)(\hat{P} - \langle\hat{P}\rangle) + (\hat{P} - \langle\hat{P}\rangle)(\hat{V} - \langle\hat{V}\rangle)\right)
\end{aligned}
$$

as an exact expansion around the background state with expectation values $\langle\hat{V}\rangle$ and $\langle\hat{P}\rangle$. Ehrenfest equations produce

$$\frac{d\langle \hat{V} \rangle}{d\varphi} = \langle \hat{V} \rangle, \quad \frac{d\langle \hat{P} \rangle}{d\varphi} = -\langle \hat{P} \rangle \tag{5.26}$$

free of quantum back-reaction thanks to the absence of cubic terms in the Hamiltonian. For moments of second order, using for instance

$$\frac{d}{d\varphi}(\Delta V)^2 = \frac{\langle [\hat{V}^2, \hat{H}] \rangle}{i\hbar} - 2\langle \hat{V} \rangle \frac{d\langle \hat{V} \rangle}{d\varphi} \tag{5.27}$$

we obtain in a similar way the equations

$$\frac{d(\Delta V)^2}{d\varphi} = 2(\Delta V)^2, \quad \frac{dC_{VP}}{d\varphi} = 0, \quad \frac{d(\Delta P)^2}{d\varphi} = -2(\Delta P)^2. \tag{5.28}$$

These linear equations can easily be solved:

$$\langle \hat{V} \rangle \propto e^{\varphi}, \quad \langle \hat{P} \rangle \propto e^{-\varphi}, \quad (\Delta V)^2 \propto e^{2\varphi}, \quad (\Delta P)^2 \propto e^{-2\varphi}. \tag{5.29}$$

As a consequence, relative fluctuations $\Delta V/\langle \hat{V} \rangle$ and $\Delta P/\langle \hat{P} \rangle$ are constant, and the semiclassicality that may be posed on an initial state is exactly preserved even if we get arbitrarily close to large curvature $\langle \hat{P} \rangle$.

If we use the self-adjoint dilation operator $\hat{D}$ instead of $\hat{P}$, we obtain equivalent results. The linear Hamiltonian

$$\hat{H} \propto \hat{D} = \langle \hat{D} \rangle + (\hat{D} - \langle \hat{D} \rangle)$$

provides equations of motion

$$\frac{d\langle \hat{V} \rangle}{d\varphi} = \langle \hat{V} \rangle, \quad \frac{d\langle \hat{D} \rangle}{d\varphi} = 0 \tag{5.30}$$

$$\frac{d(\Delta V)^2}{d\varphi} = 2(\Delta V)^2, \quad \frac{dC_{VD}}{d\varphi} = C_{VD}, \quad \frac{d(\Delta D)^2}{d\varphi} = 0. \tag{5.31}$$

The solutions obtained before are consistent with these equations, confirming that the boundary problem associated with $V = 0$ does not play a role for the dynamics of this model. However, in the formulation using $\hat{D}$ we do not gain access to the curvature fluctuation ΔP but only to ΔD, which is a higher moment from the point of view of P.

5.4.1.1 Positive Spatial Curvature

For concreteness, we first consider the case of $k = 1$ (for a unit sphere, or $k = (\mathcal{V}_{\text{unit}}/\mathcal{V})^{2/3}$ in general) and $\Lambda = 0$, choosing the refinement scheme $x = 0$ to make the Hamiltonian linear at least in V:

$$H(V, P) = V\sqrt{P^2 + kf_0^2}. \tag{5.32}$$

We have absorbed an irrelevant constant factor of $\sqrt{16\pi G/3}$ in p_φ, rescaling our internal time. For quantum equations of motion, we will have to compute expectation values of commutators with V and P. Since the commutators will no longer be linear in basic operators, equations of motion will mix expectation values with higher moments such as fluctuations.

> For a Wheeler–DeWitt model, the refinement parameters in the basic variables and the Hamiltonian just amount to the selection of a convenient choice of canonical variables. There are no quantum-geometry effects; all values for the refinement parameters are allowed, and they have no influence on the dynamics. We keep f_0 in the equations even though it has no clear meaning in a Wheeler–DeWitt formulation since it will be useful for a comparison with the loop equations at low curvature. Its value is not important in a pure Wheeler–DeWitt context since it always cancels out for expressions in terms of a.

Following the background-state procedure, truncating the Hamiltonian at cubic order in basic operators, we obtain equations of motion for expectation values up to second order in quantum variables, including expectation values of second-degree polynomials of basic operators as source terms [9]:

$$\frac{d\langle\hat{V}\rangle}{d\varphi} = \frac{\langle\hat{V}\rangle\langle\hat{P}\rangle}{\sqrt{\langle\hat{P}\rangle^2 + f_0^2}} - \frac{3}{2}f_0^2\frac{\langle\hat{V}\rangle\langle\hat{P}\rangle}{(\langle\hat{P}\rangle^2 + f_0^2)^{5/2}}(\Delta P)^2 + f_0^2\frac{C_{VP}}{(\langle\hat{P}\rangle^2 + f_0^2)^{3/2}} \tag{5.33}$$

$$\frac{d\langle\hat{P}\rangle}{d\varphi} = -\sqrt{\langle\hat{P}\rangle^2 + f_0^2} - \frac{1}{2}f_0^2\frac{(\Delta P)^2}{(\langle\hat{P}\rangle^2 + f_0^2)^{3/2}}. \tag{5.34}$$

Here, the second-order quantum variables are the fluctuation $(\Delta P)^2 = \langle(\hat{P} - \langle\hat{P}\rangle)^2\rangle$ and the covariance $C_{VP} = \frac{1}{2}\langle\hat{V}\hat{P} + \hat{P}\hat{V}\rangle - \langle\hat{V}\rangle\langle\hat{P}\rangle$; the remaining second-order moment $(\Delta V)^2 = \langle(\hat{V} - \langle\hat{V}\rangle)^2\rangle$ does not enter these equations, but its evolution is of high interest since it determines the behavior of volume fluctuations.

In order to analyze the dynamics of expectation values, we must know how ΔP and C_{VP} evolve, which cannot simply be treated as constants. They are subject to their own equations of motion, again following from the background-state expanded Hamiltonian operator and Hamiltonian-type equations of motion such as (5.27). Expanded to quantum variables of second order, we obtain

$$\frac{d(\Delta P)^2}{d\varphi} = -2\frac{P}{\sqrt{P^2 + f_0^2}}(\Delta P)^2 \tag{5.35}$$

$$\frac{dC_{VP}}{d\varphi} = f_0^2\frac{V}{(P^2 + f_0^2)^{3/2}}(\Delta P)^2 \tag{5.36}$$

$$\frac{\mathrm{d}(\Delta V)^2}{\mathrm{d}\varphi} = 2 f_0^2 \frac{V}{(P^2 + f_0^2)^{3/2}} C_{VP} + 2 \frac{P}{\sqrt{P^2 + f_0^2}} (\Delta V)^2. \tag{5.37}$$

From the equations of motion (5.33) and (5.34) we immediately read off that there is now quantum back-reaction: the evolution of expectation values is influenced by other properties of a quantum state. To determine how significant these effects are, we estimate the quantum-correction terms for a suitable class of states. Since we are looking at the small-curvature regime with this Wheeler–DeWitt type quantization, we can expect a state to be semiclassical such that its fluctuations are small. We may for instance choose a state saturating the uncertainty relation

$$(\Delta V)^2 (\Delta P)^2 - C_{VP}^2 \geq \frac{\hbar^2}{4} \tag{5.38}$$

at least at some initial time, and then see how the state evolves and how its changing shape back-reacts on expectation values. For the orders considered, one can directly see that the uncertainty relation is preserved by evolution: a state saturating the relation once will always do so; it is a dynamical coherent state as long as higher-order moments can be ignored.

Fluctuations can be large even in a dynamical coherent state if the covariance C_{VP} is large. One might assume C_{VP} to vanish initially, but it evolves according to (5.36). The main question is thus how quickly it can grow, similarly to the free particle as in Sect. 3.1. Its rate of change is proportional to curvature fluctuations $(\Delta P)^2$, which change proportionally to themselves and moreover proportionally to $\langle \hat{P} \rangle$. Both $\langle \hat{P} \rangle$ and ΔP are small in the semiclassical small-curvature regime. For initially small curvature fluctuations, curvature fluctuations thus remain small for long times and do not raise C_{VP} too much. We expect that expectation values in a small-curvature regime, especially around the recollapse phase of the closed model considered here, follow trajectories very close to the classical ones.[1] Ignoring quantum back-reaction terms, this approximation provides the general solution

$$P_{\text{classical}}(\varphi) = P_0 \cosh(\varphi - \varphi_0) + \sqrt{P_0^2 + f_0^2} \sinh(\varphi - \varphi_0) \tag{5.39}$$

$$V_{\text{classical}}(\varphi) = V_0 \frac{\sqrt{P_0^2 + f_0^2}}{-P_0 \sinh(\varphi - \varphi_0) + \sqrt{P_0^2 + f_0^2} \cosh(\varphi - \varphi_0)}. \tag{5.40}$$

To see more of the behavior of the full state, we solve the equations of motion for quantum variables. The coupled system together with expectation values is quite

---

[1] This expectation is visible numericall as well [10], a reference which for Gaussian states also agrees with several other statements made here.

complicated, but since we know that the classical solutions provide a good approximation for the expectation values, we can use them on the right-hand sides of (5.35), (5.36) and (5.37). Choosing without loss of generality $\varphi_0 = 0$ and $P(\varphi_0) = 0$ in the general solution, picking the recollapse point as initial time, we can directly integrate the equations one by one:

$$(\Delta P)^2(\varphi) = (\Delta P)_0^2 \cosh^2(\varphi) \propto \frac{1}{V(\varphi)^2} \tag{5.41}$$

$$C_{VP}(\varphi) = (C_{VP})_0 + \frac{V_0(\Delta P)_0^2}{f_0} \frac{\sinh(\varphi)}{\cosh(\varphi)} \tag{5.42}$$

$$(\Delta V)^2(\varphi) = \frac{(\Delta V)_0^2 + 2f_0^{-1}V_0(C_{VP})_0 \tanh(\varphi) + f_0^{-2}V_0^2(\Delta P)_0^2 \tanh^2(\varphi)}{\cosh^2(\varphi)}. \tag{5.43}$$

As the first line shows, curvature fluctuations are indeed small at large volume, such that our approximation is self-consistent. It also shows that these fluctuations are symmetric around the recollapse point since they depend only on V, not on whether that volume is reached before or after collapse. This is not the case for volume fluctuations, however, which in general do not have a reflection-symmetric form in φ. From (5.43) they are reflection symmetric only if $(C_{VP})_0 = 0$, realized if the state is uncorrelated at the recollapse; see Fig. 5.1.

Given the fact that correlations evolve in time and are in general non-zero, a symmetric state is thus a very special case. With long cosmic evolution times, strong correlations may indeed build up out of some uncorrelated state which may have been realized earlier. In fact, the build-up of correlations is usually what decoherence is based on; see Sect. 3.1: a state's distribution evolves to a flat elliptical shape, where some variables take on nearly classical dependence. Taken over to quantum cosmology, one should thus expect that an evolving universe, reaching the recollapse point, should be in a highly squeezed state. In the present model, this means that volume fluctuations around the recollapse are highly asymmetric: after the recollapse they might be much smaller than before—or much larger. A complete analysis of state evolution and decoherence, even in this simple model, remains to be done to see how the recollapse should be seen from a semiclassical perspective. But it highlights the importance of keeping track of all state parameters, not just unsqueezed Gaussian states as they often present the first (and only) choice for numerical evolutions of wave packets.

So far we have solved the case of $x = 0$, which makes the Hamiltonian simpler since it is linear at least in V. However, this case of lattice refinement, if it were used in a loop quantum cosmology version of the model, is not particularly realistic: if we take our solutions seriously only up to where the curvature variable $|P|$ reaches a value of the order one, which would signal the onset of strong holonomy corrections in loop quantum cosmology, we obtain a volume ratio $V_0/V_{|P|=1} = \sqrt{1 + 1/f_0^2}$. For this ratio to be sufficiently large, allowing the recollapse volume V_0 to be significantly larger than the volume where the quantum-geometry

Fig. 5.1 Symmetric and asymmetric recollapsing states [9]. The volume expectation value is represented by the central line, fluctuations by the top and bottom ones. Top diagram: The effect of volume fluctuations is hardly visible on a large scale. Bottom diagram: Differently spreading states depending on the correlations at the recollapse point. Only for vanishing correlations is the state symmetric around the recollapse.

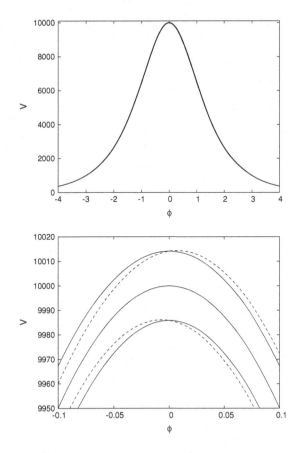

phase begins, we would have to choose f_0 extremely small. From the perspective of lattice refinement, this would be possible but not entirely natural.

We can nevertheless use our solutions to shed light on the behavior for general cases of x, without changing the qualitative properties. For that, we just need to transform non-linearly from the variables (V, P) for a desired x to new variables

$$\tilde{V} := \frac{\mathcal{V}}{G f_0} \left(\frac{(1-x) G f_0 V}{\mathcal{V}} \right)^{1/(1-x)},$$

$$\tilde{P} := P \left(\frac{\mathcal{V}}{(1-x) G f_0 V} \right)^{x/(1-x)} = P \left(\frac{\mathcal{V}}{G f_0 \tilde{V}} \right)^x. \tag{5.44}$$

Solutions for V and P will then take the form of $x = 0$-solutions found before, but with $\tilde{V}$ and $\tilde{P}$ inserted instead of V and P. Expectation values will thus refer to certain powers of the basic variables, rather than to those of V and P directly; and also quantum variables will be derived for the tilde-variables rather than the basic ones. A transformation of quantum variables based on a non-linear map of basic operators would be very complicated; in particular the orders will mix, such that fluctuations of V and P will depend on all kinds of higher moments of $\tilde{V}$ and $\tilde{P}$. But knowing the moments corresponding to some complete set of classical variables will tell us how the state behaves qualitatively, such as regarding its symmetry around the recollapse: In particular, it is clear that large asymmetries around the recollapse in general arise, irrespective of the value of x.

5.4.1.2 Negative Cosmological Constant

For a negative cosmological constant in a spatially flat universe, the Hamiltonian takes the same form as for positive spatial curvature provided we choose $x = -1/2$ instead of $x = 0$. Just some factors change and we have to put in $\Lambda < 0$, which by the previous procedure leads to solutions

$$P_{\text{classical}}(\varphi) = -\sqrt{|\Lambda|} f_0 \sinh\left(\frac{3}{2}\varphi\right) \tag{5.45}$$

$$V_{\text{classical}}(\varphi) = \frac{V_0}{\cosh\left(\frac{3}{2}\varphi\right)}, \tag{5.46}$$

again choosing the recollapse to happen at $\varphi = 0$, and

$$(\Delta P)^2(\varphi) = (\Delta P)_0^2 \cosh^2\left(\frac{3}{2}\varphi\right)$$

$$C_{VP}(\varphi) = (C_{VP})_0 + \frac{V_0 (\Delta P)_0^2}{\sqrt{|\Lambda|} f_0} \frac{\sinh\left(\frac{3}{2}\varphi\right)}{\cosh\left(\frac{3}{2}\varphi\right)}$$

$$(\Delta V)^2(\varphi) = \frac{(\Delta V)_0^2 + 2V_0 (C_{VP})_0 |\Lambda|^{-1/2} f_0^{-1} \tanh\left(\frac{3}{2}\varphi\right) + V_0^2 (\Delta P)_0^2 |\Lambda|^{-1} f_0^{-2} \tanh^2\left(\frac{3}{2}\varphi\right)}{\cosh^2\left(\frac{3}{2}\varphi\right)}.$$

$$\tag{5.47}$$

Qualitative properties are clearly the same, and compared to the value of $x = 0$ in a spatially closed universe we have the advantage that large recollapse volumes will be possible, $V_0/V_{|P|=1} = \sqrt{1 + 1/|\Lambda| f_0^2} \gg 1$, if only $|\Lambda|$ is small. This condition can be satisfied independently of f_0 and does not require fine tuning of the refinement scheme. (Notice that f_0 has different dimensions depending on the value of x). We may have to fine tune Λ, but this we are quite accustomed to, anyway.

5.4.1.3 Positive Cosmological Constant

For a positive cosmological constant, it is again the $x = -1/2$-case that makes the Hamiltonian linear in V. Equations for expectation values and quantum variables can approximately be solved as before, but here the large-volume regime is entirely different from the earlier two models. The flipped sign of the cosmological constant means that some signs in the solutions change, too, such as

$$P_{\text{classical}}(\varphi) = P_0 \cosh\left(\frac{3}{2}(\varphi - \varphi_0)\right) + \sqrt{P_0^2 - \Lambda f_0^2} \sinh\left(\frac{3}{2}(\varphi - \varphi_0)\right) \tag{5.48}$$

$$V_{\text{classical}}(\varphi) = V_0 \frac{\sqrt{P_0^2 - \Lambda f_0^2}}{P_0 \sinh\left(\frac{3}{2}(\varphi - \varphi_0)\right) + \sqrt{P_0^2 - \Lambda f_0^2} \cosh\left(\frac{3}{2}(\varphi - \varphi_0)\right)}. \quad (5.49)$$

This implies that the classical volume is unbounded from above, and in fact diverges for some finite value of φ. The unboundedness is certainly as it should be, for a free, massless scalar in a spatially flat space-time does not trigger a recollapse if there is a positive cosmological constant. Moreover, from $d\varphi/d\tau \propto p_\varphi/V(\varphi)$ (using $x = -1/2$) one can solve for proper time $\tau(\varphi)$ and verify that in this variable, unlike in internal time φ, it takes an infinite duration for the volume to diverge.

Problematic in the present scheme is, however, the fact that not only the volume diverges, as an approximation to the evolution of expectation values, but also some of the quantum variables such as ΔV or C_{VP}, while

$$(\Delta P)^2(\varphi) = (\Delta P)_0^2 \left(\cosh\left(\frac{3}{2}(\varphi - \varphi_0)\right) + \frac{P_0}{\sqrt{P_0^2 - \Lambda f_0^2}} \sinh\left(\frac{3}{2}(\varphi - \varphi_0)\right) \right)^2$$

$$(5.50)$$

remains finite. Using this solution, one can solve for the other variables, but, comparing (5.49) with (5.50), it is already clear that ΔV must diverge when V does: for infinite V the solution for $\Delta P \propto 1/V^2$ gives zero, which by the uncertainty relation requires an infinite volume fluctuation.

At this stage, the approximation breaks down: if quantum variables of second order diverge, we cannot ensure that those of higher order can be ignored in the process of solving up to second order. In fact, pushing the approximation to the next, third order and using numerical solutions indicates that several other higher moments tend to infinity, too [11]. The region of diverging volume is rather difficult to understand with the effective methods used so far, but one can nevertheless apply this scheme self-consistently to analyze the approach to large volume and see when higher moments become significant. One result, obtained up to around tenth order of the moments, is that a dynamical state quickly deviates from an initial Gaussian, but then, as in the example of Fig. 5.2, a new hierarchy of moments seems to arise.

Near the time of diverging volume one would have to use the entire collection of infinitely many moments, subject to a highly coupled dynamics. At this stage, a direct analysis of the wave function and the difference equation it satisfies becomes more useful. While analytic discussions are difficult also here, results from reduced phase-space quantization [12] (and numerical solutions) show how a state travels to diverging volume—and beyond. In fact, since these evolutions are done in terms of φ, and a self-adjoint extension of the Hamiltonian can be used for the evolution, nothing stops at the finite value of φ where the volume diverges. This is hard to interpret from the viewpoint of a classical space-time: no observer can reach this point because it would require an infinite amount of proper time. Formally, one can interpret the behavior in such a way that the self-adjoint extension of the Hamiltonian (which turns

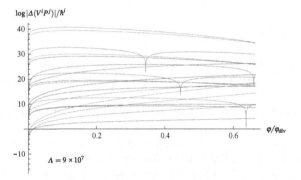

Fig. 5.2 Example for the evolution of odd-order moments, which vanish for an initial Gaussian. They rapidly deviate from the initial values, but then settle into a new hierarchy of a non-Gaussian state better adapted to the evolution [11]. (Spikes appear at zeros of the moments due to the logarithmic scale. The values of the scalar field are taken relative to the point of divergence φ_{div} where $V_{\text{classical}}(\varphi_{\text{div}}) \to \infty$).

out not to be unique in this case [13]) imposes a certain form of reflecting boundary condition at infinity, making the wave packet bounce back to finite volume. Formally, one may extend space–time; but if no observer can cross the divide, the extension lacks physical meaning. Only strong quantum back-reaction effects inducing a true recollapse at finite volume could make the extension meaningful, but such a behavior is not indicated by the higher-order analysis of [11].

Numerically, it is difficult to unravel if and in what sense the spread-out state indeed reaches infinite volume. It certainly follows the classical trajectory toward infinity very closely— as closely as one can say within the diverging quantum uncertainties. But since volume fluctuations diverge, too, the wave packet gets spread out more and more. Quantum back-reaction does ensue due to the large fluctuations and higher moments, and they yank the wave away from its classical track. One could expect a detailed analysis to show that the wave does not reach infinite volume, after all, but is held back by large moments. Then, proper time would not diverge, either, and the following collapse phase would be a true part of space–time. The low-curvature classical limit would be badly violated during the transition. However, numerically one sees, on the contrary, that the volume divergence is strengthened by the first several orders [11]. In any case, it is already clear that, rather counter-intuitively, this large-volume, small-curvature regime is one where quantum effects of a wave function are crucial to understanding the behavior of the wave function.

Diverging fluctuations in this model do not arise from strong quantum effects, but rather as an artefact of the internal time used in this specific case. The volume diverges at a finite φ, such that in a contour plot of the wave function, the region swept out by a wave packet resembles a strip parallel to the V-axis. Volume fluctuations are obtained from cross-sections of the wave function along the V-axis, and they (as well as some other moments) become large just by having a wave function moving along V. The wave function would still be considered semiclassical based on fluctuations of other quantities. All this happens at an infinite point in proper time, never reached by observers.

5.4.2 Scalar Potential

If the free, massless scalar is generalized by including a non-trivial potential, several difficulties arise. First, φ will now show some kind of non-trivial evolution, which only rarely allows it to be a good global internal time: φ may not have a monotonic relation with coordinate time (and p_φ will have zeroes during its evolution). Moreover, in the internal view the Hamiltonian $H = -p_\varphi$ now becomes time dependent via the potential $W(\varphi)$. In particular, it is no longer preserved and a dynamical state supported only on the positive part of the spectrum is not guaranteed to remain so supported at all times. Thus, dropping the absolute value in any Hamiltonian arising from p_φ is more difficult to justify. Nevertheless, for brief evolution times, where φ still serves as internal time, and for states supported at sufficiently large values of the Hamiltonian, far from the demarcation between positive and negative frequency, the approximation used so far will remain a good one.

One can demonstrate this in more detail: the second-order equation corresponding to the Hamiltonian constraint is

$$\hat{p}_\varphi^2 \psi = \frac{16\pi G}{3}(1-x)^2 \left(\widehat{V^2 P^2} - \left(\frac{8\pi G}{3}\right)^{(2+x)/(1-x)} \right.$$
$$\left. \times \left(\frac{1-x}{\mathcal{V}}\right)^{(1+2x)/(1-x)} (f_0 \hat{V})^{3/(1-x)} W(\varphi) \right) \psi. \qquad (5.51)$$

To separate the signs of frequencies, one makes use of the equation for p_φ directly, rather than its square. However, if we solve classically for $-p_\varphi = H(V, P, \varphi)$ and then quantize $H(V, P, \varphi)$, due to $[\hat{p}_\varphi, \hat{H}] \neq 0$ on kinematical states the equation

$$-\hat{p}_\varphi \psi = \hat{H} \psi \qquad (5.52)$$

is not equivalent to the second-order equation (5.51) but rather to

$$\hat{p}_\varphi^2 \psi = -\hat{p}_\varphi \hat{H} \psi = -\hat{H} \hat{p}_\varphi \psi - [\hat{p}_\varphi, \hat{H}]\psi = \hat{H}^2 \psi - [\hat{p}_\varphi, \hat{H}]\psi. \qquad (5.53)$$

A strict solution for p_φ at the operator level is much more difficult to construct in the case of a time-dependent Hamiltonian.

But for solutions one is usually interested in, the additional commutator term in (5.53) turns out to be small. The form of the Hamiltonian indicates that we have a commutator

$$[\hat{p}_\varphi, \hat{H}] \sim i\hbar \mathcal{V}^2 \left(\frac{8\pi G f_0 (1-x)}{3\mathcal{V}}\right)^{3/(1-x)} \frac{|\hat{V}|^{3/(1-x)} W'(\varphi)}{\hat{H}}$$

whose expectation value is small compared to that of $\hat{H}^2$ in (5.53) provided that H is suffiently large and the potential not too steep.

There are now two conditions to be ensured: $\mathcal{V}^2 (V/\mathcal{V})^{3/(1-x)} W(\varphi) \ll V^2 P^2 \approx H^2$ for the potential term to be perturbative, and $\hbar \mathcal{V}^2 (V/\mathcal{V})^{3/(1-x)} W'(\varphi)/H \ll H^2$

for the commutator in (5.53) to be negligible require $W \ll \mathcal{V}^{-2} H^2 (V/\mathcal{V})^{-3/(1-x)}$ and $\hbar W' \ll \mathcal{V}^{-2} H^3 (V/\mathcal{V})^{-3/(1-x)}$. For a sufficiently small and flat potential, long evolution times can be considered in our approximation, using the first-order equation and dropping absolute values.

Further, much stronger support for the treatment of effective Hamiltonians in the presence of a non-trivial potential comes from the framework of effective constraints [14, 15], to be described in more detail in Sect. 13.2. In this way one avoids the need to deparameterize, and yet an effective Hamiltonian can be derived by solving a constrained system. Sufficiently far away from regions in which p_φ is nearly zero, close to a turning point for φ in coordinate time, φ can still be used as local internal time. (As a trace of non-unitary state evolution, φ may acquire an imaginary contribution, but this has no adverse consequences for observables [16, 17].) To move through the turning point, another local internal time must be used, just as different time coordinates are usually required to coordinatize a whole manifold. Changes of local internal times can be achieved by gauge transformations of the effective constrained system.

From the analysis of cosmological-constant models, which can be interpreted as models with a constant scalar potential, it is clear that quantum back-reaction generically arises. Equations of motion take the same form as before, with $W(\varphi)$ replacing Λ. Solutions are more difficult to find due to the non-trivial time dependence of the potential, but numerically one can easily investigate the behavior for some types of potentials. It would also be of interest to summarize the effect of quantum variables in some kind of effective potential or other terms, which would no longer contain independent degrees of freedom but correct the classical equations of motion for V and P to incorporate quantum effects. This is sometimes possible in effective theories, especially if an adiabatic approximation around a ground state can be used, see Chap. 13.

In quantum cosmology we do not have a ground state, and it turns out that no other adiabatic approximation to solve for quantum variables and insert them back into the equations of motion for expectation values is possible [18]. Quantum variables thus do play a very crucial role, and they cannot be reduced to mere corrections such as an effective potential. New quantum degrees of freedom are essential, and we are dealing with a higher-dimensional effective system. This observation is especially relevant in strong quantum regimes such as those expected at small volume or high curvature. But then, we will also have to include the quantum-geometry effects of loop quantum cosmology, which we will do in the next chapter.

References

1. Ashtekar, A., Pawlowski, T., Singh, P.: Phys. Rev. D **74**, 084003 (2006). gr-qc/0607039
2. Falciano, F.T., Pinto-Neto, N., Santini, E.S.: Phys. Rev. D **76**, 083521 (2007). arXiv:0707.1088
3. Colistete, R. Jr, Fabris, J.C., Pinto-Neto, N.: Phys. Rev. D **62**, 083507 (2000). gr-qc/0005013
4. Kaminski, W., Pawlowski, T.: Phys. Rev. D **81**, 084027 (2010). arXiv:1001.2663

5. Isham, C.J. In DeWitt, B.S, Stora, R. (eds.) Relativity, Groups and Topology, 2nd edn. Lectures given at the 1983 Les Houches Summer School on Relativity, Groups and Topology (1983)
6. Klauder, J.: Int. J. Mod. Phys. D **12**, 1769 (2003). gr-qc/0305067
7. Klauder, J.: Int. J. Geom. Meth. Mod. Phys. **3**, 81 (2006). gr-qc/0507113
8. Bojowald, M.: Proc. Roy. Soc. A. **464**, 2135 (2008). arXiv:0710.4919
9. Bojowald, M., Tavakol, R.: Phys. Rev. D **78**, 023515 (2008). arXiv:0803.4484
10. Bojowald, M., Brizuela, D., Hernandez, H.H., Koop, M.J. Morales-Técotl, H.A. arXiv:1011.3022
11. Bentivegna, E., Pawlowski, T.: Phys. Rev. D **77**, 124025 (2008). arXiv:0803.4446
12. Mielczarek, J., Piechocki, W.: Phys. Rev. D **83**, 104003 (2011). arXiv:1011.3418
13. Kaminski, W., Pawlowski, T.: Phys. Rev. D **81**, 024014 (2010). arXiv:0912.0162
14. Bojowald, M., Sandhöfer, B., Skirzewski, A., Tsobanjan, A.: Rev. Math. Phys. **21**, 111 (2009). arXiv:0804.3365
15. Bojowald, M., Tsobanjan, A.: Phys. Rev. D **80**, 125008 (2009). arXiv:0906.1772
16. Bojowald, M., Höhn, P.A., Tsobanjan, A.: Class. Quantum. Grav. **28**, 035006 (2011). arXiv:1009.5953
17. Bojowald, M., Höhn, P.A., Tsobanjan, A.: Phys. Rev. D **70**, 124022 (2010). arXiv:1011.3040
18. Bojowald, M., Hernández, H., Skirzewski, A.: Phys. Rev. D. **76**, 063511 (2007). arXiv:0706.1057

Chapter 6
Harmonic Cosmology: The Universe Before the Big Bang and How Much We Can Know About It

To understand the strong quantum phase that is expected to be realized in a small and highly curved universe around the big bang, we return to the solvable model of loop quantum cosmology developed in Sect. 5.3.2. Solvability is realized for a free, massless scalar in a spatially flat isotropic universe, but formulated in the non-canonical variables V and $J = V \exp(iP)$ rather than V and P used for a Wheeler–DeWitt model. The linear Hamiltonian $H = -\sqrt{16\pi G/3}(1-x)|\frac{1}{2}iJ_-| \propto |i(J - J^*)|$ then provides solvability: the evolution of expectation values does not couple to quantum variables. (As in Sect. 5.3.1, dropping the absolute value is justified for a time-independent Hamiltonian provided initial values for a state make it be supported on a definite sign of the spectrum of iJ_-.) Equations of motion can then be solved for expectation values as well as quantum variables to determine the behavior of a whole state. With the equations, also the qualitative behavior of solutions will be insensitive to the refinement scheme as specified by the parameter x, as long at least as it is of power-law form.

6.1 Reality

We consider basic operators satisfying the linear algebra

$$[\hat{V}, \hat{J}] = \hbar \hat{J}, \quad [\hat{V}, \hat{J}^\dagger] = -\hbar \hat{J}^\dagger, \quad [\hat{J}, \hat{J}^\dagger] = -2\hbar \hat{V} \tag{6.1}$$

whose dynamics is given by a linear Hamiltonian $\hat{H} \propto i(\hat{J} - \hat{J}^\dagger)$. Just as in the Wheeler–DeWitt model before, the resulting dynamical equations will be easy to solve. But there is a new issue due to the complex-valuedness of variables and the non-canonical form of their algebra. Expectation values of these variables cannot be prescribed arbitrarily, for the relationship $\hat{J}\hat{J}^\dagger = \hat{V}^2$ which must hold with a unitary quantization of $\exp(iP)$ ordered to the right of $\hat{V}$ in $\hat{J}$ implies an equality

$$|\langle \hat{J} \rangle|^2 - \langle \hat{V} \rangle^2 = (\Delta V)^2 - (\Delta J_+)^2 + (\Delta J_-)^2 \tag{6.2}$$

M. Bojowald, *Quantum Cosmology*, Lecture Notes in Physics 835,
DOI: 10.1007/978-1-4419-8276-6_6, © Springer Science+Business Media, LLC 2011

relating expectation values to fluctuations of $\hat{V}$ and $\hat{J}_{\pm} = \hat{J} \pm \hat{J}^{\dagger}$. We will refer to this important equation, as well as to others derived for higher-order moments, as a reality condition, since it derives from the fact that the classical P entering the definition of the complex J must be real. Once (6.2) is implemented by restricting $\langle \hat{J} \rangle$, together with the simple reality condition $\langle \hat{V} \rangle \in \mathbb{R}$, we are dealing with expectation values of states in the physical Hilbert space realizing the correct adjointness relations of basic operators. The reality condition (6.2) also reduces the number of degrees of freedom contained in the expectation values back to two since the imaginary part of J can no longer be chosen independently of its real part.

> As used here, it is sometimes convenient to work with the complex J and $\bar{J}$, rather than the real or purely imaginary $J_{\pm}$. One can easily translate back and forth thanks to the linear relationship between these variables. Care should just be used in the interpretation of the variables: what may look like a fluctuation in one set, for instance $(\Delta J)^2 = \frac{1}{4}((\Delta J_+)^2 + 2\Delta(J_+ J_-) + (\Delta J_-)^2)$, actually contains a covariance in the other. As we have already seen, correlations often play special roles in semiclassical regimes via decoherence, and they do so more specifically in questions about the asymmetry of states. Despite possibly differing appearances in the variables J and $\bar{J}$, we will thus make statements precise by using only the real variables $\hat{J}_+$ and $i\hat{J}_-$ to determine what should be considered a covariance and what a fluctuation, or mixtures of both. At the present level of formulating and solving equations of motion, however, these issues can be postponed.

In these considerations, the fluctuations on the right of (6.2) are considered fixed, so that (6.2) is a reality condition for expectation values, reducing to the classical one when fluctuations and $\hbar$ vanish. The moments themselves must be subject to reality as well. In fact, analogous conditions arise at higher order which reduce the second-order moments to the right number [1]. We obtain those conditions by considering operator equations of the form

$$\hat{V}^i \hat{J}^j \hat{J}^{\dagger k}(\hat{J}\hat{J}^{\dagger} - \hat{V}^2) = 0 \quad \text{with } i + j + k > 0 \tag{6.3}$$

which directly follow from the initial operator equation and are thus implied. But expectation values in the presence of the extra factors do not agree exactly with the basic reality condition (6.2); they rather differ from it by terms involving moments of higher order, at least third. For instance, at third order, with a single extra operator in the equation, we have the third-order moments

$$\Delta(VJ\bar{J}) \equiv \langle (\hat{V} - \langle \hat{V} \rangle)(\hat{J} - \langle \hat{J} \rangle)(\hat{J}^{\dagger} - \langle \hat{J}^{\dagger} \rangle) \rangle_{\text{symm}} \tag{6.4}$$

$$\Delta(V^3) \equiv \langle (\hat{V} - \langle \hat{V} \rangle)^3 \rangle \tag{6.5}$$

appearing in the reality condition

$$\Delta(VJ\bar{J}) - \Delta(V^3) = 2\langle \hat{V} \rangle (\Delta V)^2 - 2(\text{Re}\Delta(VJ)\text{Re}\langle \hat{J} \rangle + \text{Im}\Delta(VJ)\text{Im}\langle \hat{J} \rangle)$$

that follows from the expectation value of $\hat{V}(\hat{J}\hat{J}^{\dagger} - \hat{V}^2) = 0$. Just as we use (6.2) to restrict semiclassical expectation values up to terms of order $\hbar$, we use third-order reality conditions to restrict semiclassical second-order moments. To leading order in $\hbar$,

$$\langle \hat{V} \rangle (\Delta V)^2 = \text{Re}(\langle \hat{J}^\dagger \rangle \Delta(VJ))$$
$$= \text{Re}\langle \hat{J} \rangle \text{Re}\Delta(VJ) + \text{Im}\langle \hat{J} \rangle \text{Im}\Delta(VJ), \tag{6.6}$$

as a restriction for second-order moments. Similarly, we obtain

$$\langle \hat{V} \rangle \text{Re}\Delta(VJ) = \frac{1}{2} \left(\text{Re}\langle \hat{J} \rangle \text{Re}(\Delta J)^2 + \text{Im}\langle \hat{J} \rangle \text{Im}(\Delta J)^2 + \text{Re}\langle \hat{J} \rangle \Delta(J\bar{J}) \right),$$
$$\langle \hat{V} \rangle \text{Im}\Delta(VJ) = \frac{1}{2} \left(\text{Re}\langle \hat{J} \rangle \text{Im}(\Delta J)^2 - \text{Im}\langle \hat{J} \rangle \text{Re}(\Delta J)^2 + \text{Im}\langle \hat{J} \rangle \Delta(J\bar{J}) \right) \tag{6.7}$$

from the other third-order conditions. General considerations of this type show that none of the higher-order conditions restricts second-order moments further. Thus, the six initial second-order moments (counting real and imaginary parts of $\Delta(VJ)$ and $(\Delta J)^2$ separately, while $(\Delta V)^2$ and $\Delta(J\bar{J})$ are always real) are restricted by three conditions. Three degrees of freedom remain, just as we expect it for two fluctuations and one correlation. Only the counting is more complicated for non-canonical basic operators, and must take into account all relations betweem them.

Reality conditions as derived here present an example of Casimir conditions discussed in Sect. 13.1.4. Interpreting $J\bar{J} - V^2 = 0$ as a constraint on the classical phase space, the large set of expectation-value conditions arising from (6.3) is an example of effective-constraint methods developed in [2–4]. The reality condition, although just one constraint, is of second class on the non-symplectic phase space spanned by $(V, J, \bar{J})$, using generalizations of Dirac's classification of constraints to general Poisson manifolds [5]. Thus, only conditions for the moments result but no gauge flow need be factored out.

Since all reality conditions derived from (6.3) are based on $\hat{J}\hat{J}^\dagger - \hat{V}^2 = 0$, and thus on the understanding that $\exp(iP)$ is ordered to the right of V in $\hat{J}$, some of their coefficients may change in different orderings. However, this does not affect general statements made here.

6.2 Uncertainty

Uncertainty relations follow for every pair of self-adjoint operators by a well-known application of the Schwarz inequality; see Sect. 13.1.3. For our non-canonical, partially complex basic variables in harmonic loop quantum cosmology we derive three independent inequalities for the pairs $(\hat{V}, \hat{J}_+)$, $(\hat{V}, i\hat{J}_-)$ and $(\hat{J}_+, i\hat{J}_-)$ of self-adjoint operators:

$$(\Delta V)^2 (\Delta J_+)^2 - \Delta(VJ_+)^2 \geq \hbar^2 H^2 \tag{6.8}$$

$$-(\Delta V)^2 (\Delta J_-)^2 + \Delta(VJ_-)^2 \geq \frac{1}{4}\hbar^2 \langle \hat{J}_+ \rangle^2 \tag{6.9}$$

$$-(\Delta J_+)^2(\Delta J_-)^2 + \Delta(J_+J_-)^2 \geq \hbar^2 \langle \hat{V} \rangle^2 \tag{6.10}$$

with $H = \langle \hat{H} \rangle = -\frac{1}{2}i\langle \hat{J}_- \rangle$. The second-order moments featuring in here are restricted by reality conditions as well, and just as we have seen for the number of moments, the number of uncertainty relations is reduced to the expected number (one per canonical pair) when reality is imposed.

A direct calculation shows that this is indeed the case. We first rewrite (6.6) and (6.7) in terms of $\hat{J}_\pm$ instead of $\hat{J}$ and $\hat{J}^\dagger$:

$$\langle \hat{V} \rangle (\Delta V)^2 = \frac{1}{4}(\langle \hat{J}_+ \rangle \Delta(VJ_+) + \langle \hat{J}_- \rangle \Delta(VJ_-)) \tag{6.11}$$

$$\langle \hat{V} \rangle \Delta(VJ_+) = \frac{1}{4}(\langle \hat{J}_+ \rangle (\Delta J_+)^2 + \langle \hat{J}_- \rangle \Delta(J_+J_-)) \tag{6.12}$$

$$\langle \hat{V} \rangle \Delta(VJ_-) = \frac{1}{4}(\langle \hat{J}_- \rangle (\Delta J_-)^2 + \langle \hat{J}_+ \rangle \Delta(J_+J_-)). \tag{6.13}$$

From suitable combinations we then derive

$$\langle \hat{V} \rangle^2((\Delta V)^2(\Delta J_+)^2 - \Delta(VJ_+)^2) = \frac{1}{4}\langle \hat{J}_- \rangle^2((\Delta J_+)^2(\Delta J_-)^2 - \Delta(J_+J_-)^2)$$

$$\langle \hat{V} \rangle^2(-(\Delta V)^2(\Delta J_-)^2 + \Delta(VJ_-)^2) = \frac{1}{4}\langle \hat{J}_+ \rangle^2(-(\Delta J_+)^2(\Delta J_-)^2 + \Delta(J_+J_-)^2)$$

and conclude, noting that $-i\langle \hat{J}_- \rangle = 2H$ by the constraint, that both (6.8) and (6.9) are equivalent to (6.10). Only one of the uncertainty relations is required once second-order reality is imposed, giving rise to only one independent uncertainty relation per canonical pair.

In these calculations we have ignored terms higher than second order. If higher-order terms are significant, the equivalence of second-order uncertainty relations no longer follows in the same way. In fact, when higher-order terms matter in reality conditions, one must use uncertainty relations for higher-order moments as well, not just at second order. Higher-order uncertainty relations also follow from the Schwarz inequality, but they have a more complicated form that mixes moments of different orders. For the purposes of our discussions here, second-order relations will be sufficient.

6.3 Repulsive Forces and Bouncing Cosmologies

The linear Hamiltonian provides equations of motion for expectation values which, thanks to solvability, decouple into finite sets. We will absorb a factor of $\sqrt{16\pi G/3}(1-x)$ in the internal time variable defined as $\lambda := \sqrt{16\pi G/3}(1-x)\varphi$. For expectation values we have

$$\frac{d}{d\lambda}\langle\hat{V}\rangle = \frac{1}{i\hbar}\langle[\hat{V},\hat{H}]\rangle = -\frac{1}{2}(\langle\hat{J}\rangle + \langle\hat{J}^{\dagger}\rangle) \tag{6.14}$$

$$\frac{d}{d\lambda}\langle\hat{J}\rangle = \frac{1}{i\hbar}\langle[\hat{J},\hat{H}]\rangle = -\langle\hat{V}\rangle = \frac{d}{d\lambda}\langle\hat{J}^{\dagger}\rangle. \tag{6.15}$$

and for fluctuations and correlations

$$\frac{d}{d\lambda}(\Delta V)^2 = -\Delta(VJ) - \Delta(V\bar{J}) \tag{6.16}$$

$$\frac{d}{d\lambda}(\Delta J)^2 = -2\Delta(VJ), \quad \frac{d}{d\lambda}(\Delta\bar{J})^2 = -2\Delta(V\bar{J}) \tag{6.17}$$

$$\frac{d}{d\lambda}\Delta(VJ) = -\frac{1}{2}(\Delta J)^2 - \frac{1}{2}\Delta(J\bar{J}) - (\Delta V)^2 \tag{6.18}$$

$$\frac{d}{d\lambda}\Delta(V\bar{J}) = -\frac{1}{2}(\Delta\bar{J})^2 - \frac{1}{2}\Delta(J\bar{J}) - (\Delta V)^2 \tag{6.19}$$

$$\frac{d}{d\lambda}\Delta(J\bar{J}) = -\Delta(VJ) - \Delta(V\bar{J}). \tag{6.20}$$

In particular, the right-hand side of (6.2) is constant in time, and reality conditions need be imposed only for initial expectation values. If initial values are posed in a semiclassical regime, the left-hand side of (6.2) must be of the order $\langle\hat{V}\rangle\hbar$, and thus be much smaller than the squares of expectation values themselves. In this way, the classical condition is recovered.

Both $\langle\hat{V}\rangle$ and $\langle\hat{J}\rangle$ have the same dimension as $\hbar$, and second-order moments in semiclassical states are typically of the order $\hbar$. The right combination of semiclassical behavior with dimensions results in the order of magnitude $\hbar\langle\hat{V}\rangle$ for quantum fluctuations squared and the right-hand side of (6.2). There are no free dimensionful constants in this model that could be used to provide the correct dimensions of fluctuations without reference to dynamical variables such as $\langle\hat{V}\rangle$. Indeed, the behavior $\hbar\langle\hat{V}\rangle$ for second-order moments will be shown clearly by the following discussion of dynamical coherent states.

Equations of motion of the linear model, such as (6.14) and (6.15) for expectation values and (6.16–6.20) for second-order moments, are linear and finitely coupled, making them easily solvable. The reality conditions are quadratic, but preserved in time. Thus, they need not be considered when solving equations of motion; one will just have to make sure that they are satisfied by initial values used. (Numerically one may have to monitor that reality conditions, as well as uncertainty relations, remain respected in the presence of rounding errors, which can sometimes be a subtle issue [1].) The quadratic terms they contain do not amount to quantum back-reaction of fluctuations on expectation values, and they do not complicate the solution procedure. For the expectation values, we obtain

$$\langle\hat{V}\rangle(\lambda) = \frac{1}{2}(Ae^{-\lambda} + Be^{\lambda}) \tag{6.21}$$

$$\langle \hat{J} \rangle(\lambda) = \frac{1}{2}(Ae^{-\lambda} - Be^{\lambda}) + iH \tag{6.22}$$

with two integration constants A and B, while $H = \langle \hat{H} \rangle$, the expectation value of the Hamiltonian, must equal the imaginary part of $\langle \hat{J} \rangle$. Comparing with the solutions (5.13) for a Wheeler-DeWitt quantization of the same system, we notice that there are different combinations of the exponentials. The dynamics is thus indeed different, but so far it is not clear whether the singularity would be avoided: $\langle \hat{V} \rangle$ can easily be zero at some λ provided A and B have opposite signs.

At this point, we still have to impose the reality condition (6.2), which for the specific form of solutions implies $AB = H^2 + c_1$ with another constant $c_1 := (\Delta J_+)^2 - (\Delta J_-)^2 - (\Delta V)^2$ preserved by evolution. Looking at a state which is semiclassical at least once, say at large volume, we know that the fluctuation parameter c_1 must be of the order $\hbar \langle \hat{H} \rangle$ (which is constant just like c_1).

Up to quantum corrections, we thus have $AB = H^2 + O(H\hbar) > 0$, implying that A and B must have the same sign. Choosing the positive one, and defining $A/B =: e^{2\varepsilon}$, the solution (6.21) can only be of the form $\langle \hat{V} \rangle(\lambda) = H \cosh(\lambda - \varepsilon)$ which never becomes zero: the classical as well as Wheeler–DeWitt approach to vanishing volume is replaced by a smooth bounce. Even though this is a statement about a true quantum regime, we did not have to make any assumptions about the form of the state there; all we used was the condition that it be semiclassical at large volume. In this way, as we will also see it for other questions, the solvable model allows powerful conclusions about deep quantum properties. Reality conditions are preserved dynamically; still, some moments evolve and may become more strongly quantum. But the existence of dynamical coherent states shows that the quantum behavior is not exceedingly strong in this model.

An intuitive way to understand the removal of the singularity is by an effective space-time picture, which in the case of isotropic models follows from an effective Friedmann equation. This must be an equation of motion just for $\langle \hat{V} \rangle$, a power of the scale factor, not coupled to $\langle \hat{J} \rangle$ as it is so far in (6.14) and (6.15). We can eliminate $\langle \hat{J}_- \rangle$ in terms of the scalar momentum, $p_\varphi = -H = \sqrt{4\pi G/3}(1 - x)i\langle \hat{J}_- \rangle$, which will then enter the kinetic energy density $\rho_{\text{free}} = p_\varphi^2/2a^6\mathcal{V}^2$ of the free scalar. The other variable, $\langle \hat{J}_+ \rangle$, is not independent of $\langle \hat{J}_- \rangle$ and can be obtained by using the reality condition:

$$\langle \hat{J}_+ \rangle^2 = \langle \hat{J}_- \rangle^2 + 4\langle \hat{V} \rangle^2 + 4c_1.$$

Combined, we have

$$\left(\frac{d\langle \hat{V} \rangle}{d\varphi}\right)^2 = \frac{4\pi G}{3}(1 - x)^2\langle \hat{J}_+ \rangle^2 = \frac{16\pi G}{3}(1 - x)^2\langle \hat{V} \rangle^2 - p_\varphi^2 + O(\hbar)$$

or, in terms of proper time τ appearing in the derivative denoted by a dot,

$$\left(\frac{\dot{a}}{a}\right)^2 = \left(\frac{\dot{\varphi}}{2(1-x)\langle\hat{V}\rangle}\frac{d\langle\hat{V}\rangle}{d\varphi}\right)^2 = \left(\frac{4\pi G f_0 a^{-3} p_\varphi}{3a^{2(1-x)}\mathscr{V}^2}\right)^2 \left(\frac{3\mathscr{V}^2}{4\pi G f_0^2}a^{4(1-x)} - p_\varphi^2 + O(\hbar)\right)$$

$$= \frac{4\pi G}{3}\frac{p_\varphi^2}{a^6\mathscr{V}^2}\left(1 - \frac{4\pi G f_0^2}{3\mathscr{V}^2}a^{-4(1-x)}p_\varphi^2 + O(\hbar)\right) \tag{6.23}$$

$$= \frac{8\pi G}{3}\rho_{\text{free}}\left(1 - \rho_{\text{free}}\cdot\frac{8\pi G f_0^2}{3}a^{2(1+2x)}\right) + O(\hbar). \tag{6.24}$$

(Recall $V = 3a^{2(1-x)}\mathscr{V}/8\pi G f_0(1-x)$ from (5.3).) To leading order, we thus have the Friedmann equation with the energy density of a free, massless scalar; but quantum corrections are effectively included by a term of higher order in the free scalar's kinetic energy. This term becomes important at high densities, near the critical one of

$$\rho_{\text{crit}} = \frac{3}{8\pi G f_0^2 a^{2(1+2x)}} = \frac{3}{8\pi G\gamma^2 L^2} \tag{6.25}$$

with the patch size $L = \ell_0 a$ ($\gamma\ell_0 = \mathscr{V}^{1/3}\gamma\delta(a) = f_0 a^{2x}$), where it can cancel the classical term and cause an extremum in the evolution of the scale factor.

This complete effective equation reproduces a modified Friedmann equation found first by a tree-level approximation incorporating holonomy corrections directly in the classical Hamiltonian [6, 7], which was also tested numerically [6, 7]. Thanks to solvability and the absence of quantum back-reaction, no strong extra corrections arise and the tree-level approximation turns out to be good.

Tree-level expansions have been tested numerically in several models, but only for strongly peaked states in which quantum back-reaction is weak. In this context, it is important to note that quantum back-reaction arises from moments appearing in time derivatives of equations of lower order, such as those for expectation values. Quantum back-reaction is caused by quantum fluctuations, among other moments, but is distinct from the statistical effect of quantum fluctuations. The initial meaning of quantum fluctuations is that they determine the spread of outcomes of repeated measurements done for many systems in the same state. Such an interpretation is unavailable in quantum cosmology; instead, the primary role of quantum fluctuations is as a state parameter that influences the evolution of expectation values and leads to deviations from the classical behavior, providing quantum corrections.

Keeping this in mind, even in the presence of fluctuations, deviations of $\langle\hat{V}\rangle(\varphi)$ from tree-level equations may be small, even smaller than the value of fluctuations in a state used would indicate. What is affected by quantum back-reaction is not the measurement of the volume, but the rate of change $d\langle\hat{V}\rangle/d\varphi$. As seen from (5.33) (for a Wheeler–DeWitt quantization), the rate of change of volume is affected mainly by $(\Delta P)^2$ and C_{VP}, not by $(\Delta V)^2$. Even large volume fluctuations do not change the evolution much if curvature fluctuations and correlations are sufficiently small. A comparison of deviations of $\langle\hat{V}\rangle$ from the tree-level equations with volume fluctuations ΔV, as sometimes used to justify tree-level approximations, is not meaningful because it would confuse the two different roles of fluctuations, statistical aspects and the dynamical quantum back-reaction.

Moreover, fluctuations typically grow as states are evolved. While one can always choose a suitably peaked initial state, quantum back-reaction becomes strong if one waits long enough. At this stage, tree-level equations break down and one has to include higher orders by systematic effective equations.

The second equation of motion, the one for $d\langle \hat{J}_+ \rangle/d\varphi$, can be reformulated in a similar way as a second-order equation for $\langle \hat{V} \rangle$, resembling the Raychaudhuri equation. Following the same procedure as in Sect. 5.3.1, we first derive the second-order equation

$$\frac{d^2 \langle \hat{V} \rangle}{d\varphi^2} = \frac{16\pi G}{3} (1 - x)^2 \langle \hat{V} \rangle + O(\langle \hat{V} \rangle \hbar). \tag{6.26}$$

Curiously, this is exactly the same (5.17), as obtained earlier; corrections from loop quantum cosmology arise only by the $O(\langle \hat{V} \rangle \hbar)$-terms from the reality condition, corresponding to the different factor orderings used. Holonomy corrections, therefore, do not correct the equation for $d^2 \langle \hat{V} \rangle/d\varphi^2$. However, to derive the Raychaudhuri equation for $\ddot{a}$, we use the Friedmann equation. With the modified Friedmann (6.23), we then obtain the modified Raychaudhuri equation

$$\frac{\ddot{a}}{a} = -\frac{4\pi G}{3} \rho_{\text{free}} \left(1 - (2 - x) \frac{\rho_{\text{free}}}{\rho_{\text{crit}}} \right) \tag{6.27}$$

which is corrected compared to (5.18). At high densities, near the new extremum of a, the correction implies a positive second derivative of a by time: $\ddot{a} > 0$ for $\rho_{\text{free}} > \rho_{\text{crit}}/(2 - x)$, ensuring that the high-density extremum implied by (6.23) is a minimum.

Incidentally, the phase of acceleration spans only a rather small range of densities: for $-1/2 \leq x \leq 0$, acceleration begins when ρ_{free} rises above a value between $\frac{2}{5} \rho_{\text{crit}}$ (for $x = -1/2$) and $\frac{1}{2} \rho_{\text{crit}}$ (for $x = 0$). The density at the onset of acceleration does not depend much on the value of x in the given range. In both cases, the density must be of the critical size within just one order of magnitude in order to produce acceleration from holonomy corrections.

Of course, we have already seen the presence of a minimum in the exact cosh-like solution for the free scalar model. An effective equation would not be necessary at this stage, just as it is not of much use to write down an effective action for a free quantum field theory. But the power of effective descriptions of such simple systems is that they can be used as starting points for perturbative analyses of more complicated, interacting systems. While still technically involved, such a procedure is usually much easier to perform than an outright quantization and subsequent analysis of the interacting system; and yet, as is well known from particle physics, most of the desired information can be extracted effectively and efficiently.

6.4 Quantum Big Bang

In a solvable system, things are simple and clean. In the present context we have seen that a spatially flat isotropic model sourced by a free massless scalar, in loop quantum cosmology, shows a bounce of the scale factor if the corresponding quantum state is semiclassical once (e.g. at large volume). Even if it evolves away from a semiclassical state when smaller volumes are reached, quantum properties of the free state cannot be strong enough to prevent the bounce. (In fact, in the next section we will see that a near-coherent state does remain nearly coherent throughout the bounce phase, although its fluctuations can still change noticeably).

But if the model is no longer solvable, even if it is just perturbed by another contribution to its energy density by a term small compared to the kinetic one, quantum back-reaction ensues. States may evolve much more strongly over the required cosmological periods of time, even if local changes remain perturbatively small. If there is a non-constant scalar potential, φ will not always be a global internal time and is unlikely to serve as time all the way between large volume and the bounce. One would patch together several effective space-time regions, each described in terms of φ as internal time but linked in different phases where φ would have turning points with respect to coordinate time (and p_φ vanishes). Physically, the semiclassical state at large volume is supposed to arise via decoherence at intermediate times. These complicated processes, possible only by interactions with an environment of many degrees of freedom, must now be wound backwards if the big-bang state is to be understood. In all these constructions, the solvable dynamics would severely be disturbed, implying that considerable changes must be taken into account for an analysis that could be called reliable and robust. What we need is a general effective equation, formulated for arbitrary states and in the presence of an unconstrained potential. We will see that conclusions from such an equation can in fact be drawn, and it provides the basis for specific analyses of concrete matter models or special classes of states.

If there is a matter potential not large compared to the kinetic term, perturbative techniques can be used as in the preceding chapter. Our classical φ-Hamiltonian now is

$$H := \sqrt{\frac{16\pi G}{3}}(1-x)\sqrt{V^2 P^2 - \frac{3\mathscr{V}^2}{8\pi G(1-x)^2}\left(\frac{8\pi G(1-x)f_0}{3\mathscr{V}}V\right)^{3/(1-x)}} \, W(\varphi).$$

$$(6.28)$$

After doing the loop replacement of VP by $-\frac{1}{2}iJ_-$, we expand $H = \sum_{k=0}^{\infty} H_k$ with

$$H_k := -i\sqrt{\frac{4\pi G}{3}}(1-x)J_-\binom{\frac{1}{2}}{k}\left(\frac{3\mathscr{V}^2}{2\pi G(1-x)^2}\left(\frac{8\pi G(1-x)f_0}{3\mathscr{V}}V\right)^{3/(1-x)}\frac{W(\varphi)}{J_-^2}\right)^k.$$

We are clearly dealing with a non-linear Hamiltonian, which implies quantum back-reaction terms. They can be derived from a background-state expansion around expectation values, or by the effective methods of Chap. 13 based on Poisson geometry, giving equations of motion such as [8]

$$\frac{d\langle\hat{V}\rangle}{d\varphi} = -i\langle\hat{J}_+\rangle\left(\frac{\partial H}{\partial J_-} + \sum_{n=2}^{\infty}\sum_{a=0}^{n}\frac{\partial^{n+1}H}{\partial V^a\partial J_-^{n-a+1}}\frac{\Delta(V^a J_-^{n-a})}{a!(n-a)!}\right)$$
$$-i\sum_{n=2}^{\infty}\sum_{a=0}^{n}\frac{\partial^n H}{\partial V^a\partial J_-^{n-a}}\frac{\Delta(V^a J_-^{n-a-1}J_+)}{a!(n-a-1)!}. \tag{6.29}$$

The second line arises from the fact that commutators of non-canonical basic operators are not mere constants, or that expectation values and moments do not have vanishing Poisson brackets for non-canonical variables. (In this sense, the more interacting nature of the loop-quantized model, which does not allow a harmonic formulation in canonical variables, is responsible for effects based on these terms. One consequence discussed below is the asymmetry of fluctuations before and after the bounce.) Quantum corrections, as always, arise from coupling terms of expectation values and moments $\Delta(\cdots)$ as defined in (5.23).

Although there are more terms than before in the solvable model, formally even infinitely many ones, we can follow the previous route to an effective Friedmann equation [8, 9]: The Hamiltonian $H_Q = \sum_k\langle\hat{H}_k\rangle$ is no longer linear in $\langle\hat{J}_-\rangle$, but when equated with $-p_\varphi$ provides a polynomial equation to a given perturbative order: For $x = -1/2$ and up to $k = 1$ for instance,

$$H_Q = -\sqrt{3\pi G}i\langle\hat{J}_-\rangle\left(1 - \frac{4\pi G f_0^2}{3}\frac{\langle\hat{V}\rangle^2}{(i\langle\hat{J}_-\rangle)^2}(1+\varepsilon_1)W(\varphi)\right) = -p_\varphi \tag{6.30}$$

where

$$\varepsilon_1 = \sum_{n=2}^{\infty}(-1)^n\left(\frac{\Delta(J_-^n)}{\langle\hat{J}_-\rangle^n} - 2\frac{\Delta(V J_-^{n-1})}{\langle\hat{V}\rangle\langle\hat{J}_-\rangle^{n-1}} + \frac{\Delta(V^2 J_-^{n-2})}{\langle\hat{V}\rangle^2\langle\hat{J}_-\rangle^{n-2}}\right)$$

is the first correction from quantum back-reaction (obtained as the relative-moment expansion of V^2/J_-, or of $V^{3/(1-x)}/J_-$ for general x).

Perturbatively, we can solve this polynomial equation to find $i\langle\hat{J}_-\rangle$ in terms of p_φ:

$$i\langle\hat{J}_-\rangle = \frac{p_\varphi}{\sqrt{3\pi G}}\left(1 + \frac{1}{4}\frac{W(\varphi)(1+\varepsilon_1)a^6\varphi'^2}{p_\varphi^2}\right) = \frac{p_\varphi}{\sqrt{3\pi G}}\left(1 + \frac{1}{8}\frac{\rho_{\text{pot}}}{\rho_{\text{kin}}}(1+\varepsilon_1)\right). \tag{6.31}$$

The reality condition, which does not change by interaction terms, again provides $\langle\hat{J}_+\rangle$ in terms of $\langle\hat{J}_-\rangle$, $\langle\hat{V}\rangle$ and quantum terms:

$$\frac{\langle\hat{J}_+\rangle}{2\langle\hat{V}\rangle} = \pm\sqrt{1 - \left(\frac{\langle\hat{J}_-\rangle}{i(2\langle\hat{V}\rangle)}\right)^2 - \varepsilon_0}$$

with $\varepsilon_0 = ((\Delta J_+)^2 - (\Delta J_-)^2 - (\Delta V)^2)/\langle\hat{V}\rangle^2$.

Then, in turn, we insert $\langle \hat{J} \rangle_-$ in terms of p_φ, providing

$$\langle \hat{J}_+ \rangle = \pm 2 \langle \hat{V} \rangle \sqrt{1 - \rho_Q / \rho_{\text{crit}}} \qquad (6.32)$$

with

$$\rho_Q := \rho + \varepsilon_0 \rho_{\text{crit}} + W \sum_{k=0}^{\infty} \varepsilon_{k+1} \left(\frac{W a^6 \mathscr{V}^2}{p_\varphi^2} \right)^k \qquad (6.33)$$

and the same value of ρ_{crit} as in the solvable case. Higher orders in k provide their own correction parameters as relative moment expansions of $V^{3k/(1-x)} J_-^{1-2k}$. At this stage, we notice that it is not the classical energy density ρ that appears in the quantum-geometry correction, but a quantum-corrected version ρ_Q. This quantity allows us to generalize the regularity result of the free model to an arbitrary interacting one: Since $\langle \hat{J}_+ \rangle$ must be real as one of the reality conditions, the quantum density ρ_Q cannot be larger than ρ_{crit} thanks to (6.32). (Depending on the value of x, ρ_{crit} may be a-dependent. Specific realizations of the boundedness condition then depend on the history of the scale factor.)

In contrast to the solvable model, however, reaching the upper bound for the density does not automatically imply a bounce for expectation values. Extra terms in the (6.29) of φ-motion for $\langle \hat{V} \rangle$ from quantum back-reaction mean that there are terms in the effective Friedmann equation

$$\left(\frac{\dot{a}}{a} \right)^2 = \frac{8 \pi G}{3} \left(\rho \left(1 - \frac{\rho_Q}{\rho_{\text{crit}}} \right) \pm \frac{1}{2} \sqrt{1 - \frac{\rho_Q}{\rho_{\text{crit}}}} \, \eta W + \frac{\mathscr{V}^2 a^6 W^2}{2 p_\varphi^2} \eta^2 \right) \qquad (6.34)$$

with general quantum corrections $\eta = \sum_k \eta_{k+1} (\mathscr{V}^2 a^6 W / p_\varphi^2)^k$ where, e.g.,

$$\eta_1 = \sum_{n=2}^{\infty} (-1)^n \left(n \frac{\Delta(J_-^{n-1} J_+)}{\langle \hat{J}_- \rangle^n} - 2(n-1) \frac{\Delta(V J_-^{n-2} J_+)}{\langle \hat{V} \rangle \langle \hat{J}_- \rangle^{n-1}} + (n-2) \frac{\Delta(V^2 J_-^{n-3} J_+)}{\langle \hat{V} \rangle^2 \langle \hat{J}_- \rangle^{n-2}} \right).$$
$$(6.35)$$

If $\eta \neq 0$, the point where the maximal energy density is reached does not correspond to a bounce; and whether or not there is a bounce at all depends on the specific behavior of the state. As seen from the terms in the corrections, it is only quantum correlations that could prevent the bounce. If a state is uncorrelated at the critical density, even if it is highly fluctuating, a bounce will still ensue.

What is not so clear, however, is whether the universe will indeed bounce back to large volume or, if several zeros of $\dot{a}$ become possible due to the moment dynamics, get stuck in oscillations at small volume. Such models would resemble emergent or oscillatory scenarios as in [10–15]. Another possibility is for the wave function, which initially was sharply peaked at large volume, to split up into several packets. One of them may bounce, but others may not do so or even tunnel through to the minisuperspace region where the triad is inverted. (Reflection symmetry may sometimes be treated as a large gauge transformation, requiring wave functions to be (anti-)symmetric and not leaving any freedom for tunneled wave functions. But due to the presence of fermions violating parity, this assumption of reflections as large gauge transformations is not realistic.) Even though one packet would describe a bouncing universe, the whole state would imply a much stronger quantum phase. Only decoherence scenarios could tell how the transition would be perceived in terms of measurable quantities. Which scenarios are possible is still to be determined by a systematic analysis of many specific cases, using the coupled dynamics of expectation values and the moments.

6.5 Dynamical Coherent States

Much of the interesting behavior of cosmic quantum states is determined by their leading moments: fluctuations and correlations. While fluctuations are realized in any quantum state, specific studies often assume simple uncorrelated Gaussian states. This may not capture all properties reliably, and so correlations themselves must be included and analyzed—especially since many processes such as decoherence rely on the build-up of quantum correlations. Examples for the role of correlations seen so far in isotropic quantum cosmology are the asymmetry of states around the recollapse of a closed universe (Sect. 5.4.1.1), and changes to as well as a potential prevention of a bounce in loop quantum cosmology beyond the harmonic model. For the would-be bounce, one also must know what kinds of quantum states one can expect in general, and we thus return to the solvable model and analyze its dynamical coherent states in more detail [16]. Possible implications for singularities will be discussed more generally in the following chapter.

A solvable model is of the most highly controlled form, and dynamical coherent states are the most highly controlled ones within the model. By definition, such states saturate the uncertainty relation at all times, which means that many of their properties must be preserved during evolution. Well-known examples are coherent states of the harmonic oscillator, which retain a constant shape while they follow the classical trajectory. However, there are also squeezed states, dynamically coherent as well, but with oscillating fluctuations. Their shape does not remain constant in time, but changes in a way controlled by their correlation. Similarly, what we have already seen in quantum cosmology clearly shows that we have to understand correlations of quantum states near the big bang, and see how large they and their roles could be.

"Stable" coherent states can be defined more generally [17] for a large class of anharmonic systems. Like the harmonic oscillator, these systems have families of states $|z\rangle$ labeled by a complex number z, such that $|z\rangle(t) = |z + \omega t\rangle$ with a real ω. For the harmonic oscillator, the specific expression of z in terms of state parameters depends only on expectation values; since no moments change as z evolves to $z + \omega t$, the shape of the state is dynamically preserved. For anharmonic systems, however, the expression of z in terms of state parameters depends on some of the moments. As a consequence, the shape of the state may change considerably in time. These states are not dynamically coherent, and semiclassicality can easily be lost when a state evolves.

Back in the solvable model, we have a finite set of equations of motion (6.16)–(6.20) for second-order moments, which are linear and can be solved straightforwardly:

$$(\Delta V)^2(\lambda) = \frac{1}{2}(c_3 e^{-2\lambda} + c_4 e^{2\lambda}) - \frac{1}{4}(c_1 + c_2)$$

$$(\Delta J)^2(\lambda) = \frac{1}{2}(c_3 e^{-2\lambda} + c_4 e^{2\lambda}) + \frac{1}{4}(3c_2 - c_1) - i(c_5 e^{\lambda} - c_6 e^{-\lambda})$$

$$(\Delta \bar{J})^2(\lambda) = \frac{1}{2}(c_3 e^{-2\lambda} + c_4 e^{2\lambda}) + \frac{1}{4}(3c_2 - c_1) + i(c_5 e^{\lambda} - c_6 e^{-\lambda})$$

$$\Delta(VJ)(\lambda) = \frac{1}{2}(c_3 e^{-2\lambda} - c_4 e^{2\lambda}) + \frac{i}{2}(c_5 e^{\lambda} + c_6 e^{-\lambda})$$

$$\Delta(V\bar{J})(\lambda) = \frac{1}{2}(c_3 e^{-2\lambda} - c_4 e^{2\lambda}) - \frac{i}{2}(c_5 e^{\lambda} + c_6 e^{-\lambda})$$

$$\Delta(J\bar{J})(\lambda) = \frac{1}{2}(c_3 e^{-2\lambda} + c_4 e^{2\lambda}) + \frac{1}{4}(3c_1 - c_2).$$

We have kept the constant $c_1 = -(\Delta V)^2 + \Delta(J\bar{J})$ introduced earlier, and added five more integration constants c_i. From the reality condition with the parameters in our expectation-value solution $\langle \hat{V} \rangle(\lambda) = A \cosh(\lambda)$, we have $c_1 = A^2 - H^2$, and using $\hat{H} = -\frac{1}{2}i(\hat{J} - \hat{J}^\dagger)$ we obtain

$$(\Delta H)^2 = -\frac{1}{4}\left((\Delta J)^2 - 2\Delta(J\bar{J}) + (\Delta \bar{J})^2\right) = \frac{1}{2}(c_1 - c_2). \qquad (6.36)$$

These six constants cannot be chosen arbitrarily because the three uncertainty relations (6.8), (6.9) and (6.10), or only one of them if the second-order reality conditions (6.6) and (6.7) are imposed as well, must be satisfied for the variables. Dynamical coherent states saturate these inequalities at all times λ, imposing four conditions on the integration constants:

$$4c_3 c_4 = H^2 \hbar^2 + \frac{1}{4}(c_1 + c_2)^2 \qquad (6.37)$$

$$(c_1 - c_2)c_3 - c_6^2 = \frac{1}{4}A^2 \hbar^2 = (c_1 - c_2)c_4 - c_5^2 \qquad (6.38)$$

$$4c_5 c_6 = A^2 \hbar^2 + c_2^2 - c_1^2. \qquad (6.39)$$

As a first application of these equations, we determine bounds on the asymmetry, before and after the free bounce, of fluctuations of a state. From the solutions and uncertainty constraints on the parameters, we have [18]

$$\left| \lim_{\lambda \to -\infty} \frac{(\Delta V)^2}{\langle \hat{V} \rangle^2} - \lim_{\lambda \to \infty} \frac{(\Delta V)^2}{\langle \hat{V} \rangle^2} \right| = 2 \frac{|c_3 - c_4|}{A^2}$$

$$= 4\frac{H}{A} \sqrt{\left(1 - \frac{H^2}{A^2} + \frac{1}{4}\frac{\hbar^2}{A^2}\right) \frac{(\Delta H)^2}{A^2} - \frac{1}{4}\frac{\hbar^2}{A^2} + \left(\frac{H^2}{A^2} - 1\right) \frac{(\Delta H)^4}{A^4}}$$

$$\sim 4\frac{H}{A} \sqrt{\left(1 - \frac{H^2}{A^2}\right) \frac{(\Delta H)^2}{A^2} - \frac{1}{4}\frac{\hbar^2}{A^2}} = 4\frac{H}{A} \sqrt{\frac{(\Delta H)^4}{A^4} + \frac{c_1 + c_2}{2} \frac{(\Delta H)^2}{A^4} - \frac{1}{4}\frac{\hbar^2}{A^2}}$$

$$\tag{6.40}$$

making it clear that states in general are not symmetric. A better measure than the difference of relative volume fluctuations is the ratio of relative fluctuations before and after the bounce themselves, since it is independent of their absolute size. Solving the equations for c_4 and dividing by it, we find

$$\left| 1 - \frac{(\Delta V)_-^2}{(\Delta V)_+^2} \right| = \frac{|c_4 - c_3|}{c_4}$$

$$= \frac{2\delta H / A}{\delta^2 / 2(\Delta H)^2 \pm \delta H / A + \frac{1}{2}(\Delta H)^2 H^2 / A^2 + \frac{1}{8}A^2 \hbar^2 / (\Delta H)^2}$$

$$\tag{6.41}$$

where

$$\delta := \sqrt{\left(\frac{H^2}{A^2} - 1\right)(\Delta H)^4 + \left(A^2 - H^2 + \frac{1}{4}\hbar^2\right)(\Delta H)^2 - \frac{1}{4}A^2 \hbar^2}$$

$$\sim \sqrt{(A^2 - H^2)(\Delta H)^2 - \frac{1}{4}A^2 \hbar^2}.$$

(Note that $\langle \hat{V} \rangle_+ = \langle \hat{V} \rangle_-$, and thus $(\Delta V)_- / (\Delta V)_+$ is identical to the ratio of relative volume fluctuations $\Delta V / \langle \hat{V} \rangle$.)

This asymmetry parameter would vanish for a symmetric state, but one can easily see that it can be as large as of the order of ten. Even if we set $c_1 + c_2 = 0$, absolutely minimizing $c_3 c_4$ in the saturation equation (6.37), no strong restriction arises. Thus, generic states are non-symmetric around the bounce. Also here, as seen in Sect. 5.4.1.1 for recollapses, one can verify that a correlation parameter controls the asymmetry.

These equations are consistent with second-order reality conditions. If we insert the explicit solutions into (6.6) and (6.7) and compare coefficients of different powers of e^λ, we obtain three independent equations

$$c_5 = \frac{A}{2H}\left(2c_4 - \frac{1}{2}(c_1 + c_2)\right), \quad c_6 = \frac{A}{2H}\left(2c_3 - \frac{1}{2}(c_1 + c_2)\right) \tag{6.42}$$

$$c_5 + c_6 = \frac{H}{A}(c_1 - c_2). \tag{6.43}$$

We first use these equations to eliminate c_5 and c_6 from the sum

$$c_3 + c_4 = \frac{H}{A}(c_5 + c_6) + \frac{1}{2}(c_1 + c_2) = \frac{H^2}{A^2}(c_1 - c_2) + \frac{1}{2}(c_1 + c_2). \quad (6.44)$$

The remaining constants on the right-hand side are related to state parameters via

$$c_1 + c_2 = 2(A^2 - H^2 - (\Delta H)^2), \quad c_1 - c_2 = 2(\Delta H)^2 \quad (6.45)$$

using (6.36). For the asymmetry we are interested in the absolute value of the difference of c_3 and c_4, which can be obtained from the sum using (6.37):

$$(c_3 - c_4)^2 = (c_3 + c_4)^2 - 4c_3 c_4 = \frac{H^4}{A^4}(c_1 - c_2)^2 + \frac{H^2}{A^2}(c_1^2 - c_2^2) - H^2 \hbar^2$$
$$= 4H^2 \left(\left(1 - \frac{H^2}{A^2}\right)(\Delta H)^2 - \frac{\hbar^2}{4} + \left(\frac{H^2}{A^2} - 1\right)\frac{(\Delta H)^4}{A^2} \right). \quad (6.46)$$

This equation is equivalent to (6.40).

At this stage, we have reproduced the asymmetry derived from all three uncertainty relations in a different way, using reality conditions and only one of the uncertainty relations. Since we already know that second-order reality conditions reduce the number of uncertainty relations to just one, this result is not surprising. However, the rederivation allows a powerful generalization of the asymmetry formula to all semiclassical states, not just dynamically coherent ones. Reality conditions of the form used are valid provided only that moments of order higher than second are subdominant, which is the most general definition of semiclassical states. The preceding derivation remains valid if we change the equality in (6.37) to an inequality once we depart from dynamical coherent states. In this way, the last formula changes to the inequality

$$(c_3 - c_4)^2 \leq 4H^2 \left(\left(1 - \frac{H^2}{A^2}\right)(\Delta H)^2 - \frac{\hbar^2}{4} + \left(\frac{H^2}{A^2} - 1\right)\frac{(\Delta H)^4}{A^2} \right) \quad (6.47)$$

providing the asymmetry for all semiclassical states. The inequality even applies more generally to states which may not be semiclassical as long as moments of order three or higher are suppressed compared to second-order moments. Fluctuations and correlations, on the other hand, need not be restricted for the validity of the equations used and can be large.

The relative asymmetry (6.41) is more difficult to reproduce for classes of states that are not dynamically coherent because the presence of an inequality for $(c_3 - c_4)^2$ rather than an equality makes it impossible to combine the sum and difference of c_3 and c_4 in such a way to bound $|c_4 - c_3|/c_4$ from above. Equation (6.41) is thus the key result for which dynamical coherent states are important. It has one consequence which is more difficult to see by other means. As mentioned, it implies factors of order up to ten between the fluctuations which may not seem large, but one should

Fig. 6.1 Asymmetry
parameter $|1 - \Delta_+/\Delta_-|$
from (6.41) depending on the
ratio of initial values A and
H. Different curves
correspond to different
values of H, the steepness
increasing with H

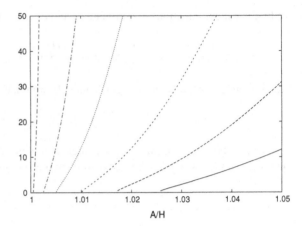

keep in mind that they are realized for highly controlled dynamical coherent states in a solvable model. Moreover, in non-solvable models quantum back-reaction means that correlation terms in the equations of motion in general significantly alter the evolution such that states spread out even more. Based on general principles, without restricting the class of states by further means, no strong statement about the behavior near or before the bounce can be made. What is more, the asymmetry behavior even of dynamical coherent states in the solvable model is very sensitive to initial values. Figure 6.1 shows examples of asymmetry parameters for different amounts of matter, showing that it reacts very sensitively to small changes in initial conditions A and H. Especially for a large $H = -p_\varphi$, which is required to ensure kinetic domination, the sensitivity is very pronounced.

Even if the solvable model could be taken as an approximation to the real universe, observations would not determine state parameters relevant at the bounce in precisely enough a manner. As illustrated by this example of cosmic forgetfulness [19], it is practically impossible to draw conclusions about full properties of the state of the universe before the big bang, as it presents itself in loop quantum cosmology. Only some parameters, such as expectation values, can be extrapolated directly without high sensitivity to initial values.

In the context of these discussions, it is important to realize that deterministic behavior is not questioned. This means that one can always choose an extremely sharply peaked initial state of precisely known properties and claim that its fluctuations do not change significantly. Cosmic forgetfulness rather refers to how precisely such information could be extracted in realistic terms with imperfect knowledge of state properties at any given time, for which the sensitivity is crucial.

As one would expect, restrictions on parameters obtained by effective equations are also consistent with relations that have been derived in a rather different way for wave-function evolution [20]. Results obtained in this way are even more general because no restriction on higher-order moments is required, as long as one expresses asymmetry equations in terms of $\Delta \log V$ rather than $\Delta V/\langle \hat{V} \rangle$. (For a semiclassical state, $\Delta \log V$ reduces to $\Delta V/\langle \hat{V} \rangle$, as can be seen by Taylor expanding the volume around its expectation value in the integral that defines ΔV. If the volume fluctuation is small, only the leading term of the Taylor expansion need be considered.) As already mentioned, the sensitivity of the asymmetry toward state parameters, and thus the basis of cosmic forgetfulness, can be seen easily only for the much higher controlled dynamical coherent states or specific classes of states.

One can see the reason for the existence of (perhaps surprisingly) sharp bounds on the spreading of states over extremely long times in the role of the covariance, combined with the fact that the model considered is harmonic. One can show that it is the covariance which controls the asymmetry [16]. If one then assumes that $\Delta \log |p_\varphi|$ is given and bounded at large volume, where a semiclasical state should be realized, the harmonic dynamics with $|p_\varphi| \sim |VP|$ at large volume immediately produces a bound on correlations for given fluctuations independent of the uncertainty relation. Moreover, the explicitly known dynamics shows that the covariance is constant until large curvature is reached just before the bounce point. The initial semiclassicality in this way controls the asymmetry in the bounce phase, explaining why sharp relations for the asymmetry are available.

Even though there are highly controlled bounds on relative fluctuations (and thus semiclassicality) in this solvable model, the asymmetry of fluctuations themselves may still be significant. The reason for this is that the difference (6.40) of relative volume fluctuations is bounded by relative matter fluctuations. (The fluctuation ΔH refers to the scalar momentum.) The more interesting relative change of volume fluctuations (6.41), which is insensitive to the actual size of volume fluctuations, is obtained by dividing relative matter fluctuations by relative volume fluctuations. In general, matter fluctuations and volume fluctuations are independent quantities; it is even reasonable to assume that matter behaves "more quantum" than geometry. (This assumption is realized also in quantum field theory on curved spacetimes.) Relative matter fluctuations should then be significantly larger than relative volume fluctuations, and the asymmetry (6.41) can easily be larger than one.

The issue of the asymmetry has appeared controversial in some part of the literature. Initially the question arose from claims in [7] saying that states have symmetric fluctuations, based on numerical evidence. However, the methods of [7] used for the plots shown most often, with very symmetric fluctuations, contain a desqueezing procedure for the initial Gaussian states explicitly mentioned in Sect. V.B.2 of that paper. Later on, an effective analysis [16, 21] showed the importance of squeezing for the asymmetry (which was already visible in [7], although not interpreted in this way). Especially the issue of cosmic forgetfulness [19] has stimulated several follow-up papers, some disputing its relevance and countering it with "cosmic recall" [22]. In this context, one should note that the sizes provided for asymmetry bounds in these different treatments were always consistent with one another, but often interpreted in diametrically opposite ways. Curiously, especially the weakest bounds found in this context have been attributed the strongest meaning [22]. In that article, an asymmetry bound was derived for semiclassical states which is linear in fluctuations. Since the asymmetry is defined for quadratic fluctuations and fluctuations are small for semiclassical states, a linear fluctuation bound is automatically smaller than quadratic fluctuations, even if no difference is taken to compute the asymmetry [23]. A significant improvement of the bounds was then given by the new scattering methods of [20], which for semiclassical states is equivalent to the bounds presented here based on effective techniques. However, in spite of the equivalence regarding the assymmetry, the results of [20] cannot address the issue of cosmic forgetfulness because they do not discuss the sensitivity to initial values.

In this context, see also the discussion in Sect. 7.4.

6.6 Lessons for Effective Actions

In this chapter, we have analyzed a harmonic system of loop quantum cosmology for which effective equations, describing the evolution of expectation values and moments of a state, can be analyzed exactly. As in this example, effective techniques for canonical systems such as canonical quantum gravity primarily provide effective Hamiltonians or constraints and effective equations of motion. It would be of interest to find an equivalent effective action, but this is not straightforward. Nevertheless, what we have seen based on effective equations and solvable models provides some insights into the general properties of a general effective action that might correspond to loop quantum cosmology.

There have been attempts to find effective actions in a backward way, by trying to reproduce effective equations of motion of harmonic systems, using $f(R)$ theories [24] or Lovelock theories [25]. None of the features pointed out here are realized in those actions. These attempts do not present a systematic derivation of an effective action; they rather try to extend the known effective dynamics from one point in superspace (the exactly solvable isotropic model) into an infinite-dimensional neighborhood in superspace. Without any control, for instance by the consideration of consistency requirements resulting from anomaly-freedom (Chap. 10), the resulting action principles are too ambiguous to be relevant for cosmological evaluations. Related aspects have been discussed in [17], pointing out in particular the nonuniqueness of effective actions obtained by simply matching the isotropic equations they provide with modified equations of isotropic quantum cosmology.

In general non-solvable models, quantum back-reaction means that the quantum system has more dynamically interacting degrees of freedom than the classical one. At the level of an effective action, this can only be modeled by higher time derivatives as they naturally arise from higher-curvature terms. The quantum action is then non-local in time, just as it must ultimately be non-local in space due to the presence of spatially integrated holonomies and fluxes as basic variables. Later, in our discussion of inhomogeneities, we will however show that higher-curvature terms cannot be the only contributions to an effective action, with interesting consequences for potential cosmological observations; see Sect. 10.3.

We can also infer that the form of the effective action must be matter dependent; it cannot be independently formulated for vacuum quantum gravity. After all, solvability is realized for a particular matter ingredient—a free, massless scalar— and this case implies the absence of quantum back-reaction. In other words, the free system underlying quantum gravity, corresponding to the harmonic oscillator for quantum field theory, is realized not in vacuum but with this specific kind of matter. Since quantum back-reaction at the level of an effective action is reflected in the higher-derivative terms, the presence of higher-derivative terms must depend on the matter ingredients, another indication that quantum corrections cannot just be of the usual higher-curvature type which would be matter independent. From this conclusion, one may also expect some kind of unification even though so far there is no hint for this in the fundamental formulation of loop quantum gravity. Quantum-correction terms in an effective action that faithfully reflects the solvability properties seen in models must provide some balance between gravitational and matter contributions.

The models analyzed so far by effective means only included holonomy effects as quantum-geometry corrections, in addition to quantum back-reaction. Another effect is the one from inverse-triad corrections, which changes the dynamics of homogeneous models but does not add much new to the general picture of effective descriptions. At the inhomogeneous level, however, we will see further implications in a later chapter.

Both types of quantum-geometry corrections, when formulated in terms of the coordinate-dependent connection component $\tilde{c}$ or the scale factor, depend on the lattice spacing ℓ_0 of an underlying state. Higher powers of the connection, as they may contribute to a higher-curvature effective action, come in the form $(\ell_0 \tilde{c})^n$, which can be written as $(\gamma v(a)^{1/3} \dot{a}/a)^n$ in terms of the fundamental patch volume $v(a)$. Coefficients of higher-order terms and their dimensions are thus provided not directly by the Planck length but by the state-dependent $v(a)$; as usual, effective actions depend on the state. The form of the state must be known for a specific effective action, and no unique one can result in the absence of a distinguished state such as the vacuum for the low-energy effective action; see Sect. 13.1.5. In particular, $v(a)$ need not be constant but may change as the universe expands. (Of course, one may Taylor expand $v(a)$ as a function of a and rearrange the higher-order corrections in terms of constant coefficients.) In this way, the elementary discreteness and its refinement enters effective descriptions and can have a bearing on potential observations. We will see this more clearly in the context of inhomogeneities, Sect. 10.1.2.3.

References

1. Bojowald, M., Mulryne, D., Nelson, W., Tavakol, R.: Phys. Rev. D **82**, 124055 (2010). arXiv:1004.3979
2. Bojowald, M., Sandhöfer, B., Skirzewski, A., Tsobanjan, A.: Rev. Math. Phys. **21**, 111 (2009). arXiv:0804.3365
3. Bojowald, M., Tsobanjan, A.: Phys. Rev. D **80**, 125008 (2009). arXiv:0906.1772
4. Bojowald, M., Tsobanjan, A.: Class. Quantum Grav. **27**, 145004 (2010). arXiv:0911.4950
5. Bojowald, M., Strobl, T.: Rev. Math. Phys. **15**, 663 (2003). hep-th/0112074
6. Singh, P., Vandersloot, K.: Phys. Rev. D **72**, 084004 (2005). gr-qc/0507029
7. Ashtekar, A., Pawlowski, T., Singh, P.: Phys. Rev. D **73**, 124038 (2006). gr-qc/0604013
8. Bojowald, M.: Gen. Rel. Grav. **40**, 2659 (2008). arXiv:0801.4001
9. Bojowald, M.: Phys. Rev. Lett. **100**, 221301 (2008). arXiv:0805.1192
10. Ellis, G.F.R., Maartens, R.: Class. Quantum Grav. **21**, 223 (2004). gr-qc/0211082
11. Ellis, G.F.R., Murugan, J., Tsagas, C.G.: Class. Quant. Grav. **21**, 233 (2004). gr-qc/0307112
12. Mulryne, D.J., Tavakol, R., Lidsey, J.E., Ellis, G.F.R.: Phys. Rev. D **71**, 123512 (2005). astro-ph/0502589
13. Bojowald, M.: Nature **436**, 920 (2005)
14. Parisi, L., Bruni, M., Maartens, R., Vandersloot, K.: Class. Quantum Grav. **24**, 6243 (2007). arXiv:0706.4431
15. Lidsey, J.E., Mulryne, D.J., Nunes, N.J., Tavakol, R.: Phys. Rev. D **70**, 063521 (2004). gr-qc/0406042

16. Bojowald, M.: Phys. Rev. D **75**, 123512 (2007). gr-qc/0703144
17. Gazeau, JP., Klauder, J.: J. Phys. A Math. Gen. **32**, 123 (1999)
18. Bojowald, M.: Proc. Roy. Soc. A **464**, 2135 (2008). arXiv:0710.4919
19. Bojowald, M.: Nat. Phys. **3**(8), 523 (2007)
20. Kaminski, W. Pawlowski, T.: Phys. Rev. D **81**, 084027 (2010). arXiv:1001.2663
21. Bojowald M.: Phys Rev D **75**, 081301(R) (2007). gr-qc/0608100
22. Corichi, A., Singh, P.: Phys. Rev. Lett. **100**, 161302 (2008). arXiv:0710.4543
23. Bojowald, M.: Phys. Rev. Lett. **101**, 209001 (2008). arXiv:0811.2790
24. Olmo, G.J., Singh, P.: JCAP **0901**, 030 (2009). arXiv:0806.2783
25. Date, G., Sengupta, S.: Class. Quantum Grav. **26**, 105002 (2009). arXiv:0811.4023

Chapter 7
What Does It Mean for a Singularity to be Resolved?

We have now seen and studied in quite some detail a general mechanism by which loop quantum cosmology can resolve singularities, based on the fundamental difference equation, and a very specific one of an effective bounce in a solvable model. In such a situation, and also in comparison with Wheeler–DeWitt quantizations, the question arises what it should mean, in general, for a singularity to be resolved.

The general issue is rather messy due to the presence of several different statements even in the classical determination of singularities. One may use curvature divergence as a physical condition known from many explicit solutions, but one difficult to handle at a general level. Strict theorems mainly make use of the comparatively weak and rather different notion of geodesic incompleteness. When potential resolutions of singularities are to be discussed, the first question to ask is which one of the classical criteria for a singularity one should focus on.

Another important condition to be considered is the genericness of resolution mechanisms. Singularities can be avoided even classically by clever choices of initial values [1]. The classical singularity problem does not state that all realistic solutions must develop singularities; the problem is rather caused by the fact that no mechanism is known that could avoid singularities generically. Genericness is also the most important condition for quantum cosmological resolutions of singularities; also here it is not that difficult to avoid specific types of singularities since the freedom, for instance in violating energy conditions, is much higher than classically.

7.1 Density Bounds

We have already encountered bounds for the quantum density ρ_Q in the context of the quantum Friedmann equation; see the remarks after (6.32). Since there are quantum corrections from fluctuations and higher moments in the expression (6.33) for the quantum density, the physical matter density that follows from the quantum matter Hamiltonian may still be unbounded in non-semiclassical states. For stronger results about density bounds, the matter density must be constrained by additional means, probably involving the dynamics of specific models.

M. Bojowald, *Quantum Cosmology*, Lecture Notes in Physics 835,
DOI: 10.1007/978-1-4419-8276-6_7, © Springer Science+Business Media, LLC 2011

An alternative result exists for models with a cosmological constant in addition to the free, massless scalar. These models are not harmonic, but have one advantage compared to general matter systems because they are still deparameterizable by φ. If one considers the spectrum of a density operator defined by quantizing the matter energy divided by the volume with inverse-volume techniques of loop quantum gravity, on upper bound for density values can be obtained [2].

> Although a rigorous derivation is important, the existence of such a bound is not altogether surprising [3, 4]. The loop modification of c by $\sin(\delta c)/\delta$ replaces the unbounded c by a bounded function. Unless quantum back-reaction is strong, the matter density is then bounded by the Friedmann equation. The non-trivial aspect of loop quantum cosmology is not to bound the density in isotropic models, but to construct consistent embeddings of such models in a full quantum system that includes inhomogeneities. Such an embedding has not yet been achieved completely consistently, but the relationship between loop quantum cosmology and loop quantum gravity is showing the available possibilities.

Upper bounds for the spectrum of density operators can be interpreted as one contribution to the resolution of singularities. However, they also show the limitations of some of the current results obtained in loop quantum cosmology. In isotropic models one can easily construct density operators, once methods for inverse-triad quantizations are known following [5]. But in inhomogeneous situations or the full theory there is no well-defined way to obtain density operators in loop quantum gravity; only quantities of density weight zero can be quantized. One may thus quantize the matter Hamiltonian $\hat{H}$ [6] and the volume $\hat{V}$ [7, 8] corresponding to some finite region. Their expectation values in a given state then provide a measure for the density $\langle\hat{H}\rangle/\langle\hat{V}\rangle$. One can use the same construction in isotropic models, which would bring the results closer to those of the full theory. An alternative expression for the matter density results, one that differs from the expectation value $\langle\widehat{HV^{-1}}\rangle$ of an isotropic density operator. The difference is given by moments of the state, and is strong in a highly quantum state. If the moments of the state are large, the physical matter density may be unbounded even if there is an upper bound on the spectrum of density operators. To decide whether actual densities are bounded, one must bring the moments and the shape of states under better control. No general, state-independent bounds as in [2] can be expected. Not many results in this direction are known, but it has already been demonstrated numerically that the bounds of [2] can be breached if the shape of states is taken into account [9].

7.2 Bounces

A more specific scenario for singularity resolution is a turn-around of a from collapse to expansion, or a bounce [10], keeping the scale factor away from zero and energy densities finite.

Example 7.1. (Fluctuation-triggered bounce) In a Wheeler–DeWitt treatment of the model analyzed in Example 2.1, the requirement of self-adjointness in the face of a boundary of the a-axis requires special fall-off conditions for wave functions. The wave function is always suported on $a > 0$, which implies that the expectation value of $\hat{a}$ cannot become zero. In contrast to the classical solutions, a bounce or at least a mechanism to keep $\langle \hat{a} \rangle$ away from the singular $a = 0$ is indicated.

In order to see the role of the boundary more clearly, we use effective equations for the system quantized with the non-canonical variables V and $D = VP$, $\{V, D\} = V$. As discussed in Sect. 2.2, group-theoretical quantization then leads to self-adjoint basic operators on the phase space $\mathbb{R}^+ \times \mathbb{R}$. In Example 2.1 we reformulated the system in Schrödinger form in the presence of dust, for which the Hamiltonian, when quantized by our non-canonical operators, can be taken to be of the self-adjoint form $\hat{H} = \hat{V}^{-1}\hat{D}^2\hat{V}^{-1}$. (By contrast, $\hat{H} = \hat{P}^2$ is not self-adjoint on the positive half-plane.) In the affine variables, the Hamiltonian is no longer quadratic and requires a choice of factor ordering, here done in a simple (but non-unique) symmetric way. The dynamics is no longer of free-particle form, resolving the problem of wave packets crossing the boundary seen in Example 2.1.

In order to determine the role of fluctuations at small volume, we perform a background-state expansion and then compute an expectation value of the Hamiltonian, or use the methods of Chap. 13. To second order, we obtain the quantum Hamiltonian

$$H_Q = \frac{\langle \hat{D} \rangle^2}{\langle \hat{V} \rangle^2} + \frac{1}{\langle \hat{V} \rangle^2}(\Delta D)^2 - 4\frac{\langle \hat{D} \rangle}{\langle \hat{V} \rangle^3}\Delta(VD) + 3\frac{\langle \hat{D} \rangle^2}{\langle \hat{V} \rangle^4}(\Delta V)^2. \tag{7.1}$$

Fluctuation terms clearly change the Hamiltonian and, when large, can affect the dynamics. For instance, for a state that remains unsqueezed Gaussian at small scales we have $\Delta(VD) = 0$ and $(\Delta D)^2 = \langle \hat{V} \rangle^2 \hbar^2/4(\Delta V)^2$. (The uncertainty relation, saturated for a Gaussian, follows as in the discussion of harmonic cosmology; see Chap. 13 for details.) For such states, we can write

$$H_Q = P^2 \left(1 + 3\frac{(\Delta V)^2}{\langle \hat{V} \rangle^2}\right) + \frac{1}{4}\frac{\hbar^2}{(\Delta V)^2} = p_T \tag{7.2}$$

with the dust momentum p_T equal to the Hamiltonian in this deparameterization. We have transformed back to the conventional curvature parameter P, now defined as $\langle \hat{D} \rangle/\langle \hat{V} \rangle$. Since there is no explicit T-dependence, p_T is conserved. Classically, it equals P^2 and is positive; thus, P or the time derivative of the scale factor cannot vanish and there is no bounce. With the fluctuation terms, there is a chance for P to vanish at non-zero p_T, in some cases corresponding to an extremum of a, provided fluctuations are significant. Fluctuations are dynamical, satisfying the equation $\mathrm{d}(\Delta V)^2/\mathrm{d}T = 4\langle \hat{D} \rangle(\Delta(VD)_2(\Delta V)^2/\langle \hat{V} \rangle)$, and the general behavior requires an analysis of the whole system coupling expectation values to fluctuations and correlations. In some examples, wave-function dynamics (for instance in the Bohmian viewpoint) has shown that bounces do arise [11–13].

Effective equations in some loop quantized systems show bounces that are not solely based on quantum fluctuations, but more importantly on quantum-geometry modifications. Such equations have not been derived yet in many models because the interacting nature of any common matter system in the presence of gravity makes this rather involved. But a solvable model has been identified, and it can be used as the basis for perturbation expansions. In such situations, which may be rather special but can be analyzed concretely, evidence for bounces has been seen. Assuming that a state remains sufficiently semiclassical at high densities and that matter is dominated by kinetic energy terms, the conclusion for a bounce to happen is rather robust. This conclusion can clearly be inferred from the perturbation equations we looked at before, which can also be extended to anisotropic or inhomogeneous situations. Sometimes one appeals to asymptotic properties, indicating that kinetic terms do become asymptotically dominant near a singularity since, in contrast to the potential in the energy density, they carry an inverse power of the scale factor. Also the Belinsky–Khalatnikov–Lifshitz (BKL) conjecture [14] may be taken to indicate that homogeneous models describe the approach to a singularity generically, in which case the dynamics for long stretches of time would be determined by the simple Bianchi I model. If the free massless scalar Bianchi I model would in general have bounces, one could expect this result to apply also to more generic situations.

But the very asymptotic regime used in this line of arguments is avoided by a bounce, which would imply that the universe never gets arbitrarily close to a singularity where kinetic domination or BKL-type arguments can be used. Theorems about asymptotic properties do not provide estimates of when exactly such a regime is reached, which could then be tested at the bounce. Even if this were possible, the presence of a bounce would become dependent on initial conditions since changing them would put the asymptotic regimes at different places. For these reasons, a kinetic-driven inhomogeneous bounce relying on asymptotic arguments such as the BKL conjecture cannot be generic.

Moreover, in these considerations one would still assume the quantum gravitational state to be sufficiently semiclassical. However, quantum back-reaction is in general important and can significantly change the behavior. Then, the classical theorems about the asymptotic dynamics would be uprooted and could no longer be used—the quantum system is one of many independent dynamical variables in coupled motion which must be re-analyzed for its asymptotic properties. Even if the time derivative of the scale factor approaches zero due to the action of some quantum repulsive force mediated by discrete geometry, quantum variables may not allow it to be precisely zero. In the effective Friedmann equation (6.34), for instance, we are comparing a density term, which in the solvable model vanishes at the bounce, with a correlation term. Even if one might have reason to expect quantum correlations to be small, if they are not exactly zero they would still be significant compared to zero.

In the big-bang phase, not just expectation values but also fluctuations, correlations and higher moments of a state are important. Effective equations describe this as a higher-dimensional dynamical system where quantum variables couple to expectation values. Such systems can have properties quite different from what a low-dimensional truncation might indicate. For instance, instead of bouncing sharply the scale factor could approach small sizes and linger there with nearly vanishing $\dot{a}$. But quantum fluctuations fluctuate, and so $\dot{a}$ may never be exactly zero as required for a

Fig. 7.1 A wave packet as a function of volume (horizontal) and time, bouncing, spreading and tunneling as indicated by the contours growing to the *right* (bouncing) and *left* (tunneling) [47]. (*Dashed*: expectation value, *long-dashed curve*: zero volume)

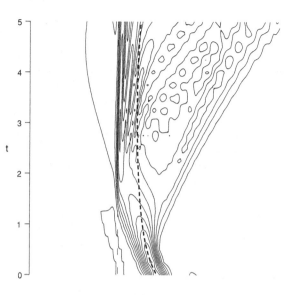

bounce; or it might become zero many times, resulting in repeated oscillations of a small universe. In principle it is even possible for a to approach the singular value zero asymptotically, despite the bound on energy density during kinetic domination. Different asymptotic behaviors resulting from the state dependence of evolution have been demonstrated in the Wheeler–DeWitt context [15, 16]. All these possibilities remain, at present, wide open. A general conclusion is much more difficult to reach than in the simple solvable model or in situations close to it.

Even if one could show the presence of a bounce generally, or at least in a sufficiently large class of models and situations, it would provide only a classical picture by an effective geometry. (In Chap. 10 we will discuss obstructions to the existence of effective line elements due to quantum-geometry effects in the presence of inhomogeneities.) Quantum effects would be included, for instance by holonomy corrections of geometry and by quantum back-reaction, but the effective bounce would only refer to the expectation value of a wave function. Repulsive forces of quantum geometry then erect a potential barrier which classical dynamics cannot penetrate. But quantum physics is rarely impressed by a barrier as long as it is of finite height. Wave functions can simply tunnel through (see Fig. 7.1), and if they reach $a = 0$ in this way one would again have to grapple with the singularity issue. Thus, the wave function must still be ensured to be non-singular in a fundamental rather than effective sense, based on the difference equation it has to obey.

At this point, we have to come back to the factor ordering in the Hamiltonian constraint, which was restricted by singularity removal based on the difference equation, but which was independently chosen in a particular form to make the free scalar model solvable. These orderings do not agree: using the solvable ordering for a fundamental singularity analysis would not allow one to evolve through $\mu = 0$, as can be seen by looking at the recurrence that follows from (5.20). The solvable ordering must be seen as an approximation, just as we had to ignore inverse-triad corrections to realize solvability. Re-ordering terms will introduce additional corrections which spoil the linear nature of the model, but which like ordinary interactions can be included perturbatively as long as quantum effects are not very strong.

Ultimately referring to the difference equation to study the singularity issue and to shed light on what came before the big bang makes us face the interpretational question of what time is in a deep quantum regime. It is unlikely to be a classically supported internal time such as φ or a, especially if deparameterizability is required to realize such a time choice. To cross quantum regimes, one might even have to switch to genuine quantum variables as time parameters which would have no classical analog [17, 18]. Covariances would indeed be a good choice since they are often monotonic when a state gets ever more squeezed. (Squeezing has in fact been related to entropy in several cosmological settings [19–25].) What the difference equation, essentially using as internal time the triad component (the scale factor squared with a sign for orientation), gives us as a picture for the transition through the classical singularity is a branch of a negatively oriented universe flipping to a positively oriented one, turning its inside out. This view is of particular interest if parity-violating matter is coupled to gravity, as this would strongly influence the transition and make the pre- and post-big bang phases differ in fundamental properties.

What the effective scenarios studied so far indicate, on the other hand, would be a bouncing trajectory with respect to some matter clock, which stays at one orientation of space but provides a minimum for volume. But this is borne out clearly only if one avoids strong quantum regimes by choosing a large matter content and large p_φ, making the universe kinetic-dominated at large volume where it can bounce easily by loop effects. In a deep quantum regime, single trajectories no longer matter; many different branches typically arise in superposition. Such a superposition can include both orientations of space, even for "times" which would all be considered as being before the big bang. Maybe there is not even a single notion of time that can explain the whole transition; time like geometry would have to emerge from the wave function. No such scenario for the emergence of time is available yet, but taken together all indications for the singularity removal in loop quantum cosmology show that this is what must be realized. Approaching the strong quantum regime with more refined techniques than are available now will open up access to this deep conceptual problem.

7.3 Quantum Cosmology

In the general setting of quantum cosmology, not using specific loop-quantization or other effects, singularity removal has been suggested and discussed in many different ways. None of these effects is as strong as the repulsive force arising from holonomy corrections in loop quantum cosmology, but subtle mechanisms sometimes exist. Due to the relative weakness, a conclusion about singularity removal here depends more strongly on interpretational issues of the wave function. See for instance [26] for a definition and analysis in the context of the consistent-histories formulation.

7.3.1 Interpretational Issues

Quantum cosmology plays a special role in quantum physics because the observer is always within the quantum system. The Copenhagen interpretation becomes inapplicable, but other options are available. Several of them have been claimed to be the only viable one in this context, most emphatically for the Bohmian viewpoint and the consistent-histories approach [27–29] (or even the many-worlds interpretation). The question of how to interpret quantum states is relevant for the singularity problem in quantum cosmology whenever arguments are based on typical quantum properties of wave functions rather than quantum-geometry effects. From this perspective, one can describe the main progress made by loop quantum cosmology as avoiding interpretational issues either by using quantum hyperbolicity for general wave functions, or by providing strong quantum-geometry effects.

In many applications of loop quantum cosmology, and in most parts of this book, one can take a pragmatic view. For observational aspects it is mainly the dynamics of expectation values and fluctuations that matters. Fluctuations, via quantum back-reaction, are then important for the dynamics, but not for the measurement problem where interpretational issues would strike with full force. It may even be possible to base the quantum-to-classical transition from fluctuations to matter perturbations in the early universe on quantum back-reaction of modes [30], but this problem still needs to be explained, perhaps by decoherence. (see also Sect. 10.4.1.)

7.3.2 Examples

If singularity resolution is to be based on aspects of wave functions, specific mechanisms require further input in addition to just a quantization scheme, as it has been done in the traditional proposals for initial conditions of a wave function by Vilenkin [31] and Hartle–Hawking [32]. This approach accepts the presence of a classically counterintuitive regime around the singularity, and replaces it with quantum notions such as tunneling or signature change, or outright quantum potentials [33]. Such schemes have primarily been formulated in isotropic models, and one often encounters difficulties when one tries to extend them to more general situations [34, 35]. Also future singularities have been discussed in this spirit [36]. In loop quantum cosmology, restrictions for states arise via dynamical initial conditions [37, 38].

Sometimes it is argued that quantization of a cosmological model with the positive scale factor a as the configuration variable resolves the singularity in the sense that a-expectation values of regular wave functions cannot be zero: any normalizable state must be supported at non-vanishing values of a, providing contributions to the expectation value which cannot completely cancel since no negative a are allowed. A minimum for a in the sense of expectation values may thus arise at small volume and strong curvature (where one may not trust Wheeler–DeWitt quantizations but rather prefer a loop quantization which then removes zero as the minimum for the basic geometrical variable p). The problem with this statement is that it depends on interpretational issues of the wave function as well as on implicit assumptions on the physical inner product. No explicit geometrical picture of how the singularity is resolved arises in this way, but it has been realized in some cases; see e.g. [39].

More specifically referring to quantum dynamics, quantum back-reaction provides additional terms to an effective Friedmann equation which should be imortant near $a = 0$. Some cases where a bounce at small volume ensues are indeed known [11–13, 15], not derived by the methods used in the preceding chapters but by explicit wave functions or a Bohmian formulation of quantum cosmology; see also the example at the beginning of this chapter. In this way, geometrical pictures of singularity resolution result which qualitatively can be compared with those of loop quantum cosmology. But the robustness in this case is much less investigated. In fact, at least the solvable Wheeler–DeWitt model analyzed before, which is free of quantum back-reaction, does not have a bounce. Whether or not singularities are removed by such a mechanism alone would thus depend on the matter type.

7.3.3 Dependence on Ambiguities

Any mechanism to avoid singularities must happen or at least mainly apply in strong quantum regimes. It may thus be sensitive to ambiguities such as factor ordering choices which have no classical analog. As an example, we look at the volume quantization in loop quantum cosmology, which has an influence not just on the dynamical approach to a singularity (the volume operator features prominently in the dynamics) but also on the identification of homogeneous singularities as states annihilated by the volume.

In analyzing inverse-triad corrections, here and in cosmological applications of the corrected perturbation equations of Chap. 10, we mostly use the behavior for larger values of μ where correction functions (3.59) drop off to the classical value one, as in Fig. 3.3. On very small scales, the approach to zero at $\mu = 0$ is characteristic for operators with U(1)-holonomies as they appear in homogeneous models or in the perturbative treatment. In particular, as we have seen explicitly in isotropic models the volume operator $\hat{V}$ and gauge-covariant combinations of commutators such as $\mathrm{tr}(\tau^i \hat{h}[\hat{h}^{-1}, \hat{V}])$ commute. It is thus meaningful to speak of the (eigen-)value of inverse volume on zero-volume eigenstates. For non-Abelian holonomies such as those for SU(2) in the full theory, the operators become non-commuting [40]. The inverse volume at zero-volume eigenstates thus becomes unsharp and one can at most make statements about expectation values rather than eigenvalues, which again requires more information on suitable states. Then, the expectation values are not expected to become sharply zero at zero volume, as calculations indeed show [41] (using a kinematical notion of coherent states). In addition, also here quantization ambiguities matter: We can write volume itself, and not just inverse volume, through Poisson brackets such as [40]

$$
V = \int d^3x \left(\frac{\varepsilon^{abc}\varepsilon_{ijk}}{6(10\pi\gamma G/3)^3} \int d^3y_1 \{A_a^i(x), |\det e(y_1)|^{5/6}\} \right.
$$

$$
\left. \times \int d^3y_2 \{A_b^j(x), |\det e(y_2)|^{5/6}\} \int d^3y_3 \{A_c^j(x), |\det e(y_3)|^{5/6}\} \right)^2 .
$$

After regularization, splitting the integration into sums over small patches of coordinate size ℓ_0^3 with volume contributions $V_v \approx \ell_0^3 |\det(e_a^i)|$, we obtain

$$
V_v = \ell_0^6 \left(\frac{|\det e|}{\sqrt{V_v}} \right)^2 = \ell_0^6 \left(\varepsilon^{abc}\varepsilon_{ijk} \frac{e_a^i}{V_v^{1/6}} \frac{e_b^j}{V_v^{1/6}} \frac{e_c^k}{V_v^{1/6}} \right)^2
$$

$$
= \left(\frac{\varepsilon^{abc}\varepsilon_{ijk}}{6(10\pi\gamma G/3)^3} \ell_0^3 \{A_a^i, V_v^{5/6}\}\{A_b^j, V_v^{5/6}\}\{A_c^j, V_v^{5/6}\} \right)^2
$$

whose quantization, making use of commutators, differs from the original volume operator of loop quantum gravity. If non-Abelian holonomies are used, the new volume operator does not commute with the full volume operator of [7] or [8]. This clearly shows that the usual quantization ambiguity in writing inverse-triad expressions also applies to what is considered the relevant geometrical volume. (Related ambiguities for flux operators have been discussed in [42].) For geometrical properties one may not only consider the original volume operator constructed directly from fluxes, but any operator having volume as the classical limit. In order to find zero-volume states to be related to classical singularities, the general fundamental dynamics of the form (4.15) indicates that operators constructed through commutators with the original volume operator are more relevant than the volume operator constructed directly from fluxes [40]. Thus, as one example of the relevance of quantization ambiguities for questions of singularity removal, specific volume eigenstates have to be used with great care in applications with non-Abelian holonomies.

7.4 Negative Attitude

Singularity removal in general terms may depend sensitively on specific properties of the quantum state of the universe which is not under observational control. This problem applies to the question whether singularities are removed, but also to how specifically resolution would be achieved. In such a situation, a "negative attitude" is useful, where one tries to quantify limitations to what can actually be said in detail, and to find out which possibilities remain within the bounds. An example is the analysis of asymmetries of states before and after a bounce of the solvable model, as already discussed in detail. Continuing such an analysis in this and other models will show clearly how much the small-volume behavior of quantum cosmology can be elucidated, and which questions must remain open. The derivations of strong symmetry bounds in highly specific or even solvable models, by contrast, is much less relevant because it does not show how the behavior may be restricted in realistic situations.

Several results in loop quantum cosmology can be seen to be investigated along the viewpoint of negative attitude. For instance, singularities in tree-level equations have been identified in scalar models, which are not big-bang but sudden future singularities [43]. While the analysis based on tree-level equations would still have to be justified by a detailed derivation of effective equations with the potential used, as a negative result the statement is of interest: it shows the limitations of tree-level equations and what quantum back-reaction terms would have to provide to achieve non-singular behavior in that case. Also, extending c^2 in the isotropic Hamiltonian constraint to almost-periodic functions in more complicated ways than normally done can lead to new types of singularities of diverging Hubble parameter even before the high-density regime of the holonomy-induced bounce is reached [44]. Such studies show how generic the singularity-resolution mechanisms of loop quantum cosmology are.

Similarly, possible cases of the large-volume behavior have been studied with a parameterized form for the growth of quantum fluctuations. The model used was actually the solvable one, where explicit solutions for the behavior of fluctuations are available, but with a different factor ordering. What was found by the parameterization was that the behavior of fluctuations can be very important for the large-volume behavior, too, even in the absence of a positive cosmological constant as we discussed it earlier. If the wrong behavior of fluctuations is used, even a flat model can lead to a recollapse at large volume in disagreement with the expected classical limit [45, 46]. This is true even for a behavior of fluctuations that would still allow one to interpret states as semiclassical. The analysis underlines the importance of considering the exact state properties for effective equations, rather than picking a certain form of dynamical fluctuations or correlations. All quantum variables must be evolved from their initial values onwards to ensure that the correct dynamics, also of expectation values, is captured.

The distinguishing feature of those examples is that they show possible limitations to proposed singularity-avoidance mechanisms, rather than providing more examples of the same special scenario such as a bounce. As the theory is further developed, instances of possible breaches, rather than superficial uniqueness claims of simple mechanisms, will be very valuable to probe its general behavior.

References

1. Senovilla, J.M.M.: Phys. Rev. Lett. **64**, 2219 (1990)
2. Kaminski, W., Lewandowski, J., Pawlowski, T.: Class. Quantum. Grav. **26**, 035012 (2009)
3. Haro, J., Elizalde, E.: arXiv:0901.2861

4. Helling, R.: arXiv:0912.3011
5. Thiemann, T.: Class. Quantum. Grav. **15**, 839 (1998). gr-qc/9606089
6. Thiemann, T.: Class. Quantum. Grav. **15**, 1281 (1998). gr-qc/9705019
7. Rovelli, C., Smolin, L.: Nucl. Phys. B **442**, 593 (1995). gr-qc/9411005. Erratum: Nucl. Phys. B 456:753
8. Ashtekar, A., Lewandowski, J.: Adv. Theor. Math. Phys. **1**, 388 (1998). gr-qc/9711031
9. Bojowald, M., Mulryne, D., Nelson, W., Tavakol, R.: Phys. Rev. D **82**, 124055 (2010). arXiv:1004.3979
10. Novello, M., Bergliaffa, S.E.P.: Phys. Rep. **463**, 127 (2008)
11. Alvarenga, F.G., Fabris, J.C., Lemos, N.A., Monerat, G.A.: Gen. Rel. Grav. **34**, 651 (2002). gr-qc/0106051
12. Acacio de Barros, J., Pinto-Neto, N., Sagiaro-Leal, M.A.: Phys. Lett. A **241**, 229 (1998). gr-qc/9710084
13. Pinto-Neto, N.: Phys. Rev. D **79**, 083514 (2009). arXiv:0904.4454
14. Belinskii, V.A., Khalatnikov, I.M., Lifschitz, E.M.: Adv. Phys. **13**, 639 (1982)
15. Colistete, R. Jr, Fabris, J.C., Pinto-Neto, N.: Phys. Rev. D **62**, 083507 (2000). gr-qc/0005013
16. Falciano, F.T., Pinto-Neto, N., Santini, E.S.: Phys. Rev. D **76**, 083521 (2007). arXiv:0707.1088
17. Bojowald, M., Tavakol, R.: Phys. Rev. D **78**, 023515 (2008). arXiv:0803.4484
18. Bojowald, M.: A Momentous Arrow of Time. Springer, Berlin (2011)
19. Gasperini, M., Giovannini, M.: Class. Quantum. Grav. **10**, L133 (1993)
20. Kruczenski, M., Oxman, L.E., Zaldarriaga, M.: Class. Quantum. Grav. **11**, 2317 (1994)
21. Koks, D., Matacz, A., Hu, B.L.: Phys. Rev. D **55**, 5917 (1997). Erratum: Koks, D., Matacz, A., Hu, B.L. (1997). Phys. Rev. D 56:5281.
22. Kim, S.P., Kim, S.W.: Phys. Rev. D **49**, R1679 (1994)
23. Kim, S.P., Kim, S.W.: Phys. Rev. D **51**, 4254 (1995)
24. Kim, S.P., Kim, S.W.: Nuovo. Cim. B **115**, 1039 (2000)
25. Kiefer, C., Polarski, D., Starobinsky, A.A.: Phys. Rev. D **62**, 043518 (2000)
26. Craig, D.A., Singh, P.: Phys. Rev. D **82**, 123526 (2010). arXiv:1006.3837
27. Griffiths, R.B.: Phys. Rev. Lett. **70**, 2201 (1993)
28. Omnés, R.: Rev. Mod. Phys. **64**, 339 (1992)
29. Gell-Mann, M., Hartle, J.B.: Phys. Rev. D **47**, 3345 (1993). gr-qc/9210010
30. Bojowald, M., Skirzewski, A.: Adv. Sci. Lett. **1**, 92 (2008). arXiv:0808.0701
31. Vilenkin, A.: Phys. Rev. D **30**, 509 (1984)
32. Hartle, J.B., Hawking, S.W.: Phys. Rev. D **28**, 2960 (1983)
33. Conradi, H.D., Zeh, H.D.: Phys. Lett. A **154**, 321 (1991)
34. del Campo, S., Vilenkin, A.: Phys. Lett. B **224**, 45 (1989)
35. Hawking, S.W., Luttrell, J.C.: Phys. Lett. B **143**, 83 (1984)
36. Kamenshchik, A., Kiefer, C., Sandhöfer, B.: Phys. Rev. D **76**, 064032 (2007). arXiv:0705.1688
37. Bojowald, M.: Phys. Rev. Lett. **87**, 121301 (2001). gr-qc/0104072
38. Bojowald, M.: Gen. Rel. Grav. **35**, 1877 (2003). gr-qc/0305069
39. Amemiya, F., Koike, T.: Phys. Rev. D **80**, 103507 (2009). arXiv:0910.4256
40. Bojowald, M.: Class. Quantum. Grav. **23**, 987 (2006). gr-qc/0508118
41. Brunnemann, J., Thiemann, T.: Class. Quantum. Grav. **23**, 1429 (2006). gr-qc/0505033
42. Giesel, K., Thiemann, T.: Class. Quantum. Grav. **23**, 5667 (2006). gr-qc/0507036
43. Cailleteau, T., Cardoso, A., Vandersloot, K., Wands, D.: Phys. Rev. Lett. **101**, 251302 (2008). arXiv:0808.0190
44. Mielczarek, J., Szydłowski, M.: Phys. Rev. D **77**, 124008 (2008). arXiv:0801.1073
45. Ding, Y., Ma, Y., Yang, J.: Phys. Rev. Lett. **102**, 051301 (2009). arXiv:0808.0990
46. Yang, J., Ding, Y., Ma, Y.: Phys. Lett. B **682**, 1 (2009). arXiv:0904.4379
47. Bojowald, M., Singh, P., Skirzewski, A.: Phys. Rev. D **70**, 124022 (2004). gr-qc/0408094

Part III
Beyond Isotropic Models

Isotropic models are special already in their classical dynamics and show properties, in particular regarding singularities, that are not generically realized in anisotropic cases. For instance, the typical approach to a homogeneous singularity is not isotropic but has two directions contracting while the third one expands. This generic behavior is captured in Bianchi models with their homogeneous but not isotropic geometries. The BKL scenario suggests that these models describe even the general classical approach to a space like singularity. Other examples for important phenomena not captured in isotropic models are black holes and their singularities.

Techniques must thus be extended from isotropic models to much more general situations, while still keeping constructions and evaluations manageable. If this can be achieved, many new opportunities to apply quantum gravity ensue: On a technical level, results obtained in isotropic models can be checked for their robustness. Physically, black holes and collapse models can be analyzed, for instance regarding the end state of Hawking evaporation; and back in a cosmological context perturbative inhomogeneities can be evolved and possibly compared with observations.

Part III
Beyond Isotropic Models

Chapter 8
Anisotropy

Models can be useful to analyze a complicated general theory. To that end, they must be simple enough to be tractable but at the same time be able to capture crucial properties. Symmetry reduction is a popular tool to do this with general relativity, and it also applies to quantum gravity. In loop quantum gravity, already the kinematical framework can show several technical difficulties, mainly due to the non-Abelian nature which makes operators such as that for volume quite involved. Such issues have to be faced before one even tries to describe and solve the dynamics, and so already a kinematical reduction must be undertaken with care.

8.1 Constructing Models

In the full setting of loop quantum gravity we only have the connection representation; no triad representation is available because not all flux operators commute with one another. They are, after all, SU(2)-derivatives with the well-known angular-momentum algebra for all pairs of fluxes corresponding to intersecting surfaces. This sense of non-commutativity [1] is not very strong, and it is not immediately clear whether there might be implications for the structure of space–time. (In Chap. 10 we will see rather indirect implications on the structure of space–time from properties of flux operators, but not from their non-commutativity.) Nevertheless, there are technical implications at the level of kinematical representations of the theory. In particular, it is natural in such a situation to choose the connection representation as the starting point also for symmetry reduction, which is in fact facilitated by the availability of detailed classifications of invariant connections on principal fiber bundles; see Sect. 3.2.3.1 The structure of invariant objects in the space of all connections is well understood and can be used for quantum reduction.

The first step is to find the reduced form of holonomies. Once this is known, one can, by analogy with (3.35), construct the reduced kinematical Hilbert space as the one generated by holonomies out of the constant state. A symmetric sector of

M. Bojowald, *Quantum Cosmology*, Lecture Notes in Physics 835,
DOI: 10.1007/978-1-4419-8276-6_8, © Springer Science+Business Media, LLC 2011

the kinematical setting of loop quantum gravity follows, derived by acting with a reduced set of holonomies on the basic state. In a second step, one would use the canonical structure to represent triad components. This step, however, is not always trivial since triads, after reduction, are not necessarily represented by pure derivative operators. In this way, the non-Abelian structure of the full theory may feature also in models. The simplest models are those that can be formulated in Abelian terms, either directly for the triads by virtue of their invariant form, or for a new set of variables with the same geometric information.

Invariant connections are subject to a set of conditions implementing the symmetry or other properties (such as polarization conditions). Once these conditions are solved for to parameterize a space of invariant connections, a sufficiently large set of functions must be found which separates these connections (and so has different values for at least two such functions when evaluated in two different connections) and which can be realized as resulting from matrix elements of certain holonomies of the connections. These conditions do not specify the function space completely; for instance, one can choose holonomies along different kinds of curves to result in different classes of separating functions. Among the possible cases one would then, as always, pick one which provides the strongest simplifications in the algebraic behavior (such as straight edges along symmetry generators for homogeneous models). To some degree, this is analogous to choosing adapted coordinates when discussing symmetric space–times; in arbitrary coordinates even a highly symmetric solution can look complicated and hide important properties. In models of quantum gravity, however, there is an additional uncertainty concerning whether the general behavior is already captured by one choice of basic functions. In contrast to the choice of coordinates there is, after all, no simple transformation that would guarantee independence of, say, the choice of curves. Different sectors might thus arise from choosing different types of curves, and differences in the resulting dynamics should be studied to see how general conclusions can be drawn.

For isotropy, for instance, we have a single component c of invariant connections in a fixed SU(2)-gauge. Taking holonomies for straight edges along generators of translations on the homogeneous space, a complete set of matrix elements is of the form $\{e^{i\mu c} : \mu \in \mathbb{R}\}$ as used earlier [2]. The resulting function space turns out to be that of almost-periodic functions: all continuous functions on the Bohr compactification of the real line. Choosing a set $\{e^{inc} : n \in \mathbb{Z}\}$, on the other hand, would not have separated connections (although it would have come from holonomies via edges of a fixed coordinate length). Strictly periodic functions do not provide a complete kinematical quantization, even though, as we already saw, they are sufficient to capture the crucial behavior of isotropic quantum dynamics [3].

One could as well choose different edges in this or other examples, not resulting in almost-periodic functions; see [4]. Among those possibilities, the quantization based on the Bohr compactification is arguably the simplest one, providing an Abelianized version of the model since the Bohr compactification of $\mathbb{R}$, just like the real line itself, is an Abelian group.

Moreover, holonomies evaluated along more general edges are asymptotically almost-periodic [5] for large c. The behavior near classical singularities (small μ) can thus be expected to be captured well by almost-periodic functions, while general lattice refinement in any case removes the almost-periodicity of states when one evolves to larger volume.

The choice of curves, or other ingredients, is thus an extra step in addition to implementing the symmetry of connections. Most of the work done so far in loop quantum cosmology is based on the use of straight curves, or integral curves of the symmetry generators, and the simple almost-periodic structures they imply. As in this case, ideal would always be a formulation of models in Abelian form, where a complete set of commuting momenta exists and invariant triad components can be represented just as derivative operators by the invariant connection components. This is not always possible even in symmetric models, but non-Abelian features may nevertheless be tractable more generally in such settings. In many models, it can also happen that momenta of the invariant connection components are not identical with invariant triad components, especially if a canonical transformation is involved to bring the invariant space of connections into a quantizable form. In such a case, even if momenta do not agree with triad components, one could call quantizations of the momenta flux operators since they would be part of a basic algebra analogous to the holonomy-flux algebra in the full theory. In this way, non-Abelian traces of triad operators can show up in models even if they allow commuting momenta conjugate to the connections. Here, possibilities to understand implications of the non-Abelianness arise: While fluxes commute, simplifying calculations, triad operators may be more complicated and show non-Abelian features explicitly. In particular, calculations and constructions can in such a case often be simplified using not the connection representation but the flux representation.

Once the kinematics is formulated, constraint operators are to be constructed. The diffeomorphism constraint can usually be dealt with just as in the full theory, and the main task is to construct the Hamiltonian constraint operator. At this point, the relationship of the class of functions used to separate invariant connections to holonomies along certain edges becomes important. For a correct implementation of lattice refinement, which is the crucial ingredient in the dynamics given by a Hamiltonian constraint, one must know the relation between quantizations of connection components showing up in the classical constraint and holonomies as they can be used to quantize them. Generic properties of an underlying lattice structure then tell one how parameters of holonomies must depend on geometrical properties such as the size and shapes of regions considered. Since this information goes beyond what can be *derived* in a pure mini- or midi-superspace model (even though it can certainly be *formulated* completely in such a setting), no unique dynamics can result. But suitably parameterized, the possible choices can be analyzed well and further restricted by the phenomenology or other aspects they imply.

The following sections and later chapters will provide several examples for what has been discussed here in general terms.

8.2 Bianchi Cosmology

Bianchi models are based on homogeneous spaces with a 3-dimensional symmetry group acting transitively. The most general anisotropic spatial metric of diagonal form that obeys these symmetries can be written as

$$h_{ab} = \sum_{I=1}^{3} h_I \omega_a^I \omega_b^I,$$

where the ω_a^I are left-invariant 1-forms on the homogeneous space. A homogeneous spatial manifold can be identified with the group space up to possible discrete identifications. For any given Lie group with elements parameterized as $G(x^a)$ with as many group coordinates x^a as the group has dimensions, left-invariant 1-forms can explicitly be calculated via the Maurer–Cartan form $G^{-1}\partial_a G = T_I \omega_a^I$, where T_I are Lie-algebra generators. (See e.g., [6] for more details relevant for this section.)

8.2.1 Bianchi Dynamics

The form of the ω_a^I follows just from the symmetry type and does not contain dynamical information, which instead is completely contained in the metric coefficients h_I. They are spatial constants but may be functions of time, with momenta $\pi^I \propto \dot{h}_I$. The dynamics on the phase space is controlled by the Hamiltonian constraint

$$NC_{\text{grav}} = \frac{1}{2}\left(h_1^2(\pi^1)^2 + h_2^2(\pi^2)^2 + h_3^2(\pi^3)^2\right) - h_1 h_2 \pi^1 \pi^2 \qquad (8.1)$$

$$-h_1 h_3 \pi^1 \pi^3 - h_2 h_3 \pi^2 \pi^3 - \frac{\det h}{(16\pi G)^2}{}^{(3)}R = 0 \qquad (8.2)$$

for a lapse $N = \sqrt{\det h}/16\pi G$, with the spatial Ricci scalar

$${}^{(3)}R = -\frac{1}{2}\left(\frac{n^1 h_1}{h_2 h_3} + \frac{n^2 h_2}{h_1 h_3} + \frac{n^3 h_3}{h_1 h_2} - 2\frac{n^1 n^2}{h_3} - 2\frac{n^1 n^3}{h_2} - 2\frac{n^2 n^3}{h_1}\right)$$

in terms of Bianchi parameters $n^I \in \{0, \pm 1\}$ determining the symmetry type: Structure constants of the symmetry group (of class A) are $C_{JK}^I = \varepsilon_{JKI} n^{(I)}$. (For instance, $n^I = 0$ for the Bianchi I model with group $\mathbb{R}^3$, and $n^I = 1$ for the Bianchi IX model with group SU(2)).

It is often useful to apply a canonical transformation

$$\alpha_I := \frac{1}{2}\log h_I, \quad \rho^I := 2h_{(I)}\pi^I \qquad (8.3)$$

and diagonalize the resulting kinetic term in the constraint by introducing Misner parameters [7]:

$$\alpha_1 =: \alpha + \beta_+ + \sqrt{3}\beta_-, \quad \alpha_2 =: \alpha + \beta_+ - \sqrt{3}\beta_-, \quad \alpha_3 =: \alpha - 2\beta_+ \tag{8.4}$$

for the metric variables and

$$\rho^1 =: \frac{1}{3}p_\alpha + \frac{1}{6}p_+ + \frac{1}{2\sqrt{3}}p_-, \quad \rho^2 =: \frac{1}{3}p_\alpha + \frac{1}{6}p_+ - \frac{1}{2\sqrt{3}}p_-, \quad \rho^3 =: \frac{1}{3}p_\alpha - \frac{1}{3}p_+$$

for the momenta. The constraint then takes the form

$$N C_{\text{grav}} = \frac{1}{24}(-p_\alpha^2 + p_+^2 + p_-^2) - \frac{e^{6\alpha}}{(16\pi G)^2} {}^{(3)}R \tag{8.5}$$

with the anisotropy potential

$$
{}^{(3)}R = -\frac{1}{2}e^{-2\alpha}\left(n^1 e^{4(\beta_+ + \sqrt{3}\beta_-)} + n^2 e^{4(\beta_+ - \sqrt{3}\beta_-)} + n^3 e^{-8\beta_+}\right.
$$
$$
\left. -2n^1 n^2 e^{4\beta_+} - 2n^1 n^3 e^{-2(\beta_+ - \sqrt{3}\beta_-)} - 2n^2 n^3 e^{-2(\beta_+ + \sqrt{3}\beta_-)}\right). \tag{8.6}
$$

Example 8.1 (Kasner model) The Bianchi I model is obtained for the choice $n^I = 0$, thus ${}^{(3)}R = 0$. Equations of motion

$$\dot{\rho}^I = \{\rho^I, N C_{\text{grav}}\} = 0$$

imply that all ρ^I are constant in time, and

$$\dot{\alpha}_I = \{\alpha_I, N C_{\text{grav}}\} = \frac{1}{4}(2\rho^I - \rho^1 - \rho^2 - \rho^3) =: \tilde{v}^I.$$

These coefficients enter the spatial metric components

$$h_I = e^{2\alpha_I} = h_I^{(0)} e^{2\tilde{v}^I t}.$$

Solutions $\alpha_I(t) = \tilde{v}^I t + \alpha_I^{(0)}$ are subject to the constraint

$$0 = N C_{\text{grav}} = \sum_I \dot{\alpha}_I^2 - \left(\sum_I \dot{\alpha}_I\right)^2 = \sum_I (\tilde{v}^I)^2 - \left(\sum_I \tilde{v}^I\right)^2.$$

Since the Hamiltonian constraint was used with a lapse function $N \propto \sqrt{\det h} \propto \exp((\tilde{v}^1 + \tilde{v}^2 + \tilde{v}^3)t)$, we transform to proper time by $\tau = \int^\tau N(t)dt \propto \exp((\tilde{v}^1 + \tilde{v}^2 + \tilde{v}^3)t)$ and write the line element as the Kasner solution

$$ds^2 = -d\tau^2 + \tau^{2v^1}(dx^1)^2 + \tau^{2v^2}(dx^2)^2 + \tau^{2v^3}(dx^3)^2$$

with Kasner exponents

$$v^I = \frac{\tilde{v}^I}{\sum_J \tilde{v}^J}$$

Due to their definition and the constraint, the exponents must satisfy

$$\sum_I v^I = 1 = \sum_I (v^I)^2.$$

For all solutions of this form, one v^I is negative, two are positive. The behavior of expansion or contraction is thus very different from isotropic. The volume $V \propto \tau^{v^1 v^2 v^3}$ is always monotonic and vanishes at $\tau = 0$, a singularity.

8.2.2 Connection Variables and Holonomies

Invariant connections of Bianchi type are of the form $A_a^i = \tilde{c}_I^i \omega_a^I$ with spatial constants $\tilde{c}_I^i$. Taking straight lines e_I along vector fields $X_I^a = \dot{e}_I^a$ dual to the ω_a^I ($X_I^a \omega_a^J = \delta_I^J$), we obtain holonomies

$$h_{e_I} = \exp\left(\int_{e_I} A_a^i \tau_i X_I^a \mathrm{d}s \right) = \exp(\ell_0^{(I)} \tau_i \tilde{c}_I^i) \tag{8.7}$$

with the length parameters $\ell_0^I = \int_{e_I} \mathrm{d}s$ as before. All three holonomies h_{e_I} in the three spatial directions can be arbitrary elements of SU(2), fully independent of one another. They certainly do not commute in general, such that the full non-Abelian nature has been reduced in no way. Triads would still be represented as non-commuting derivative operators

$$\hat{J}_I^i = \mathrm{tr}((h_I \tau^i)^T \partial/\partial h_I) \tag{8.8}$$

on the three copies of SU(2); see (3.39). The volume operator is constructed from $\varepsilon^{IJK} \varepsilon_{ijk} \hat{J}_I^i \hat{J}_J^j \hat{J}_K^k$, whose action is equivalent to that of the full volume operator on the six-vertex obtained as the intersection of three closed loops. In this case, the volume operator is complicated and cannot explicitly be diagonalized completely; nor can the Hamiltonian constraint be implemented explicitly.

In the homogeneous setting a functional derivative operator by connection components sees both endpoints of edges at the same time, such that the h_{e_I}, from the point of view of flux operators, correspond to three closed edges acted on by SU(2)-derivatives at a vertex where they all intersect. The full volume operator is sensitive to the way holonomies are contracted in a vertex, and possible contractions are dictated by the gauge behavior of holonomies. In this way, the interpretation of homogeneous vertices follows by using the gauge transformations $h_{e_I} \mapsto g^{-1} h_{e_I} g$ with the same g for all holonomies and, thanks to homogeneity, without a distinction between gauge transformations at the starting and endpoint of the edge. In the sense of a spin network, gauge-invariant contractions $C^{B_1 B_2 B_3}_{A_1 A_2 A_3} \rho_{j_1}(h_{e_1})^{A_1}{}_{B_1} \rho_{j_2}(h_{e_2})^{A_2}{}_{B_2} \rho_{j_3}(h_{e_3})^{A_3}{}_{B_3}$ of all three holonomies must then be such that

$$C^{B_1 B_2 B_3}_{A_1 A_2 A_3} \rho_{j_1}(g^{-1})^{A_1}{}_{C_1} \rho_{j_2}(g^{-1})^{A_2}{}_{C_2} \rho_{j_3}(g^{-1})^{A_3}{}_{C_3} \rho_{j_1}(g)^{D_1}{}_{B_1} \rho_{j_2}(g)^{D_2}{}_{B_2} \rho_{j_3}(g)^{D_3}{}_{B_3} = C^{D_1 D_2 D_3}_{C_1 C_2 C_3}.$$

This is the same condition as the one for a 6-valent vertex where three straight edges with spins j_1, j_2 and j_3, respectively, intersect.

8.2.2.1 Abelianization

Bianchi models of the so-called type A can be diagonalized, which means that the form $\tilde{c}^i_I = \tilde{c}_{(I)} \delta^i_I$ in terms of only three independent components $\tilde{c}_I$ will be preserved by evolution. Most of the important dynamical properties, for instance in the approach to a singularity, are shown in full detail already by such a diagonalization, which is thus of interest also for an implementation in a quantum model. Putting the diagonal form into general Bianchi holonomies results in expressions

$$h_{e_I} = \exp(\ell_0^{(I)} \tilde{c}_I \tau_I) = \cos(\ell_0^{(I)} \tilde{c}_I / 2) + 2\tau_I \sin(\ell_0^{(I)} \tilde{c}_I / 2) \tag{8.9}$$

which can be evaluated easily but still do not commute. However, these holonomies are of a restricted form, satisfying for instance

$$\mathrm{tr}(\log(h_{e_I}) \log(h_{e_J})) = \ell_0^{(I)} \ell_0^{(J)} \tilde{c}_I \tilde{c}_J \mathrm{tr}(\tau_I \tau_J) = 0$$

for $I \neq J$. Any pair of such holonomies obeys $gh = hg + h^{-1}g + hg^{-1} - \mathrm{tr}(hg)$; even though they do not commute, there is a general way to reorder them in simple terms, a manipulation that would not be possible if arbitrary SU(2)-elements would be allowed as in the non-diagonal case.

This restricted behavior indicates that a reformulation of the variables allows an Abelian representation, which is indeed the case [8]: Taking matrix elements of (8.9) shows that a complete set of functions on diagonal homogeneous connections is given by $\exp(i\ell_0^{(I)} \tilde{c}_I / 2)$, such that the Hilbert space $L^2(\bar{\mathbb{R}}^3_{\mathrm{Bohr}}, d\mu^3)$ arises simply as the triple product of the isotropic one: we use functions of three connection components $\tilde{c}_I$, almost periodic in each of them. Without the diagonalization, by contrast, we have the form

$$h_{e_I} = \exp(\ell_0^{(I)} \tilde{c}^i_I \tau_i) = \cos(\ell_0^{(I)} \tilde{C}_I / 2) + 2n^i_I \tau_i \sin(\ell_0^{(I)} \tilde{C}_I / 2)$$

with $\tilde{C}^2_I = \sum_{i=1}^3 (\tilde{c}^i_I)^2$ and $n^i_I = \tilde{c}^i_I / \tilde{C}_I$ of general holonomies. Information in the non-trivial components n^i_I, which in the diagonal case would reduce to δ^i_I, is not captured by almost-periodic functions.

A diagonal invariant triad has the form $E^a_i = \tilde{p}^{(I)} \delta^I_i X^a_I$ with components $\tilde{p}^I$ dual and conjugate to the $\tilde{c}_I$:

$$\{\tilde{c}_I, \tilde{p}^J\} = \frac{8\pi\gamma G\delta_I^J}{\mathcal{V}}$$

with the coordinate volume $\mathcal{V} = \mathcal{L}^1\mathcal{L}^2\mathcal{L}^3$ in terms of independent sizes $\mathcal{L}^I$ along the three directions. As in the isotropic case, we redefine

$$c_I := \mathcal{L}^{(I)}\tilde{c}_I, \quad p^J := \frac{1}{2}\mathcal{L}^K\mathcal{L}^L\varepsilon_{KL(J)}\tilde{p}^J = \frac{\mathcal{V}}{\mathcal{L}^{(J)}}\tilde{p}^J \tag{8.10}$$

such that

$$\{c_I, p^J\} = 8\pi\gamma G\delta_I^J. \tag{8.11}$$

Our phase space is just the co-tangent space over the triple product of the Bohr compactification of $\mathbb{R}$, such that the quantized fluxes have the simple form

$$\hat{p}^I = \frac{8\pi\gamma\ell_{\mathrm{P}}^2}{\mathrm{i}}\frac{\partial}{\partial c_I} \tag{8.12}$$

Their spectra can directly be determined as in the isotropic case:

$$\hat{p}^I|\mu_1, \mu_2, \mu_3\rangle = 4\pi\gamma\ell_{\mathrm{P}}^2\mu_I|\mu_1, \mu_2, \mu_3\rangle \tag{8.13}$$

with eigenstates

$$\langle c_1, c_2, c_3|\mu_1, \mu_2, \mu_3\rangle = \exp(\mathrm{i}(\mu_1 c_1 + \mu_2 c_2 + \mu_3 c_3)/2). \tag{8.14}$$

Also the volume operator is simple: in the diagonal case, classically $V = \sqrt{|p^1 p^2 p^3|}$ for a region of coordinate size $\mathcal{V}$, which directly gives the volume operator and its complete spectrum $(4\pi\gamma\ell_{\mathrm{P}}^2)^{3/2}|\mu_1\mu_2\mu_3|^{1/2}$. The reduction of the SU(2)-gauge leaves a residual transformation $p^I \mapsto -p^I$, which requires a corresponding symmetry property of states $\psi_{\mu_1\mu_2\mu_3}$ in the triad eigenbasis under $\mu_I \mapsto -\mu_I$.

One way to probe whether an Abelian formulation of a model is possible is to look at the reduced Gauss constraint. It has the general form $\partial_a E_i^a + \varepsilon_{ijk}A_a^j E_k^a = T_i$ where T_i is a possible source term resulting from a theory with non-vanishing torsion. Ignoring torsion for now, homogeneous models lead to variables where $\partial_a E_i^a = 0$, such that $A_I^i = X_I^a A_a^i$ and $E_i^I = \omega_a^I E_i^a$ must point in the same internal direction for fixed I; their vector product in internal space vanishes. We have implicitly used this property before since we diagonalized the connection and triad in the same frame. If this were not the case, the derivatives by connection components would not be identical with triad components but rotated against them. With a constant rotation there would be no problem; but if the rotation is phase-space dependent, containing the connection components themselves, triad operators would be complicated expressions in terms of basic operators, depending not just on the fluxes but also on connections. In this case, they may not commute and develop complicated spectra.

We will see an explicit example later on in the discussion of spherically symmetric and other midi-superspace models, where the partial derivative in the Gauss constraint no longer vanishes. Another example is a non-vanishing torsion source, as it results when fermions are coupled. In such cases, canonical transformations can sometimes be used to facilitate explicit quantizations in manageable terms.

In the presence of torsion, connections and triads cannot be diagonalized in the same basis, providing an interesting model to probe non-Abelian features. If we couple fermions to gravity the Gauss constraint changes by a source term from the fermion axial current [9, 10]:

$$G[\Lambda] := \int_\Sigma d^3x\, \Lambda^i \left(\mathcal{D}_b P_i^b - \tfrac{1}{2}\sqrt{\det h}\, J_i \right) = \int_\Sigma d^3x\, \Lambda^i \left(\mathcal{D}_b P_i^b - \pi_\xi \tau_i \xi - \pi_\chi \tau_i \chi \right)$$

(8.15)

in terms of half-densitized [11] 2-spinors χ, ξ and their momenta $\pi_\chi = i\chi^\dagger$ and $\pi_\xi = i\xi^\dagger$. Reduced to a Bianchi-type model, this constraint becomes [12]

$$\phi_I^j p_k^I \varepsilon_{ijk} = \frac{1}{2}\sqrt{|\det(p_j^I)|}\, J_i.$$

(8.16)

If we assume $\phi_I^i = c_{(I)}\Lambda_I^i$ and $p_i^I = p^{(I)}\Lambda_i^I$, diagonalized in the same basis Λ_I^i, vanishing spatial components of the fermion current would be required in contrast to the freedom allowed in the classical model. In fact, the spin connection now receives a contribution from a torsion term of the form

$$\frac{8\pi\gamma G}{4(1+\gamma^2)} \varepsilon^j{}_{kl} e_a^k J^l$$

(8.17)

which turns out to be inconsistent with the diagonal ansatz $C_{(K)}\Lambda_K^i \omega_a^K$ if $J^i \neq 0$ (using the Λ_K^i as it appears in the diagonal triad). Since the spin connection is part of the Ashtekar–Barbero connection, the latter cannot be diagonal in the triad basis. There are additional torsion terms contributing to the connection in the presence of fermions, which must be used for the canonical homogeneous variables. To be sufficiently general, we write

$$A_a^i = (\mathcal{L}^K)^{-1} c_{(K)}\Lambda_K^i \omega_a^K, \quad E_i^a = \frac{\mathcal{L}^{(K)}}{\sqrt{}} p^{(K)} T_i^K X_K^a$$

(8.18)

where T_i^I is not required to equal Λ_i^j. In fact, the relation between Λ_I^i and T_i^I is partially determined by dynamical fields as the reduced Gauss constraint shows. This provides one example where connection and triad components after reduction are no longer canonically conjugate: we have

$$\int_\Sigma d^3x\, E_i^a \mathcal{L}_t A_a^i = p^{(I)} T_i^I \mathcal{L}_t \left(c_{(I)}\Lambda_I^i \right) = p^{(I)} \mathcal{L}_t \left(c_{(I)}\Lambda_I^i T_i^I \right) - c_{(I)} p^{(I)} \Lambda_I^i \mathcal{L}_t T_i^I.$$

(8.19)

Not c_I is conjugate to p^I but $c_{(I)}\Lambda_I^i T_i^{(I)}$ is. (Euler) parameterizing T_i^I as $T(\varepsilon_I) = \exp(\varepsilon_3 T_3)\exp(\varepsilon_2 T_1)\exp(\varepsilon_1 T_3)$ using generators T_I of SO(3) and inserting this product in the Liouville form shows that the angles ε_I are canonically conjugate to functions of those in Λ_I^i. (For instance, ε_1 is conjugate to $-\text{tr}((c \cdot \Lambda)(p \cdot T(\varepsilon_1 + \pi/2, \varepsilon_2, \varepsilon_3)))$, where c and p here denote the diagonal matrices with components c_I and p^I, respectively.)

For simplicity, we continue the discussion with the case where isotropy is realized in surfaces transversal to the fermion current. This allows the ansatz

$$
\Lambda_j^J = \begin{pmatrix} 1 & 0 & 0 \\ 0 & \cos\rho & -\sin\rho \\ 0 & \sin\rho & \cos\rho \end{pmatrix}, \quad T_j^J = \begin{pmatrix} 1 & 0 & 0 \\ 0 & \cos\sigma & \sin\sigma \\ 0 & -\sin\sigma & \cos\sigma \end{pmatrix} \tag{8.20}
$$

where ρ and σ are the only non-vanishing rotation angles. The Liouville term in the action then simplifies:

$$
\begin{aligned}
\frac{1}{8\pi\gamma G}\int_\Sigma d^3x\, E_i^a \mathscr{L}_t A_a^i &= \frac{1}{8\pi\gamma G} p^{(I)}\mathscr{L}_t\left(c_{(I)}\Lambda_I^i T_i^I\right) - c_{(I)} p^{(I)}\Lambda_I^i \mathscr{L}_t T_i^I \\
&= \frac{1}{8\pi\gamma G}\left(\dot{c}_1 p^1 + \mathscr{L}_t(c_2\cos(\rho-\sigma))p^2 + \mathscr{L}_t(c_3\cos(\rho-\sigma))p^3 \right. \\
&\qquad\qquad \left. - \dot{\sigma}(c_2 p^2 + c_3 p^3)\sin(\rho-\sigma)\right) \\
&= \frac{1}{8\pi\gamma G}\left(\dot{c}_1 p^1 + \dot{\bar{c}}_2 p^2 + \dot{\bar{c}}_3 p^3 + \dot{\sigma} p_\sigma\right),
\end{aligned}
\tag{8.21}
$$

where we introduced

$$
\bar{c}_2 = c_2\cos(\rho-\sigma), \quad \bar{c}_3 = c_3\cos(\rho-\sigma), \quad p_\sigma = -(c_2 p^2 + c_3 p^3)\sin(\rho-\sigma). \tag{8.22}
$$

(The variable p_σ is fixed in terms of the fermion current via the Gauss constraint.) In these components, the symplectic structure is

$$
\{c_1, p^1\} = 8\pi\gamma G, \quad \{\bar{c}_2, p^2\} = 8\pi\gamma G, \quad \{\bar{c}_3, p^3\} = 8\pi\gamma G, \quad \{\sigma, p_\sigma\} = 8\pi\gamma G. \tag{8.23}
$$

The kinematical degree of freedom σ is due to torsion, but will be removed after solving the Gauss constraint (which would be trivial for the same symmetry type in the absence of torsion). This Poisson algebra can easily be quantized by loop techniques. For c_1, $\bar{c}_2$ and $\bar{c}_3$ a Bohr representation arises, while the angle σ is simply represented on U(1).

At this stage, we see that some connection components have to be modified in (8.22) to keep them conjugate to the triad. Alternatively, we could have chosen to keep the connection components unchanged but transform the triad components to new forms. Then, triads and fluxes would not agree, and triads (unlike fluxes) may not commute in a resulting quantum representation. These issues are thus related to the non-Abelian nature of the full theory, which is modeled in this setting. It turns out to be more convenient to transform the connection components, however, since this amounts simply to an implicit subtraction of torsion terms [12]: Writing

$$
\bar{c}_2 = c_2\cos(\rho-\sigma) = c_2\Lambda_2^i T_i^2 = \phi_2^i T_i^2
$$

and recalling that T_i^I gives the direction of E_i^a, we can interpret $\bar{c}_2$ as a component

$$
(\mathscr{L}^{(2)-1(K)})^-\bar{c}_2 = \mathscr{L}^1\mathscr{L}^3 A_a^i E_i^b \frac{X_2^a \omega_b^2}{p^2}
$$

of the projection of A_a^i onto E_i^a. This projection can be seen to remove exactly the torsion contribution to extrinsic curvature contained in A_a^i. The projection transversal to E_i^a, and thus the torsion contribution in separation, is realized by the new variable p_σ. We will provide more details about the dynamical implications of fermions in Sect. 8.2.4.3

For another method to deal with non-Abelian features in non-diagonal Bianchi models, see [13].

8.2.3 Dynamics and Refinement

For the dynamics, we start with the classical expression

$$
\begin{aligned}
H = \frac{1}{8\pi G} &\left\{ \left[(c_2\Gamma_3 + c_3\Gamma_2 - \Gamma_2\Gamma_3)(1 + \gamma^{-2}) - n^1 c_1 - \gamma^{-2} c_2 c_3 \right] \sqrt{\left| \frac{p^2 p^3}{p^1} \right|} \right. \\
&+ \left[(c_1\Gamma_3 + c_3\Gamma_1 - \Gamma_1\Gamma_3)(1 + \gamma^{-2}) - n^2 c_2 - \gamma^{-2} c_1 c_3 \right] \sqrt{\left| \frac{p^1 p^3}{p^2} \right|} \\
&+ \left. \left[(c_1\Gamma_2 + c_2\Gamma_1 - \Gamma_1\Gamma_2)(1 + \gamma^{-2}) - n^3 c_3 - \gamma^{-2} c_1 c_2 \right] \sqrt{\left| \frac{p^1 p^2}{p^3} \right|} \right\}
\end{aligned}
$$

(8.24)

of the constraint in terms of the diagonal variables, where

$$
\Gamma_I = \frac{1}{2} \left(\frac{p^K}{p^J} n^J + \frac{p^J}{p^K} n^K - \frac{p^J p^K}{(p^I)^2} n^I \right)
$$

(8.25)

are the spin-connection components and n^I again classify the Bianchi type.

In order to loop quantize the constraint, we write the general form

$$
\hat{H} = \frac{i}{8\pi^2 \gamma G^2 \hbar \delta_1 \delta_2 \delta_3} \sum_{IJK} \varepsilon^{IJK} \mathrm{tr}(h_I h_J h_I^{-1} h_J^{-1} h_K [h_K^{-1}, \hat{V}]) + \hat{H}_\Gamma
$$

(8.26)

of a constraint operator [8] as in (4.5), where $\hat{H}_\Gamma$ depends on the Bianchi model and incorporates the spin-connection terms [14]. Connection components are then expressed via the three independent holonomies $h_I = \exp(i\delta_{(I)} c_I / 2)$ with $\delta_I = \ell_0^I / \mathscr{L}^{(I)}$.

Following the same steps as in isotropic cosmology, we can take the trace, represent the constraint as a combination of shift operators, and derive a difference equation in the triad representation in terms of the independent quantum numbers μ_I. We will see examples for such difference equations in later sections, and now focus on general aspects regarding the form of models.

8.2.3.1 Lattice Refinement

In an anisotropic setting with several independent connection components, the connection-dependent part of the constraint would, compared to isotropy, allow even

more choices to represent it in terms of almost-periodic functions. The freedom can
again be reduced by realizing the expression in the form of a holonomy around a
square loop. Such a relation to holonomies is also important for implementing different refinement schemes. To that end one takes the constraint operator (or difference
equation) obtained for constant holonomy parameters δ_I in $h_I = \exp(i\delta_I c_I/2)$,
relates these parameters to geometrical sizes such as lengths or areas, and introduces
a possible dependence on the geometry. Also here, there is now much more freedom
than in isotropic models since holonomies along the three independent directions
can depend differently on the geometry. Even if one restricts attention to power-law
cases, a single parameter such as x before does not suffice. This parameter can still
be used to characterize the total volume dependence of the number of patches, but
in each direction patches may refine differently. To summarize this, we now write

$$\mathcal{N}_1(t)v_1(t)\mathcal{N}_2(t)v_2(t)\mathcal{N}_3(t)v_3(t) = \mathcal{V}\sqrt{|p^1(t)p^2(t)p^3(t)|} \qquad (8.27)$$

for the volume, which must factorize into three independent equations

$$\mathcal{N}_I(t)v_I(t) = \mathcal{L}^I\sqrt{|p^I(t)|} \qquad (8.28)$$

for the extensions of the three directions, where $\mathcal{V} = \mathcal{L}^1\mathcal{L}^2\mathcal{L}^3$. The product
$\mathcal{N}_1\mathcal{N}_2\mathcal{N}_3$ as the total number of patches can then be parameterized to be propor-
tional to a power $|p^1p^2p^3|^{-x}$ of volume, but this does not fix the individual depen-
dences of $\mathcal{N}_I(p^1, p^2, p^3)$. If dependences are such that $\mathcal{N}_I(p^I)$ depends only on the
geometry of its own direction, for instance $\mathcal{N}_I \propto |p^I|^{1/2}$ or $|p^I|^{-x}$ more generally,
the refinement resembles what we have in isotropic models, just for three directions
independently. In this case, while step sizes may not immediately be equidistant,
variables can be redefined to make them so. But in anisotropic models the generic
behavior would be one where each $\mathcal{N}_I$ depends on all the triad variables, which in
general results in a difference equation whose steps cannot all be made uniform. In
fact, it is rather natural to expect that $\mathcal{N}_I$ depends on the extension in direction I,
given by the co-triad component $|e_I| = \sqrt{|p^1p^2p^3|}/p^I$. (This non-trivial case is
required for consistent dynamics in the sense of stability [15]. At least one of the
variables can be made uniform, corresponding to the total volume as the isotropic
average of the degrees of freedom. Also at the quantum level there thus seems to
be an advantage to using Misner-type variables splitting the scale factors into the
volume and anisotropy parameters.) One thus has to deal with difference equations
on a genuinely non-uniform lattice, which poses new mathematical and numerical
problems. Methods of handling non-equidistant difference equations of this form are
being developed [16, 17].

8.2.3.2 Singularity

The general type of difference equations can be used to infer properties similar to
those of isotropic models. First, there is no singularity because wave functions are,

starting with initial and boundary values, uniquely evolved on all of minisuperspace, including configurations which from the classical perspective would lie beyond a singularity. For this conclusion, only steps in the difference equation around where one of the μ_I vanishes need be used, which is independent of the refinement scheme. Further restrictions on the refinement do, however, arise by making sure that the difference equation has good stability properties: solutions in classically allowed regions must be oscillatory rather than increase exponentially. For this, the refinement must be such that classical regions are provided with sufficiently fine lattices, while quantum regions can have significant coarseness. In contrast to isotropic models, ensuring this behavior in the whole minisuperspace is non-trivial and restricts refinements even beyond the volume dependence; see Sect. 8.3 and Sect. 11.2.1.3 for further details.

8.2.3.3 Fermions

Another interesting application of anisotropic models is the inclusion of fermions in the dynamics. The direction distinguished by the fermion current does not allow its inclusion in isotropic models, but rather simple anisotropic ones can be obtained. An advantage compared to isotropic models with a scalar source is that a fundamental fermion restricts, via Pauli's exclusion principle, the amount of matter that can be present per degree of freedom. The kinetic energy of a scalar field can be made arbitrarily large by raising its momentum, a freedom often exploited to produce bounces in kinetic-dominated regimes. A fermion field, on the other hand, has a maximum excitation level per degree of freedom, which does not allow one to raise the density arbitrarily.

Continuing with the constructions in Sect. 8.2.3 we now incorporate a matter source $H_{\text{fermion}} = \frac{3}{2}\pi G \mathscr{J}_1^2/p^2 \sqrt{|p^1|}$ in the Hamiltonian constraint, with a densitized fermion current $\mathscr{J}_1 := \sqrt{\det h} J_1$ pointing in the anisotropic direction. (At this stage, we assume the Gauss constraint to be solved; see [12] for more details.) We represent fermion states as the space of functions $f(\Theta_\alpha)$ of four independent half-densitized Grassmann-valued variables Θ_α, $\alpha = 1, \ldots, 4$, for the four components contained in the fermion fields ξ and χ. The fermionic momenta $\pi_\xi = i\xi^\dagger$ and $\pi_\chi = i\chi^\dagger$ then give rise to components $\bar{\Theta}_\alpha$, represented as $\hbar\partial/\partial\Theta_\alpha$. The densitized current component $\mathscr{J}_1 = \xi^\dagger \sigma_1 \xi + \chi^\dagger \sigma_1 \chi$ then becomes

$$\hat{\mathscr{J}}_1 = \hbar \frac{\partial}{\partial\Theta_2}\Theta_1 + \hbar\frac{\partial}{\partial\Theta_1}\Theta_2 + \hbar\frac{\partial}{\partial\Theta_4}\Theta_3 + \hbar\frac{\partial}{\partial\Theta_3}\Theta_4. \tag{8.29}$$

The operator $\frac{\partial}{\partial\Theta_2}\Theta_1 + \frac{\partial}{\partial\Theta_1}\Theta_2$ can easily be diagonalized: Each 2-spinor copy has two eigenstates $f_0(\Theta) = 1$ and $f^0(\Theta) = \Theta_1\Theta_2$ of eigenvalue zero, and one eigenstate each $f_\pm(\Theta) = \Theta_1 \pm \Theta_2$ of eigenvalue ± 1. Taking the tensor product of both 2-spinor copies ξ and χ then gives current eigenvalues zero, $\pm\hbar$ and $\pm 2\hbar$.

States in the triad representation are now given by wave functions $\psi_{\mu_1,\mu_2}(\Theta)$ with the flux eigenvalues μ_1 and μ_2 and the fermion dependence via Θ. For the vacuum model, there is a symmetry transformation flipping the sign of μ_1, or changing the orientation of space. It thus amounts to a parity transformation which does not leave the fermion fields invariant. The complete parity transformation $\hat{\Pi}$ is represented as

$$\psi_{\mu_1,\mu_2}(\Theta_1, \Theta_2, \Theta_3, \Theta_4) \xrightarrow{\hat{\Pi}} \psi_{-\mu_1,\mu_2}(\Theta_3, \Theta_4, \Theta_1, \Theta_2) \tag{8.30}$$

switching the places of the 2-spinor components.

The energy density of fermions can only be microscopic: eigenvalues of $\mathscr{J}_1$ can be at most $2\hbar$. Fermion models can thus be expected to be significantly different from kinetic-dominated scalar models, but they have not been studied in detail yet.

8.2.4 Reduction from Anisotropy to Isotropy

Anisotropic models are useful also in that they allow explicit symmetry reductions at the quantum level, providing a test-bed for the general procedure of deriving models from the full theory. Starting with a general homogeneous setting, isotropy can be introduced and then compared with the original isotropic reduction. In this way one again goes beyond the minisuperspace quantization of a single model, putting all homogeneous models in one setting. Moreover, the techniques allow perturbations around isotropic models, adding perturbative anisotropies to the dynamics.

To illustrate the reduction, we follow [18] and use an LRS Bianchi I model which has one rotational symmetry axis in addition to the homogeneous Bianchi I symmetries. Invariant connections and triads in this case are

$$A_a^i dx^a \tau_i = \tilde{A}\tau^1 dx + \tilde{A}\tau^2 dy + \tilde{C}\tau^3 dz \tag{8.31}$$

$$E_i^a \frac{\partial}{\partial x^a}\tau^i = \tilde{p}_A\tau^1\partial_x + \tilde{p}_A\tau^2\partial_y + \tilde{p}_C\tau^3\partial_z \tag{8.32}$$

and have Poisson brackets

$$\{\tilde{A}, \tilde{p}_A\} = \frac{4\pi\gamma G}{\mathscr{V}}, \quad \{\tilde{C}, \tilde{p}_C\} = \frac{8\pi\gamma G}{\mathscr{V}}. \tag{8.33}$$

As before, we factorize the coordinate volume into $\mathscr{V} = (\mathscr{L}^1)^2\mathscr{L}^3$ (with $\mathscr{L}^1 = \mathscr{L}^2$ owing to the rotational symmetry) and absorb factors into the basic variables by rescaling

$$A = \mathscr{L}^1\tilde{A}, \quad C = \mathscr{L}^3\tilde{C}, \quad p_A = \mathscr{L}^1\mathscr{L}^3\tilde{p}_A, \quad p_C = (\mathscr{L}^1)^2\tilde{p}_C. \tag{8.34}$$

The new variables are invariant under rescaling the coordinates by $x \mapsto \lambda_1 x$, $y \mapsto \lambda_1 y$, $z \mapsto \lambda_3 z$. (Notice the density weight of momenta: when we rescale coordinates, the densitized vector field $E^I\partial/\partial x^I$ transforms to $\lambda_1^2\lambda_3 E^I\partial/\partial(\lambda_{(I)}x^I)$).

As in general diagonal Bianchi models, there is a residual gauge transformation $p_A \mapsto -p_A$ which can be fixed by requiring $p_A \geq 0$. The sign of p_C, however, does have invariant kinematical meaning as the orientation of the triad. Both triad components together determine the volume

$$V_{\text{aniso}} = \sqrt{p_A^2 |p_C|}. \tag{8.35}$$

The classical Hamiltonian constraint takes the simple form

$$H_{\text{aniso}} = -(8\pi G)^{-1} \gamma^{-2} \left(A^2 \frac{p_A}{\sqrt{|p_C|}} + 2AC\sqrt{|p_C|} \right) + H_{\text{matter}}(p_A, p_C) = 0 \tag{8.36}$$

as a reduction of the general Bianchi Hamiltonian (8.24).

For a perturbative treatment it is useful to perform a linear canonical transformation which explicitly splits the variables into isotropic ones and perturbations (a form of Misner variables for LRS Bianchi models):

$$(\bar{c}, \bar{p}) = \left(\frac{1}{3}(2A + C), \frac{1}{3}(2p_A + p_C) \right) \tag{8.37}$$

$$(\varepsilon, p_\varepsilon) = \left(\frac{1}{3}(A - C), \frac{1}{3}(p_A - p_C) \right) \tag{8.38}$$

with Poisson brackets

$$\{\bar{c}, \bar{p}\} = \frac{8\pi \gamma G}{3}, \quad \{\varepsilon, p_\varepsilon\} = \frac{4\pi \gamma G}{3}. \tag{8.39}$$

The original variables are obtained by the inverse transformation

$$(A, C) = (\bar{c} + \varepsilon, \bar{c} - 2\varepsilon), \quad (p_A, p_C) = (\bar{p} + p_\varepsilon, \bar{p} - 2p_\varepsilon). \tag{8.40}$$

For perturbations we assume $p_\varepsilon \ll \bar{p}$; in particular, the approximation will break down close to classical singularities of the isotropic type where $\bar{p} = 0$. For the connection we only assume $\varepsilon \gg 1$ regardless of the value of the isotropic $\bar{c}$.

There are different ways to introduce such a transformation; the one chosen here gives rise to a volume expression

$$V_{\text{aniso}} = \sqrt{|\bar{p}^3 - 3p_\varepsilon^2 \bar{p} - 2p_\varepsilon^3|} = |\bar{p}|^{3/2} \left(1 - \frac{3}{2} \frac{p_\varepsilon^2}{\bar{p}^2} + O(p_\varepsilon^3/\bar{p}^3) \right) \tag{8.41}$$

which agrees with the isotropic one up to terms of at least second order in p_ε. The Hamiltonian constraint becomes

$$H_{\text{aniso}} = -\frac{3}{8\pi G \gamma^2} |\bar{p}|^{-3/2} \left(\bar{c}^2 \bar{p}^2 - \varepsilon^2 \bar{p}^2 + \tfrac{1}{2} \bar{c}^2 p_\varepsilon^2 + 2\bar{c} \bar{p} \varepsilon p_\varepsilon + O(p_\varepsilon^3/\bar{p}^3) \right) + H_{\text{matter}}(\bar{p}, p_\varepsilon) \tag{8.42}$$

up to terms of third order in the perturbation.

At the non-perturbative anisotropic level, the Hilbert space and the basic holonomy-flux representation on it are constructed as in general anisotropic models. We have the space

$$\mathcal{H}_{\text{aniso}} \cong \mathcal{H}_{\text{iso}} \otimes \mathcal{H}_{\text{iso}} \tag{8.43}$$

of almost-periodic functions in two variables, with orthonormal basis

$$\langle A, C | \mu, \nu \rangle = \exp(i(\mu A + \nu C)/2), \quad \mu, \nu \in \mathbb{R}. \tag{8.44}$$

Holonomy operators act by multiplication as before,

$$h_A^{(\rho)} | \mu, \nu \rangle = | \mu + \rho, \nu \rangle, \quad h_C^{(\tau)} | \mu, \nu \rangle = | \mu, \nu + \tau \rangle \tag{8.45}$$

for

$$h_A^{(\rho)} = \exp(i\rho A/2), \quad h_C^{(\tau)} = \exp(i\tau C/2), \tag{8.46}$$

and fluxes are

$$\hat{p}_A | \mu, \nu \rangle = 2\pi \gamma \ell_{\text{P}}^2 \mu | \mu, \nu \rangle, \quad \hat{p}_C | \mu, \nu \rangle = 4\pi \gamma \ell_{\text{P}}^2 \nu | \mu, \nu \rangle. \tag{8.47}$$

Gauge invariance of states under the residual transformation can be ensured by working only with states $\sum_{\mu,\nu} \psi_{\mu,\nu} | \mu, \nu \rangle$ where $\psi_{\mu,\nu}$ is symmetric in μ. In terms of the triad operators, we have the volume operator $\hat{V} = \sqrt{\hat{p}_A^2 | \hat{p}_C |}$.

8.2.4.1 Isotropic Distributions

The relationship between isotropic and anisotropic variables is now to be formulated at the quantum level. Isotropic states are not contained in the anisotropic Hilbert space, for they are not normalizable from that perspective. They can instead be realized as distributions. To do so explicitly, we select the dense subspaces Cyl_{iso} and $\text{Cyl}_{\text{aniso}}$ of cylindrical functions in the two Hilbert spaces. They are finite linear combinations of the basis states $| \mu \rangle$ and $| \mu, \nu \rangle$, respectively. Distributional states are linear functionals on the cylindrical subspaces. Taking all spaces together, we write the two Gel'fand triples $\text{Cyl}_{\text{iso}} \subset \mathcal{H}_{\text{iso}} \subset \text{Cyl}_{\text{iso}}^\star$ and $\text{Cyl}_{\text{aniso}} \subset \mathcal{H}_{\text{aniso}} \subset \text{Cyl}_{\text{aniso}}^\star$.

Isotropic states [19] in the anisotropic model can now be introduced as distributions supported only on isotropic connections with $\varepsilon = 0$. Such states must indeed be distributional: they are supported on a set of measure zero in the space of anisotropic connections. Explicitly, they can be implemented by an antilinear map

$$\sigma : \text{Cyl}_{\text{iso}} \to \text{Cyl}_{\text{aniso}}^\star, \quad | \mu \rangle \mapsto \langle \mu | \tag{8.48}$$

such that

$$\sigma(|\mu\rangle)[|\rho,\tau\rangle] = (\mu|\rho,\tau) = \langle\mu|\rho,\tau\rangle|_{A=C=c} \quad \text{for all } |\rho,\tau\rangle$$

implements the distributional restriction to isotropy $A = C = c$ by using

$$\langle A, C|\rho,\tau\rangle|_{A=C=c} = \langle c, c|\rho,\tau\rangle = \exp(\mathrm{i}(\rho+\tau)c/2).$$

We compute $(\mu|$ by expanding $\sigma(|\mu\rangle) =: \sum_{\kappa,\lambda} \sigma_{\kappa,\lambda}(\mu)\langle\kappa,\lambda|$:

$$\sigma_{\rho,\tau}(\mu) = \sigma(|\mu\rangle)[|\rho,\tau\rangle] = \int e^{-\mathrm{i}\mu c/2} e^{\mathrm{i}(\rho+\tau)c/2} \mathrm{d}\mu_H(c) = \delta_{\mu,\rho+\tau}.$$

Thus,

$$(\mu| = \sigma(|\mu\rangle)) = \sum_{\rho,\tau} \delta_{\mu,\rho+\tau}\langle\rho,\tau| = \sum_{\rho}\langle\rho,\mu-\rho| \tag{8.49}$$

summing over all real numbers, which for a distribution in $\mathrm{Cyl}^*_{\mathrm{aniso}}$ is well-defined.

To confirm the symmetry of the states, we evaluate $(\mu|\rho,\tau) = \delta_{\mu,\rho+\tau}$: the distribution is non-zero only if the averaged label $\rho + \tau$ equals the isotropic one, in eigenvalues

$$\tfrac{1}{2}\gamma\ell_P^2(\rho+\tau) = 2(\hat{p}_A)_\rho + (\hat{p}_C)_\tau = 3(\hat{\tilde{p}})_{\rho+\tau}$$

as required for triad operators quantizing the second relation (8.40). Fluxes are thus isotropic in the states. Moreover, $(\mu|h_A^{(\rho)} h_C^{(\rho)-1} = (\mu|$: the dual action of $\exp(\mathrm{i}\rho A/2)\exp(-\mathrm{i}\rho C/2) = \exp(3\mathrm{i}\rho\varepsilon/2)$ acts trivially on symmetric distributions, in accordance with the fact that they are supported only on isotropic connections with $\varepsilon = 0$.

In this simple, finite-dimensional case the definition of a symmetric state can easily be seen to amount to the following procedure: Take an isotropic state in the connection representation and multiply it with a δ-distribution supported at $\varepsilon = 0$ on the Bohr compactification of the real line, $\delta(A - C) = \sum_\rho e^{\mathrm{i}\rho(A-C)/2}$,

$$e^{\mathrm{i}\mu C/2}\delta(A - C) = e^{\mathrm{i}\mu C/2}\sum_\rho e^{\mathrm{i}\rho(A-C)/2} = \sum_\rho e^{\mathrm{i}(\rho A+(\mu-\rho)C)/2} = \sigma(|\mu\rangle)(|A, C\rangle).$$

The result is a symmetric distributional state in the anisotropic model according to (8.49). Our original definition (8.48) generalizes this concept to more general systems where explicit constructions of delta-distributions on subspaces of invariant connections would be too complicated.

Unlike holonomies, not all operators map a symmetric state to another symmetric one by their dual action. This prevents us from directly defining all operators for the isotropic model, such as the Hamiltonian constraint, by deriving them from the dual action of anisotropic operators. But suitable operators do exist for which a reduction can be done, and they suffice to derive the basic operators of the model and the holonomy-flux algebra. For flux operators, one can easily see that $2\hat{p}_A + \hat{p}_C = 3\hat{\bar{p}}$ is the only one mapping an isotropic state $(\mu|$ to another isotropic state:

$$\sigma(|\mu\rangle)\hat{p} = \tfrac{1}{3}(\mu|(2\hat{p}_A + \hat{p}_C) = \tfrac{4}{3}\pi\gamma\ell_P^2\mu(\mu| = \sigma(\hat{\bar{p}}|\mu)). \qquad (8.50)$$

It also agrees with the isotropic flux operator $\hat{\bar{p}}$ defined in the isotropic model: the actions of $\hat{p}$ and σ commute. For holonomies, arbitrary products of h_A and h_C map an isotropic state to another one, which follows from the fact that $h_A h_C^{-1}$ acts as the identity while the remaining factor simply amounts to an isotropic holonomy operator. Holonomy operators form a closed algebra with the isotropic flux operator $\hat{\bar{p}} = \tfrac{1}{3}(2\hat{p}_A + \hat{p}_C)$:

$$[h_A^{(\rho)}h_C^{(\tau)}, \hat{\bar{p}}]|\mu, \nu\rangle = -\tfrac{4}{3}\pi\gamma\ell_P^2(\rho + \tau)|\mu + \rho, \nu + \tau\rangle = -\tfrac{4}{3}\pi\gamma\ell_P^2(\rho + \tau)h_A^{(\rho)}h_C^{(\tau)}|\mu, \nu\rangle, \qquad (8.51)$$

thus $[h_A^{(\rho)}h_C^{(\tau)}, \hat{\bar{p}}] = -\tfrac{4}{3}\pi\gamma\ell_P^2(\rho + \tau)h_A^{(\rho)}h_C^{(\tau)}$. The anisotropy operator $\hat{p}_\varepsilon :=$ $\tfrac{1}{3}(\hat{p}_A - \hat{p}_C)$ could also be included in a closed algebra due to

$$[h_A^{(\rho)}h_C^{(\tau)}, \hat{p}_\varepsilon] = -\tfrac{1}{6}\gamma\ell_P^2\left(\tfrac{1}{2}\rho - \tau\right)h_A^{(\rho)}h_C^{(\tau)} \qquad (8.52)$$

(after all, the anisotropic operators form a closed algebra) but it does not map an isotropic distribution to another such distribution.

Operators isolated so far represent many of the elements of the classical basic algebra, split into averaged isotropic and perturbative anisotropic ones. The unique flux $\hat{\bar{p}}$ fixing isotropic distributions corresponds to the average flux. Holonomy products fixing any isotropic state represent the connection perturbation ε. While these operators are intrinsically defined, making use only of the symmetry properties, no unique quantization corresponding to $\bar{c}$ can be obtained in this way. Even at the algebraic level of operators, where the distributional nature of symmetric states does not matter, a reduced model is not simply a subspace of a less symmetric one. Instead, it also requires a factorization procedure: We can act with all holonomies on symmetric states since they leave this space invariant. Thus, the unique averaged flux together with all anisotropic holonomies defines the reduced basic algebra. On any given symmetric state, many anisotropic holonomies have the same action. A unique representation thus requires factoring out equivalent actions.

This is fully analogous to the classical situation where the splitting of anisotropic variables A and C into an isotropic average $\bar{c}$ and an anisotropy is not unique. The situation here is thus equivalent to classical averaging problems, which for inhomogeneities play important roles in cosmology. Classically, a given $\bar{p}$, corresponding to our unique flux operator mapping isotropic states to isotropic ones, determines a form of ε (corresponding to the uniquely defined holonomy operator $h_A h_C^{-1}$ fixing isotropic states) as a linear combination of A and C by requiring $\{\bar{p}, \varepsilon\} = 0$. A unique form for $\bar{c}$ and p_ε can then be obtained only with an additional choice. (We did this classically by requiring the volume (8.41) to receive corrections only to second order.) One can for instance define $\hat{p}_\varepsilon = \tfrac{1}{3}(\hat{p}_A - \hat{p}_C)$ as above and then require that (8.52) vanish, analogously to $\{\bar{c}, p_\varepsilon\} = 0$. This prescription yields $h_A^2 h_C$

as a specific choice of the isotropic holonomy operator in addition to the flux $\hat{\bar{p}}$. The pair indeed forms a subalgebra of the anisotropic operator algebra, isomorphic to the isotropic algebra and mapping isotropic states to isotropic states.

This example illustrates the constructions of symmetric models within a fuller setting. Models cannot be directly embedded in the full theory, but their basic algebra and quantum representation can be derived from the full one: We use the classical relations between symmetric and non-symmetric variables to define a distinguished subalgebra of the non-symmetric holonomy-flux algebra $\mathscr{A}_{\text{aniso}}$. Using the fact that the non-symmetric representation is cyclic on $|0, 0\rangle$ such that the subspace $\mathscr{A}_{\text{aniso}}|0, 0\rangle$ is dense in $\mathscr{H}_{\text{aniso}}$, we generate the representation of the reduced model by acting only with the symmetric subalgebra on the cyclic state. In this way we define the reduced state space, which is equipped with an inner product by requiring holonomy operators to be unitary and fluxes to be self-adjoint. Upon completion, this inner-product space defines the Hilbert space of the isotropic model and its quantum representation in agreement with the loop quantization of the classically reduced model. Key properties of holonomies and fluxes, such as the spatial discreteness they imply, then descend to the models. Characteristic features of the representation are inherited from those of the less symmetric system or even the full theory. From the basic algebra one can construct more complicated operators such as the Hamiltonian constraint by following the constructions done in the full theory.

The constructions in the anisotropic setting can also be used to illustrate difficulties expected for a reduction of models from the full theory: (i) The Hamiltonian constraint is constructed by analogy, not derived, even though its characteristic features known from the full theory are implemented. (ii) The averaging problem is relevant since one must know not only the classical embedding of symmetric configurations in non-symmetric ones but also a projection or factorization map in the opposite direction, here used to single out the isotropic degree of freedom $\bar{c}$. This problem is easily tractable for isotropic models within anisotropic ones, but much more complicated for inhomogeneities even classically [20, 21]. Once the reduced algebra and Hamiltonian constraint have been constructed, the development of perturbation theory is merely a technical issue. For isotropic models within LRS Bianchi I space–times, this has been done in [18] (see the next subsection), for isotropic models in Bianchi IX space–times in [22].

8.2.4.2 Anisotropic Perturbations

For small connection components, the nature of the fundamental evolution equation as a difference equation does not matter much and a Wheeler–DeWitt limit can be taken. This is routinely done to confirm the continuum, small-curvature limit of loop quantum cosmology [23]. (See also [24].) In the present context of uninhibited isotropic curvature, we intend to treat the average isotropic component $\bar{c}$ still in its quantum configuration space $\bar{\mathbb{R}}_{\text{Bohr}}$, while ε is a small perturbation for which only a part of the quantum configuration space near $\varepsilon = 0$ should matter, blind to the Bohr-compactified nature. A wave packet peaked at a small value of ε, for instance, would not be sensitive to the whole configuration space. The behavior of the perturbation can thus be formulated on the ordinary Schrödinger Hilbert space $\mathcal{H}_S = L^2(\mathbb{R}, d\varepsilon)$, with a dynamical equation in the triad representation expected to be of difference type for $\bar{p}$, but differential for p_ε.

States: Accordingly, we assume that the wave function does not vary rapidly when p_ε changes. In an approximate sense, an operator for ε, rather than just its holonomy, then exists. With the Schrödinger Hilbert space $\mathcal{H}_S$ for anisotropies, we define the perturbative Hilbert space

$$\mathcal{H}_{\text{pert}} := \mathcal{H}_{\text{iso}} \otimes \mathcal{H}_S \qquad (8.53)$$

and realize its dense subset $\text{Cyl}_{\text{iso}} \otimes \text{Cyl}_S$ as a subspace of the dual $\text{Cyl}^\star_{\text{aniso}}$. In the Schrödinger Hilbert space we have to choose a suitable dense set Cyl_S, which for our purposes will be the set of all functions that are products of a polynomial and a Gaussian. Schrödinger states in Cyl_S can be interpreted as distributions on the Bohr Hilbert space [25] using the antilinear map

$$\pi : \text{Cyl}_S \to \text{Cyl}^\star_{\text{Bohr}}, \pi(\psi)[|\rho\rangle] := \int e^{i\rho\varepsilon}\overline{\psi(\varepsilon)}d\varepsilon \quad \text{for } \psi \in \mathcal{H}_S.$$

Dual actions of basic operators on those states are given by

$$(\hat{p}_\varepsilon \pi(\psi))[|\rho\rangle] = \pi(\psi)[\hat{p}_\varepsilon^\dagger|\rho\rangle] = \tfrac{4}{3}\pi\gamma\ell_P^2 \int \rho e^{i\rho\varepsilon}\overline{\psi(\varepsilon)}d\varepsilon$$

$$= \tfrac{4}{3}\pi\gamma\ell_P^2 \int e^{i\rho\varepsilon} \cdot i\frac{d}{d\varepsilon}\overline{\psi(\varepsilon)}d\varepsilon = \pi\left(-\tfrac{4}{3}i\pi\gamma\ell_P^2 d\psi/d\varepsilon\right)[|\rho\rangle]$$

$$(8.54)$$

and

$$(e^{i\tau\varepsilon}\pi(\psi))[|\rho\rangle] = \psi[e^{-i\tau\varepsilon}|\rho\rangle] = \int e^{i(\rho-\tau)\varepsilon}\overline{\psi(\varepsilon)}d\varepsilon = \int e^{i\rho\varepsilon}\overline{e^{i\tau\varepsilon}\psi(\varepsilon)}d\varepsilon$$

$$= \pi(e^{i\tau\varepsilon}\psi)[|\rho\rangle]. \qquad (8.55)$$

The momentum operator is a derivative operator on the Schrödinger as well as the Bohr Hilbert space, while $(e^{i\tau\varepsilon}\pi(\psi))(\varepsilon) = \pi(e^{i\tau\varepsilon}\psi)(\varepsilon)$. On the Schrödinger Hilbert space, unlike the Bohr Hilbert space, we can now take the derivative with respect to τ. By going to the dual action on the image of π, we obtain a simple multiplication operator for ε, well-defined with domain $\pi(\text{Cyl}_S)$.

While the perturbative space does not have a canonical isotropic subspace, any fixed state $\Psi \in \text{Cyl}_S$, understood as the map $\mathbb{C} \to \text{Cyl}_S, 1 \mapsto \Psi$, provides a mapping $\text{id} \otimes \Psi : \text{Cyl}_{\text{iso}} \to \text{Cyl}_{\text{pert}}$ from isotropic cylindrical states to perturbative cylindrical states. For states in the image, anisotropies are small in mean value, but not eliminated exactly, if the state is chosen to have significant support only for small perturbations. There are now two ways to implement symmetries: the strict one by σ and the perturbative one by $\pi_\Psi := (\star \otimes \pi) \circ (\text{id} \otimes \Psi)$, both as maps from Cyl_{iso} to $\text{Cyl}^\star_{\text{aniso}}$:

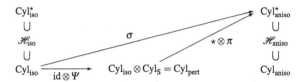

The $\star$ here denotes the antilinear dualization of isotropic states.

On the perturbative Hilbert space, averaged operators can be implemented as before. In addition, we now have non-trivial operators for perturbations. Any anisotropic operator $\hat{O}$ acting on $\mathscr{H}_{\text{aniso}}$ has a dual action on $\star \otimes \pi(\text{Cyl}_{\text{pert}})$, but does not necessarily fix this subspace of $\text{Cyl}_{\text{aniso}}^{\star}$. It does so perturbatively when we expand it as a sum of operators in the perturbative sector. With an unsqueezed Gaussian for Ψ, for instance, we have perturbative states of the form

$$\psi(A, C) = e^{i\bar{v}\bar{c}/2} e^{-(\varepsilon-\varepsilon_0)^2/4\sigma^2} e^{3i\varepsilon p_\varepsilon^0/4\pi\gamma\ell_P^2}$$

where $\bar{c} = \frac{1}{3}(2A + C)$ and $\varepsilon = \frac{1}{3}(A - C)$ are understood as functions of A and C. By the chain rule, we then have flux operators

$$\hat{p}_A\psi = -4\pi i\gamma\ell_P^2\frac{\partial}{\partial A}\psi = -\frac{4}{3}\pi i\gamma\ell_P^2\left(2\frac{\partial}{\partial\bar{c}} + \frac{\partial}{\partial\varepsilon}\right)\psi = \left(\frac{4}{3}\pi\gamma\ell_P^2\bar{v} + \hat{p}_\varepsilon\right)\psi \tag{8.56}$$

and

$$\hat{p}_C\psi = -8\pi i\gamma\ell_P^2\frac{\partial}{\partial C}\psi = -\frac{8}{3}\pi i\gamma\ell_P^2\left(\frac{\partial}{\partial\bar{c}} - \frac{\partial}{\partial\varepsilon}\right)\psi = \left(\frac{4}{3}\pi\gamma\ell_P^2\bar{v} - 2\hat{p}_\varepsilon\right)\psi. \tag{8.57}$$

Composite operators. Expansions can be used for composite operators, for instance when eigenvalues are already known. If an anisotropic operator $\hat{O}$ has eigenstates $|\mu, \nu\rangle$, we take the eigenvalues $O_{\mu,\nu}$ and insert, following (8.56), (8.57),

$$\begin{aligned}\mu &= \tfrac{2}{3}\bar{v} + p_\varepsilon/2\pi\gamma\ell_P^2 = \tfrac{2}{3}\bar{v} + P, \\ \nu &= \tfrac{1}{3}\bar{v} - p_\varepsilon/2\pi\gamma\ell_P^2 = \tfrac{1}{3}\bar{v} - P\end{aligned} \tag{8.58}$$

with $P := p_\varepsilon/2\pi\gamma\ell_P^2$. This procedure yields a function $O(\bar{v}, P)$ which we expand in the perturbation $P/\bar{v}$,

$$O(\bar{v}, P) = \sum_k O_{\text{iso}}^{(k)}(\bar{v})P^k. \tag{8.59}$$

Note that P itself need not be small compared to one by our assumptions, which would mean $p_\varepsilon \ll \ell_P^2$. We have, however, used $P \ll \bar{v}$ for anisotropies small compared to the isotropic average, such that each $O_{\text{iso}}^{(k)}(\bar{v})$ must drop off at least as $\bar{v}^{-k}$. For any fixed k, the values $O_{\text{iso}}^{(k)}(\bar{v})$, interpreted as eigenvalues, define an isotropic operator $\hat{O}_{\text{iso}}^{(k)} = \sum_{\bar{v}} O_{\text{iso}}^{(k)}(\bar{v})|\bar{v}\rangle\langle\bar{v}|$ such that $\hat{O}_{\text{iso}}^{(k)}|\bar{v}\rangle := O_{\text{iso}}^{(k)}(\bar{v})|\bar{v}\rangle$. Thus, we obtain the expansion

$$\hat{O} \sim \sum_k \hat{O}_{\text{iso}}^{(k)} \otimes \hat{P}^k \tag{8.60}$$

acting on $\mathscr{H}_{\text{pert}}$.

Such expansions will also have to be applied to operators implementing lattice refinement. For instance, the C-holonomy, taken not as a basic operator but as one with refinement $\mathscr{N}_3(p_A, p_C)$ in its argument, has the action

$$h_C^{(\mathscr{N}_3)}|\mu, v\rangle = |\mu, v + \mathscr{N}_3(\mu, v)^{-1}\rangle$$

on anisotropic states. We expand

$$\mathscr{N}_3(\mu, v) = \mathscr{N}_3\left(\tfrac{2}{3}\bar{v} + P, \tfrac{1}{3}\bar{v} - P\right) = \mathscr{N}_3\left(\tfrac{2}{3}\bar{v}, \tfrac{1}{3}\bar{v}\right) + P\left(\partial_\mu\mathscr{N}_3 - \partial_v\mathscr{N}_3\right) + \cdots$$

and write the refinement in terms of isotropic discrete labels only, but with higher-order terms in the P-expansion:

$$h_C^{(\mathscr{N}_3)}|\mu, v\rangle = |\mu, v + \mathscr{N}_3\left(\tfrac{2}{3}\bar{v}, \tfrac{1}{3}\bar{v}\right)^{-1} - P(\partial_\mu\mathscr{N}_3 - \partial_v\mathscr{N}_3) + \cdots\rangle.$$

A similar equation is obtained for $h_A^{(\mathscr{N}_1)}$. As a result, the action of

$$(h_A^{(\mathscr{N}_1)})^{-1}h_C^{(\mathscr{N}_3)}|\mu, v\rangle = |\mu - \mathscr{N}_1\left(\tfrac{2}{3}\bar{v}, \tfrac{1}{3}\bar{v}\right)^{-1} + \cdots, v + \mathscr{N}_3\left(\tfrac{2}{3}\bar{v}, \tfrac{1}{3}\bar{v}\right)^{-1} + \cdots\rangle$$

on isotropic distributions is the identity only for isotropic refinements, $\mathscr{N}_1 = \mathscr{N}_3$.

For operators not having $|\mu, v\rangle$ as eigenstates, an expansion is not always possible. To continue with the example of lattice-refining holonomies, we first prepare the expansion by writing, for instance, $h_C^{(\mathscr{N}_3)} = \exp(iC/2\mathscr{N}_3) = \exp(i\bar{c}/2\mathscr{N}_3)\exp(-i\varepsilon/\mathscr{N}_3)$. The ε-expansion can now easily be done, but the first exponential may depend on P via $\mathscr{N}_3$ whose expansion is non-trivial. We may write

$$\exp(i\bar{c}/2\mathscr{N}_3) = \exp(i\bar{c}/2\mathscr{N}_3(2\bar{v}/3, \bar{v}/3)) - \frac{1}{2}i\bar{c}P(\partial_\mu\mathscr{N}_3 - \partial_v\mathscr{N}_3) + \cdots$$

to expand in P and write an operator in a form analogous to (8.60), but the expansion coefficients are not all almost-periodic in $\bar{c}$ as they should if the operator were to act on Cyl_{pert}. Lattice-refining holonomies can be expanded around the isotropic model if and only if the refinement functions are of the form $\mathscr{N}_I(\mu + v) = \mathscr{N}_I(\bar{v})$ depending only on the isotropic discrete scale. Then, no coefficients with non-almost periodic functions in $\bar{c}$ arise. The condition for a reduction and an expansion to exist, requiring anisotropic refinement functions to depend only on the isotropic discreteness scale, is consistent with the absence of refinement in Wheeler–DeWitt models, whose representation is used here for the anisotropy parameter. Anisotropic refinement cannot fully be modeled perturbatively.

Hamiltonian constraint. The main example for an operator to be expanded is the Hamiltonian constraint. It will provide the form of a difference-differential equation, because the isotropic average is Bohr quantized, while the perturbation is Schrödinger quantized. The form resembles that of isotropic matter models where for instance a kinetic scalar Hamiltonian can simply be quantized to a second-order derivative as used before. However, the structure of the equation for perturbative anisotropies turns out to be crucially different from what is obtained in isotropic matter models (4.15): Now even highest-order terms in the difference operator contain operator-valued coefficients, of the form

$$\hat{A}_{\bar{\nu}+4}\psi_{\bar{\nu}+4}(\varepsilon) + \hat{B}_{\bar{\nu}}\psi_{\bar{\nu}}(\varepsilon) + \hat{C}_{\bar{\nu}-4}\psi_{\bar{\nu}-4}(\varepsilon) = 0. \tag{8.61}$$

In order to derive the form of the operator coefficients and some of their crucial properties we now perform the anisotropic expansion as already defined in general terms.

Following the general construction of Hamiltonian constraint operators in loop quantum gravity—(4.5) in the full theory or in model systems as in (8.26)—we begin with

$$\begin{aligned}\hat{H} =&\, 8\delta^{-3} \sin\left(\tfrac{1}{2}\delta A\right) \cos\left(\tfrac{1}{2}\delta A\right) \sin\left(\tfrac{1}{2}\delta C\right) \cos\left(\tfrac{1}{2}\delta C\right) \hat{O}_A \\ &+ 4\delta^{-3} \sin^2\left(\tfrac{1}{2}\delta A\right) \cos^2\left(\tfrac{1}{2}\delta A\right) \hat{O}_C\end{aligned} \tag{8.62}$$

with

$$\hat{O}_A := \frac{\mathrm{i}}{2(\pi\gamma)^{3/2}\ell_{\mathrm{P}}^2} \left(\sin\left(\tfrac{1}{2}\delta A\right) \hat{V} \cos\left(\tfrac{1}{2}\delta A\right) - \cos\left(\tfrac{1}{2}\delta A\right) \hat{V} \sin\left(\tfrac{1}{2}\delta A\right) \right) \tag{8.63}$$

$$\hat{O}_C := \frac{\mathrm{i}}{2(\pi\gamma)^{3/2}\ell_{\mathrm{P}}^2} \left(\sin\left(\tfrac{1}{2}\delta C\right) \hat{V} \cos\left(\tfrac{1}{2}\delta C\right) - \cos\left(\tfrac{1}{2}\delta C\right) \hat{V} \sin\left(\tfrac{1}{2}\delta C\right) \right). \tag{8.64}$$

The numerical coefficient is chosen mainly for simplicity of the following equations, and including a factor $\ell_{\mathrm{P}}^{-2} = (G\hbar)^{-1}$ as it results from quantizations of the Poisson brackets $\{A, V\}$ and $\{C, V\}$, respectively, as used to represent the inverse determinant of the triad. In all these expressions, we use the same δ for both sets of independent variables since we are interested in a comparison with isotropy, and we ignore lattice refinement in order to highlight the structure of the resulting equations more clearly.

The operators $\hat{O}_A$ and $\hat{O}_C$ are diagonal in the triad eigenbasis, which allows us to perform their expansion via the eigenvalues

$$(\hat{O}_A)_{\mu,\nu} = 2\delta\ell_{\mathrm{P}}\sqrt{|\nu|}, \quad (\hat{O}_C)_{\mu,\nu} = \ell_{\mathrm{P}}\mu\left(\sqrt{|\nu+\delta|} - \sqrt{|\nu-\delta|}\right). \tag{8.65}$$

Inserting the perturbed $\mu = \tfrac{2}{3}\bar{\nu} + P$ and $\nu = \tfrac{1}{3}\bar{\nu} - P$, we obtain

$$O_A(\bar{v}, P) = \frac{2}{\sqrt{3}}\delta\ell_P\left(\sqrt{|\bar{v}|} - \frac{3}{2}\frac{1}{\sqrt{|\bar{v}|}}P - \frac{9}{8}\frac{1}{|\bar{v}|^{3/2}}P^2 + \cdots\right) \quad (8.66)$$

$$O_C(\bar{v}, P) = \frac{2}{\sqrt{3}}\delta\ell_P\,\mathrm{sgn}(\bar{v})\left(2|\bar{v}|\Delta_{3\delta}\sqrt{|\bar{v}|} + 3\left(\Delta_{3\delta}\sqrt{|\bar{v}|} - |\bar{v}|\Delta_{3\delta}\frac{1}{\sqrt{|\bar{v}|}}\right)P\right.$$
$$\left. - \frac{9}{4}\left(2|\bar{v}|\Delta_{3\delta}\frac{1}{\sqrt{|\bar{v}|}} + |\bar{v}|\Delta_{3\delta}\frac{1}{|\bar{v}|^{3/2}}\right)P^2 + \cdots\right) \quad (8.67)$$

with

$$\Delta_\delta f(\bar{v}) := \frac{1}{2\delta}(f(\bar{v} + \delta) - f(\bar{v} - \delta)). \quad (8.68)$$

The holonomy contributions can be expanded by ordinary Taylor expansions of sine and cosine with $A = \bar{c} + \varepsilon$ and $C = \bar{c} - 2\epsilon$:

$$4\sin\left(\tfrac{1}{2}\delta A\right)\cos\left(\tfrac{1}{2}\delta A\right)\sin\left(\tfrac{1}{2}\delta C\right)\cos\left(\tfrac{1}{2}\delta C\right) = \sin^2(\delta\bar{c}) - \delta\sin(\delta\bar{c})\cos(\delta\bar{c})\varepsilon$$
$$- \delta^2\left(2 + 3\sin^2(\delta\bar{c})\right)\varepsilon^2 + \cdots \quad (8.69)$$

$$4\sin^2\left(\tfrac{1}{2}\delta A\right)\cos^2\left(\tfrac{1}{2}\delta A\right) = \sin^2(\delta\bar{c}) + 2\delta\sin(\delta\bar{c})\cos(\delta\bar{c})\varepsilon$$
$$+ \delta^2\left(1 - 3\sin^2(\delta\bar{c})\right)\varepsilon^2 + \cdots. \quad (8.70)$$

Combining all terms present in (8.62) and expanding in P as well as ε, we finally obtain

$$\frac{4\ell_P}{\sqrt{3}}\frac{\sin^2(\delta\bar{c})}{\delta^2}\left(\sqrt{|\bar{v}|} + \bar{v}\Delta_{3\delta}\sqrt{|\bar{v}|}\right)$$
$$- 2\sqrt{3}\ell_P\frac{\sin^2(\delta\bar{c})}{\delta^2}\left(\frac{1}{\sqrt{|\bar{v}|}} - \mathrm{sgn}(\bar{v})\left(\Delta_{3\delta}\sqrt{|\bar{v}|} - \frac{1}{3}|\bar{v}|\Delta_{3\delta}\frac{1}{\sqrt{|\bar{v}|}}\right)\right)P$$
$$- \frac{4\ell_P}{\sqrt{3}}\frac{\sin(\delta\bar{c})\cos(\delta\bar{c})}{\delta}\left(\sqrt{|\bar{v}|} - 2\bar{v}\Delta_{3\delta}\sqrt{|\bar{v}|}\right)\varepsilon$$
$$- \frac{3\sqrt{3}\ell_P}{2}\frac{\sin^2(\delta\bar{c})}{\delta^2}\left(\frac{1}{|\bar{v}|^{3/2}} + \mathrm{sgn}(\bar{v})\left(2\Delta_{3\delta}\frac{1}{\sqrt{|\bar{v}|}} + |\bar{v}|\Delta_{3\delta}\frac{1}{|\bar{v}|^{3/2}}\right)\right)P^2$$
$$- 2\sqrt{3}\ell_P\frac{\sin(\delta\bar{c})\cos(\delta\bar{c})}{\delta}\left(\frac{1}{\sqrt{|\bar{v}|}} - 2\mathrm{sgn}(\bar{v})\left(\Delta_{3\delta}\sqrt{|\bar{v}|} - |\bar{v}|\Delta_{3\delta}\frac{1}{\sqrt{|\bar{v}|}}\right)\right)\varepsilon P$$
$$- \frac{4\ell_P}{\sqrt{3}}\left((2 + 3\sin^2(\delta\bar{c}))\sqrt{|\bar{v}|} - (1 - 3\sin^2(\delta\bar{c}))\bar{v}\Delta_{3\delta}\sqrt{|\bar{v}|}\right)\varepsilon^2 + \cdots$$
$$(8.71)$$

By a tedious but straightforward calculation one turns this expansion into a difference equation by representing the sines and cosines on isotropic triad eigenstates, while ε and P become operators in the Schrödinger Hilbert space for the perturbative anisotropy; the operator form of $\hat{A}_{\bar{\nu}}$, $\hat{B}_{\bar{\nu}}$ and $\hat{C}_{\bar{\nu}}$ in (8.61) follows. To leading order for large $\bar{\nu} \gg \delta$ (for which the action of $\Delta_{3\delta}$ reduces to a derivative by $\bar{\nu}$) the linear perturbation terms drop out, as in the classical expansion. For the recurrence one must invert the operator coefficients, which is not guaranteed to be possible for all states. A breakdown may signal singularities in the perturbative formulation.

8.2.4.3 Singularities

As usual, the perturbative expansion has a limited range of validity. Generically, one should expect it to break down in particular close to a classical singularity where anisotropies usually grow very large. Nevertheless, there are special initial conditions for which anisotropies would remain small rather close to a classical singularity, and so it is of interest to see if the singularity removal based on quantum hyperbolicity of the loop quantized anisotropic model can also be seen perturbatively. As already noted, the recurrence is crucially different from that of non-perturbative models in that operators, not just complex coefficients, must be inverted.

It turns out that large values of $\bar{\nu}$ ensure the invertibility of differential operators in the dynamical equation, but this does not happen at small $\bar{\nu}$ where the $\hat{P}$-terms in (8.71) are more relevant. Some of the coefficients even diverge, for instance $\Delta_{3\delta}|\bar{\nu}|^{-3/2}$ for $\bar{\nu} = 3\delta$, and the perturbed equation itself becomes ill-defined. In fact, such a breakdown where coefficients start to diverge happens even if $\bar{\nu}$ is not zero. Although such values of $\bar{\nu}$ are small and of the order one, one can still arrange states for which $P \ll \bar{\nu}$ remains satisfied. There are thus initial conditions for which the perturbative quantum dynamics breaks down before the perturbation assumption does: the perturbed model does not remove the singularity, even though it is a perturbation of the non-singular isotropic model within the non-singular anisotropic one. While those states may be very special, their existence shows that the generic singularity-removal mechanism of quantum hyperbolicity cannot be realized perturbatively.

The analysis provides a cautionary conclusion: perturbative treatments of quantum space–times near singularities do not always provide a reliable result. Perturbations may appear singular even though the underlying non-perturbative treatment makes singularities disappear, and also the opposite behavior is conceivable. Perturbation theory can, however, be used well to analyze the approach to a classical singularity and signal any deviation from classical behavior; or it can be used for evolution away from a singularity, as in observational cosmology.

8.3 Black Hole Models Inside the Horizon

Anisotropic homogeneous models can be used for non-rotating black holes. The Schwarzschild space–time

$$ds^2 = -\left(1 - \frac{2M}{r}\right)dt^2 + \frac{1}{(1 - 2M/r)}dr^2 + r^2 d\Omega^2 \qquad (8.72)$$

is spherically symmetric, static outside the horizon at $r = 2M$, and homogeneous inside the horizon. Inside the horizon, it is not of any of the Bianchi types but rather of Kantowski–Sachs form: a geometry that is homogeneous with one rotational axis, not implementable as an LRS Bianchi model. It presents an interesting system for loop quantization because it can be used to shed light on black-hole singularities. And the presence of a horizon poses additional questions for consistency, not just regarding the usual causality aspects but also for lattice refinement: near the horizon, the size of any homogeneous region becomes very small such that a refinement model, as used earlier, for which $\mathcal{N} \propto V$, provides only a few patches even though the near-horizon region for large black holes is supposed to be macroscopic. We will discuss this issue after presenting the technical formulation of the model.

8.3.1 Canonical Formulation

We now have invariant variables of the form [26, 27]

$$A_a^i \tau_i dx^a = \tilde{c}\tau_3 dx + (\tilde{a}\tau_1 + \tilde{b}\tau_2)d\vartheta + (-\tilde{b}\tau_1 + \tilde{a}\tau_2)\sin\vartheta d\varphi + \tau_3 \cos\vartheta d\varphi \qquad (8.73)$$

$$E_i^a \tau^i \frac{\partial}{\partial x^a} = \tilde{p}_c \tau_3 \sin\vartheta \frac{\partial}{\partial x} + (\tilde{p}_a \tau_1 + \tilde{p}_b \tau_2)\sin\vartheta \frac{\partial}{\partial \vartheta} + (-\tilde{p}_b \tau_1 + \tilde{p}_a \tau_2)\frac{\partial}{\partial \varphi}. \qquad (8.74)$$

Our coordinates (x, ϑ, φ) are adapted to the symmetry, where we denote the radial variable as x to indicate that it need not be the area radius r (for which the angular part of the metric would have a coefficient r^2). In fact, in the interior of (8.72) with $r < 2M$ the non-angular spatial coordinate is $x = t$: the angular metric component is not x^2. For a general densitized triad of the symmetric form, we have the spatial line element

$$ds^2 = \frac{\tilde{p}_a^2 + \tilde{p}_b^2}{|\tilde{p}_c|}dx^2 + |\tilde{p}_c|d\Omega^2 \qquad (8.75)$$

obtained from $q^{ab} = E_i^a E_i^b / |\det(E_j^c)|$. For comparisons of quantum and classical behaviors, we will also use the co-triad

$$e_a^i \tau_i dx^a = e_c \tau_3 dx + (e_a \tau_1 + e_b \tau_2)d\vartheta + (-e_b \tau_1 + e_a \tau_2)\sin\vartheta d\varphi \qquad (8.76)$$

with components

$$e_a = \frac{\sqrt{|\tilde{p}_c|}\tilde{p}_a}{\sqrt{\tilde{p}_a^2 + \tilde{p}_b^2}}, \quad e_b = \frac{\sqrt{|\tilde{p}_c|}\tilde{p}_b}{\sqrt{\tilde{p}_a^2 + \tilde{p}_b^2}}, \quad e_c = \frac{\mathrm{sgn}\,\tilde{p}_c \sqrt{\tilde{p}_a^2 + \tilde{p}_b^2}}{\sqrt{|\tilde{p}_c|}}. \qquad (8.77)$$

The phase space is spanned by the spatial constants $(\tilde{a}, \tilde{b}, \tilde{c}, \tilde{p}_a, \tilde{p}_b, \tilde{p}_c) \in \mathbb{R}^6$, which have non-vanishing Poisson brackets

$$\{\tilde{a}, \tilde{p}_a\} = \gamma G/\mathcal{L}, \quad \{\tilde{b}, \tilde{p}_b\} = \gamma G/\mathcal{L}, \quad \{\tilde{c}, \tilde{p}_c\} = 2\gamma G/\mathcal{L}$$

Here, we have fixed the orbital coordinate area of size 4π, and $\mathcal{L}$ is the size of a coordinate interval along x used in integrating out the fields in

$$\frac{1}{8\pi\gamma G} \int \mathrm{d}^3 x \, \dot{A}_a^i E_i^a = \frac{\mathcal{L}}{2\gamma G}\dot{\tilde{c}}\tilde{p}_c + \frac{\mathcal{L}}{\gamma G}\dot{\tilde{b}}\tilde{p}_b + \frac{\mathcal{L}}{\gamma G}\dot{\tilde{a}}\tilde{p}_a$$

to derive the reduced symplectic structure. The SU(2)-gauge transformations rotating a general triad are partially fixed to U(1) by demanding the x-component of E_i^a to point in the internal τ_3-direction in (8.74). The U(1)-gauge freedom allows one to set $\tilde{a} = 0 = \tilde{p}_a$, still leaving, as in the Bianchi I LRS model of the preceding section, a discrete residual gauge freedom $(\tilde{b}, \tilde{p}_b) \mapsto (-\tilde{b}, -\tilde{p}_b)$. The remaining variables can be rescaled as

$$(b, c) := (\tilde{b}, \mathcal{L}\tilde{c}), \quad (p_b, p_c) := (\mathcal{L}\tilde{p}_b, \tilde{p}_c). \tag{8.78}$$

to make the canonical structure $\mathcal{L}$-independent:

$$\{b, p_b\} = \gamma G, \quad \{c, p_c\} = 2\gamma G. \tag{8.79}$$

All this mimics what we had done for isotropic cosmology, just replacing the coordinate volume $\mathcal{V}$ used earlier by $\mathcal{L}$. Also as before, this parameter will play a role in lattice refinement. In particular, there are only three $\mathcal{L}$-independent quantities: b, p_c and c/p_b. A proper discussion of lattice refinement is necessary to allow for a dependence of observable expressions on all four phase-space variables, in particular on p_b by itself as it is suggested for instance by inverse-triad corrections. Without lattice refinement, one would be misled into believing that inverse-triad corrections, or any quantum correction depending on p_b but not on c, could not exist.

8.3.2 Loop Quantization

To express the elementary variables through holonomies it suffices to choose curves along the x-direction of coordinate length $\ell_0^x = \tau\mathcal{L}$ and along ϑ of coordinate length $\ell_0^\vartheta = \mu$ since this captures all information in the two connection components,

$$h_x^{(\tau)}(A) = \exp \int_0^{\tau\mathcal{L}} \mathrm{d}x \tilde{c}\tau_3 = \cos\frac{\tau c}{2} + 2\tau_3 \sin\frac{\tau c}{2} \tag{8.80}$$

$$h_\vartheta^{(\mu)}(A) = \exp \int_0^\mu \mathrm{d}\vartheta \tilde{b}\tau_2 = \cos\frac{\mu b}{2} + 2\tau_2 \sin\frac{\mu b}{2}. \tag{8.81}$$

As in general anisotropic models, the quantum Hilbert space is based on cylindrical states depending on the connection through countably many holonomies, which can always be written as almost-periodic functions $f(b, c) = \sum_{\mu, \tau} f_{\mu, \tau} \exp(i(\mu b + \tau c)/2)$ of two variables. These form the set of functions on the double product of the Bohr compactification of the real line, a compact Abelian group. Its Haar measure defines the inner product of the (non-separable) Hilbert space, where states

$$\langle b, c | \mu, \tau \rangle = e^{i(\mu b + \tau c)/2} \qquad \mu, \tau \in \mathbb{R} \tag{8.82}$$

form an orthonormal basis. Holonomies simply act by multiplication on these states, while densitized-triad components become derivative operators

$$\hat{p}_b = -i\gamma \ell_P^2 \frac{\partial}{\partial b}, \qquad \hat{p}_c = -2i\gamma \ell_P^2 \frac{\partial}{\partial c}. \tag{8.83}$$

They act as

$$\hat{p}_b |\mu, \tau\rangle = \tfrac{1}{2} \gamma \ell_P^2 \mu |\mu, \tau\rangle, \qquad \hat{p}_c |\mu, \tau\rangle = \gamma \ell_P^2 \tau |\mu, \tau\rangle, \tag{8.84}$$

immediately showing their eigenvalues.

The Hamiltonian constraint, now to be quantized, requires care due to the presence of intrinsic curvature (or a non-vanishing spin connection in the homogeneous slicing). It can be written as

$$C_{\text{grav}} = \frac{1}{\gamma^2} \int d^3 x \, \varepsilon_{ijk} (-{}^0F_{ab}^k + \gamma^2 \Omega_{ab}^k) \frac{E^{ai} E^{bj}}{\sqrt{|\det E|}} \tag{8.85}$$

where $\Omega_{ab}^k \tau_k dx^a \wedge dx^b = -\sin\vartheta \, \tau_3 d\vartheta \wedge d\varphi$ is the intrinsic curvature of two-spheres, while ${}^0F_{ab}^k$ is the curvature computed from A_a^i ignoring the spin-connection term $\sin\vartheta \, \tau_3 d\varphi$ in (8.73). (The factors of γ ensure that (8.85) is the Lorentzian constraint, which in homogeneous models can be formulated without explicit reference to extrinsic curvature.) We replace the inverse determinant of E_i^a by a Poisson bracket, following [28],

$$\varepsilon_{ijk} \tau^i \frac{E^{aj} E^{bk}}{\sqrt{|\det E|}} = \frac{-1}{4\pi\gamma G} \sum_{K \in \{x, \vartheta, \varphi\}} \frac{1}{\mathscr{L}^K} \varepsilon^{abc} \omega_c^K h_K^{(\delta_K)} \{h_K^{(\delta_K)-1}, V\} \tag{8.86}$$

with edge lengths $\ell_0^x = \delta_x \mathscr{L}$ and $\ell_0^{\vartheta/\varphi} = \delta_\vartheta$, and left-invariant one-forms ω_c^K on the symmetry group manifold. We use independent δ-parameters for the different directions. So far, they are treated as constants and could take the same value, but they are mainly place-holders for lattice refinement, for which in general different behaviors are realized for the independent directions; see below.

For curvature components ${}^0F_{ab}^k$ we use a holonomy around a closed loop

$${}^0F_{ab}^i(x) \tau_i = \frac{\omega_a^I \omega_b^J}{\mathscr{A}_{(IJ)}} (h_{IJ}^{(\delta)} - 1) + O((b^2 + c^2)^{3/2} \sqrt{\mathscr{A}}) \tag{8.87}$$

with

$$h^{(\delta)}_{IJ} = h^{(\delta_I)}_I h^{(\delta_J)}_J (h^{(\delta_I)}_I)^{-1} (h^{(\delta_J)}_J)^{-1} \tag{8.88}$$

and $\mathscr{A}_{IJ}$ being the coordinate area of the loop, using the corresponding combinations of $\mathscr{L}^I$. Putting all factors together and replacing Poisson brackets by commutators, we have the Hamiltonian constraint operator

$$\hat{C}^{(\delta)} = 2i(\gamma^3 \delta_x \delta_\vartheta^2 \ell_{\mathrm{P}}^2)^{-1} \mathrm{tr}\Bigg(\sum_{IJK} \epsilon^{IJK} \hat{h}^{(\delta_I)}_I \hat{h}^{(\delta_J)}_J \hat{h}^{(\delta_I)-1}_I \hat{h}^{(\delta_J)-1}_J \hat{h}^{(\delta_K)}_K [\hat{h}^{(\delta_K)-1}_K, \hat{V}]$$

$$+ 2\gamma^2 \delta_\vartheta^2 \tau_\vartheta \hat{h}^{(\delta_x)}_x [\hat{h}^{(\delta_x)-1}_x, \hat{V}] \Bigg)$$

$$= 4i(\gamma^3 \delta_x \delta_\vartheta^2 \ell_{\mathrm{P}}^2)^{-1} \Bigg(8 \sin\frac{\delta_\vartheta b}{2} \cos\frac{\delta_\vartheta b}{2} \sin\frac{\delta_x c}{2} \cos\frac{\delta_x c}{2}$$

$$\times \left(\sin\frac{\delta_\vartheta b}{2} \hat{V} \cos\frac{\delta_\vartheta b}{2} - \cos\frac{\delta_\vartheta b}{2} \hat{V} \sin\frac{\delta_\vartheta b}{2} \right)$$

$$+ \left(4\sin^2\frac{\delta_\vartheta b}{2} \cos^2\frac{\delta_\vartheta b}{2} + \gamma^2 \delta_\vartheta^2 \right) \left(\sin\frac{\delta_x c}{2} \hat{V} \cos\frac{\delta_x c}{2} - \cos\frac{\delta_x c}{2} \hat{V} \sin\frac{\delta_x c}{2} \right) \Bigg) \tag{8.89}$$

which acts as

$$\hat{C}^{(\delta)}|\mu, \tau\rangle = (2\gamma^3 \delta_x \delta_\vartheta^2 \ell_{\mathrm{P}}^2)^{-1} \Big(2(V_{\mu+\delta_\vartheta,\tau} - V_{\mu-\delta_\vartheta,\tau})(|\mu + 2\delta_\vartheta, \tau + 2\delta_x\rangle$$

$$- |\mu + 2\delta_\vartheta, \tau - 2\delta_x\rangle - |\mu - 2\delta_\vartheta, \tau + 2\delta_x\rangle + |\mu - 2\delta_\vartheta, \tau - 2\delta_x\rangle)$$

$$+ (V_{\mu,\tau+\delta_x} - V_{\mu,\tau-\delta_x})(|\mu + 4\delta_\vartheta, \tau\rangle - 2(1 + 2\gamma^2 \delta_\vartheta^2)|\mu, \tau\rangle + |\mu - 4\delta_\vartheta, \tau\rangle)) \Big)$$

on basis states. This operator may be ordered symmetrically, defining $\hat{C}^{(\delta)}_{\mathrm{symm}} := \frac{1}{2}(\hat{C}^{(\delta)} + \hat{C}^{(\delta)\dagger})$, whose action is

$$\hat{C}^{(\delta)}_{\mathrm{symm}}|\mu, \tau\rangle = (2\gamma^3 \delta_x \delta_\vartheta^2 \ell_{\mathrm{P}}^2)^{-1} \big((V_{\mu+\delta_\vartheta,\tau} - V_{\mu-\delta_\vartheta,\tau} + V_{\mu+3\delta_\vartheta,\tau+2\delta_x} - V_{\mu+\delta_\vartheta,\tau+2\delta_x})$$

$$\times |\mu + 2\delta_\vartheta, \tau + 2\delta_x\rangle$$

$$- (V_{\mu+\delta_\vartheta,\tau} - V_{\mu-\delta_\vartheta,\tau} + V_{\mu+3\delta_\vartheta,\tau-2\delta_x} - V_{\mu+\delta_\vartheta,\tau-2\delta_x})|\mu + 2\delta_\vartheta, \tau - 2\delta_x\rangle$$

$$- (V_{\mu+\delta_\vartheta,\tau} - V_{\mu-\delta_\vartheta,\tau} + V_{\mu-\delta_\vartheta,\tau+2\delta_x} - V_{\mu-3\delta_\vartheta,\tau+2\delta_x})|\mu - 2\delta_\vartheta, \tau + 2\delta_x\rangle$$

$$+ (V_{\mu+\delta_\vartheta,\tau} - V_{\mu-\delta_\vartheta,\tau} + V_{\mu-\delta_\vartheta,\tau-2\delta_x} - V_{\mu-3\delta_\vartheta,\tau-2\delta_x})|\mu - 2\delta_\vartheta, \tau - 2\delta_x\rangle$$

$$+ \frac{1}{2}(V_{\mu,\tau+\delta_x} - V_{\mu,\tau-\delta_x} + V_{\mu+4\delta_\vartheta,\tau+\delta_x} - V_{\mu+4\delta_\vartheta,\tau-\delta_x})|\mu + 4\delta_\vartheta, \tau\rangle$$

$$- 2(1 + 2\gamma^2 \delta_\vartheta^2)(V_{\mu,\tau+\delta_x} - V_{\mu,\tau-\delta_x})|\mu, \tau\rangle$$

$$+ \frac{1}{2}(V_{\mu,\tau+\delta_x} - V_{\mu,\tau-\delta_x} + V_{\mu-4\delta_\vartheta,\tau+\delta_x} - V_{\mu-4\delta_\vartheta,\tau-\delta_x})|\mu - 4\delta_\vartheta, \tau\rangle \big). \tag{8.90}$$

Transforming this operator to the triad representation obtained as coefficients of a wave function $|\psi\rangle = \sum_{\mu,\tau} \psi_{\mu,\tau}|\mu, \tau\rangle$ in the triad eigenbasis and using the volume eigenvalues

$$V_{\mu,\tau} = 4\pi\sqrt{|(\hat{p}_c)_{\mu,\tau}|}(\hat{p}_b)_{\mu,\tau} = 2\pi(\gamma\ell_{\mathrm{P}}^2)^{3/2}\sqrt{|\tau|}\mu,$$

a difference equation

$$\frac{\gamma^{3/2}\delta_x\delta_\vartheta^2}{\pi\ell_P}(\hat{C}_{\text{symm}}^{(\delta)}|\psi\rangle)_{\mu,\tau} = 2\delta_\vartheta\left(\sqrt{|\tau+2\delta_x|}+\sqrt{|\tau|}\right)\left(\psi_{\mu+2\delta_\vartheta,\tau+2\delta_x}-\psi_{\mu-2\delta_\vartheta,\tau+2\delta_x}\right)$$
$$+\left(\sqrt{|\tau+\delta_x|}-\sqrt{|\tau-\delta_x|}\right)((\mu+2\delta_\vartheta)\psi_{\mu+4\delta_\vartheta,\tau}$$
$$-2(1+2\gamma^2\delta_\vartheta^2)\mu\psi_{\mu,\tau}+(\mu-2\delta_\vartheta)\psi_{\mu-4\delta_\vartheta,\tau})$$
$$+2\delta(\sqrt{|\tau-2\delta_x|}+\sqrt{|\tau|})\left(\psi_{\mu-2\delta_\vartheta,\tau-2\delta_x}-\psi_{\mu+2\delta_\vartheta,\tau-2\delta_x}\right)=0$$
$$(8.91)$$

results for physical states. (For small μ the equation has to be specialized further due to the remaining gauge freedom; see [26].)

8.3.3 Tree-Level Equations

For a first analysis of the consequences of quantum-geometry corrections one may modify the classical equations by holonomy (or inverse-triad) terms suggested by the loop-quantized Hamiltonian constraint. In this way, ignoring quantum back-reaction in effective equations, one derives the tree-level approximation to loop quantum cosmology. Even with this simplification, a complete analysis of global space–times is complicated, not the least because of the considerable ambiguities in formulating detailed quantum-geometry corrections for homogeneous models, and then extending them to the spherically symmetric exterior outside the horizon. Investigations are still ongoing, even while some results have already emerged. For instance, possible non-singular space–times and extended horizons, based on holonomy or discretization corrections, have been constructed [29–35], deriving consequences for instance for Hawking evaporation in [36]. Black-hole collapse models with corrections motivated by the behavior of inverse triads are studied in [37–43].

These considerations must await a fully consistent, anomaly-free formulation of inhomogeneities before reliable constructions of global space–times can be achieved. So far, consistent deformations of spherically symmetric models have been found only in models with inverse-triad corrections [44–46] or a subclass of holonomy corrections [47]; see also the next chapter. The equations of [32] are based on a consistent set of discretizations, but ones obtained after a partial gauge-fixing.

8.3.4 Lattice Refinement

For lattice refinement, let us now assume that we have a lattice with $\mathcal{N}$ vertices in a form adapted to the symmetry: there are $\mathcal{N}_x$ vertices along the x-direction (whose triad component p_c gives rise to the label τ) and $\mathcal{N}_\vartheta^2$ vertices in spherical orbits of the symmetry group (whose triad component p_b gives rise to the label μ). Thus, $\mathcal{N} = \mathcal{N}_x\mathcal{N}_\vartheta^2$. Since holonomies in such a lattice setting are computed along single links, rather than through all of space (or the whole cell of size $\mathcal{L}$), basic ones are

$h_x = \exp(\ell_0^x \tilde{c}\tau_3)$ and $h_\vartheta = \exp(\ell_0^\vartheta \tilde{b}\tau_2)$. Edge lengths are related to the number of vertices in each direction by $\ell_0^x = \mathscr{L}/\mathscr{N}_x$ and $\ell_0^\vartheta = 1/\mathscr{N}_\vartheta$, or $\delta_x = 1/\mathscr{N}_x$, $\delta_\vartheta = 1/\mathscr{N}_\vartheta$. With the rescaled connection components $c = \mathscr{L}\tilde{c}$ and $b = \tilde{b}$ we have basic holonomies

$$h_x = \exp(\ell_0^x \mathscr{L}^{-1} c\tau_3) = \exp(c\tau_3/\mathscr{N}_x),$$
$$h_\vartheta = \exp(\ell_0^\vartheta b\tau_2) = \exp(b\tau_2/\mathscr{N}_\vartheta). \qquad (8.92)$$

Using this in the Hamiltonian constraint operator then gives a difference equation whose step-sizes are $1/\mathscr{N}_I$.

Now allowing a phase-space dependent $\mathscr{N}$ to implement refinement, we obtain an operator containing flux-dependent holonomies instead of basic ones, for instance $\mathscr{N}_x(\mu,\tau)h_x = \mathscr{N}_x(\mu,\tau)\exp(c\tau_3/\mathscr{N}_x(\mu,\tau))$ which reduces to an $\mathscr{N}_x$-independent connection component c in regimes where curvature is small. Keeping track of all prefactors and holonomies in the commutator as well as the closed loop, one obtains the difference equation [15]

$$C_+(\mu,\tau)\left(\psi_{\mu+2\mathscr{N}_\vartheta(\mu,\tau)^{-1},\tau+2\mathscr{N}_x(\mu,\tau)^{-1}} - \psi_{\mu-2\mathscr{N}_\vartheta(\mu,\tau)^{-1},\tau+2\mathscr{N}_x(\mu,\tau)^{-1}}\right)$$
$$+ C_0(\mu,\tau)\left((\mu+2\mathscr{N}_\vartheta(\mu,\tau)^{-1})\psi_{\mu+4\mathscr{N}_\vartheta(\mu,\tau)^{-1},\tau}\right.$$
$$-2(1+2\gamma^2\mathscr{N}_\vartheta(\mu,\tau)^{-2})\mu\psi_{\mu,\tau} + (\mu-2\mathscr{N}_\vartheta(\mu,\tau)^{-1})\psi_{\mu-4\mathscr{N}_\vartheta(\mu,\tau)^{-1},\tau}\bigg)$$
$$+ C_-(\mu,\tau)\left(\psi_{\mu-2\mathscr{N}_\vartheta(\mu,\tau)^{-1},\tau-2\mathscr{N}_x(\mu,\tau)^{-1}} - \psi_{\mu+2\mathscr{N}_\vartheta(\mu,\tau)^{-1},\tau-2\mathscr{N}_x(\mu,\tau)^{-1}}\right) = 0. \qquad (8.94)$$

with

$$C_\pm(\mu,\tau) = 2\mathscr{N}_\vartheta(\mu,\tau)^{-1}(\sqrt{|\tau \pm 2\mathscr{N}_x(\mu,\tau)^{-1}|} + \sqrt{|\tau|}) \qquad (8.95)$$

$$C_0(\mu,\tau) = \sqrt{|\tau + \mathscr{N}_x(\mu,\tau)^{-1}|} - \sqrt{|\tau - \mathscr{N}_x(\mu,\tau)^{-1}|}. \qquad (8.96)$$

(A total factor $\mathscr{N}_x\mathscr{N}_\vartheta^2$ for the number of vertices drops out because the right-hand side is zero in vacuum, but would multiply the left-hand side in the presence of a matter term.)

As in isotropic models, different refinement schemes can be analyzed. First, the volume dependence of the total number of vertices allows different choices, such as $\mathscr{N}_x\mathscr{N}_\vartheta^2 \propto V^{-2x}$ in power-law form. Even if this power is fixed, for instance by using a number of vertices proportional to volume, different options remain. If the number of vertices is proportional to transversal areas, we are led to $\mathscr{N}_x \propto \sqrt{|\tau|}$ and $\mathscr{N}_\vartheta \propto \sqrt{\mu}$ as introduced in [48]. Here, $\mathscr{N}_x$ and $\mathscr{N}_\vartheta$ are determined by eigenstates of $\hat{p}_c$ and $\hat{p}_b$, respectively. If the number of vertices in a given direction is proportional to its linear extension, we have $\mathscr{N}_\vartheta \propto \sqrt{|\tau|}$ and $\mathscr{N}_x \propto \mu/\sqrt{|\tau|}$ as determined by eigenstates of quantized co-triad components; see (8.77). Interestingly, these two cases can clearly be distinguished, and the first one be ruled out [15]: it leads to unstable evolution where wave functions would behave exponentially even in large parts of the phase space in which they should be semiclassical. (See Sect. 11.2.1.3).Here we have an example for restrictions of different refinement models. In particular, it shows that one cannot always confine attention to difference equations which are of constant step-size, or can be transformed to be of this form; for this would be the case only for the refinement model of the first type (with $\mathcal{N}_x$ a function of τ and $\mathcal{N}_\vartheta$ a function of μ only) which is ruled out.

Actually, no refinement model whose number of vertices follows a power law in its volume dependence can be consistent in the black-hole interior case. Here, the presence of a horizon becomes important, which is supposed to be a semiclassical regime of low curvature (for large black holes). However, in the phase-space coordinates corresponding to a homogeneous slicing the horizon is encountered at $\mu = 0$, where the volume collapses. Thus, the number of vertices with a power-law dependence of the volume with $x < 0$ is very small in some neighborhood of the horizon, where the homogeneous model should still be valid, and strong discreteness corrections ensue. For a consistent refinement scheme one has to ensure that the number of vertices remains large even when the volume vanishes at the horizon (but not necessarily at the singularity where the volume vanishes as well due to $\tau = 0$). Power-law forms do not result in semiclassical behavior in sufficiently large regions of the phase space. Strict prescriptions of the volume dependence of refinement, such as the case $x = -1/2$ going back to [49], or $\mathcal{N} \propto V$, are not viable.

8.3.5 Singularity

The singularity must be analyzed directly at the level of the difference equation for a few recurrence steps around $\tau = 0$, where the refinement does not matter. As in general anisotropic models, the recurrence continues through $\tau = 0$, removing the classical singularity. Some coefficients do vanish in the recurrence, implying decouplings of some values. Detailed considerations show that this requires solutions to be symmetric under $\tau \to -\tau$ [50], which is a consequence of the dynamics. (If one would add a matter term to the difference equation, this behavior would no longer be realized. But then the model would lose its interpretation as black-hole interior, although it could still be regarded as a cosmological model.) The reflection symmetry of solutions is an interesting consistency test: it implies that the behavior after the classical singularity is the same as that before. Since the exterior is static in the classical solution, the reflection symmetry is an important condition for the non-singular interior to be matchable to a static exterior. For such a matching to be done explicitly, we must consider the quantization of spherically symmetric models.

References

1. Ashtekar, A., Corichi, A., Zapata, J.: Class. Quantum Grav. **15**, 2955 (1998). grqc9806041
2. Ashtekar, A., Bojowald, M., Lewandowski, J.: Adv. Theor. Math. Phys. **7**, 233 (2003). gr-qc0304074
3. Bojowald, M.: Class. Quantum Grav. **19**, 2717 (2002). gr-qc0202077

4. Brunnemann, J., Fleischhack, C: (2007). arXiv:0709.1621
5. Brunnemann J., Koslowski T.A.: arXiv:1012.0053
6. Bojowald, M.: Canonical Gravity and Applications: Cosmology, Black Holes, and Quantum Gravity. Cambridge University Press, Cambridge (2010)
7. Misner, C.W.: Phys. Rev. Lett. **22**, 1071 (1969)
8. Bojowald, M.: Class. Quantum Grav. **20**, 2595 (2003). gr-qc0303073
9. Mercuri, S.: Phys. Rev. D **73**, 084016 (2006). gr-qc0601013
10. Bojowald, M., Das, R.: Phys. Rev. D **78**, 064009 (2008). arXiv:0710.5722
11. Thiemann, T.: Class. Quantum Grav. **15**, 1487 (1998). gr-qc9705021
12. Bojowald, M., Das, R.: Class. Quantum Grav. **25**, 195006 (2008). arXiv:0806.2821
13. Lamon, R.: arXiv:0909.2578
14. Bojowald, M., Date, G., Vandersloot, K.: Class. Quantum Grav. **21**, 1253 (2004). gr-qc0311004
15. Bojowald, M., Cartin, D., Khanna, G.: Phys. Rev. D **76**, 064018 (2007). arXiv:0704.1137
16. Sabharwal, S., Khanna, G.: Class. Quantum Grav. **25**, 085009 (2008). arXiv:0711.2086
17. Nelson, W., Sakellariadou, M.: Phys. Rev. D **78**, 024030 (2008). arXiv:0803.4483
18. Bojowald, M., Hernández, H.H., Morales-Técotl, H.A.: Class. Quantum Grav. **23**, 3491 (2006). gr-qc0511058
19. Bojowald, M., Kastrup, H.A.: Class. Quantum Grav. **17**, 3009 (2000). hep-th9907042
20. Ellis G.F.R.: Relativistic Cosmology: Its Nature, Aims and Problems, pp. 215–288. Reidel (1984)
21. Ellis G.F.R., Buchert T., gr-qc/0506106
22. Wilson-Ewing, E.: Phys. Rev. D **82**, 043508 (2010). arXiv:1005.5565
23. Bojowald, M.: Class. Quantum Grav. **18**, L109 (2001). gr-qc0105113
24. Corichi, A., Vukasinac, T., Zapata, J.A.: Phys. Rev. D **76**, 044016 (2007). arXiv:0704.0007
25. Ashtekar, A., Fairhurst, S., Willis, J.: Class.Quant. Grav. **20**, 1031 (2003). gr-qc/0207106
26. Ashtekar, A., Bojowald, M.: Class. Quantum Grav. **23**, 391 (2006). gr-qc/0509075
27. Bojowald, M.: Quantum Riemannian Geometry and Black Holes, Trends in Quantum Gravity Research, Nova Science (2006). gr-qc/0602100
28. Thiemann, T.: Class. Quantum Grav. **15**, 839 (1998). gr-qc/9606089
29. Modesto, L.: gr-qc/0612084
30. Modesto, L.: arXiv:0811.2196
31. Böhmer, C.G., Vandersloot, K.: Phys. Rev. D **76**, 104030 (2007). arXiv:0709.2129
32. Campiglia, M., Gambini, R., Pullin, J.: AIP Conf. Proc. **977**, 52 (2008). arXiv:0712.0817
33. Gambini, R., Pullin, J.: Phys. Rev. Lett. **101**, 161301 (2008). arXiv:0805.1187
34. Brown, E., Mann, R., Modesto, L.: Phys. Lett. B **695**, 376 (2011). arXiv:1006.4164
35. Peltola, A., Kunstatter, G.: Phys. Rev. D **79**, 061501 (2009). arXiv:0811.3240
36. Alesci E., Modesto L., arXiv:1101.5792
37. Husain, V., Winkler, O.: Class. Quantum Grav. **22**, L127 (2005). gr-qc/0410125
38. Husain, V., Winkler, O.: Class. Quantum Grav. **22**, L135 (2005). gr-qc/0412039
39. Husain, V., Winkler, O.: Int. J. Mod. Phys. D **14**, 2233 (2005). gr-qc/0505153
40. Husain, V., Winkler, O.: Phys. Rev. D **73**, 124007 (2006). gr-qc/0601082
41. Husain, V.: Adv. Sci. Lett. **2**, 214 (2009). arXiv:0808.0949
42. Ziprick, J., Kunstatter, G.: Phys. Rev. D **80**, 024032 (2009). arXiv:0902.3224
43. Ziprick J., Kunstatter G., arXiv:1004.0525
44. Bojowald, M., Harada, T., Tibrewala, R.: Phys. Rev. D **78**, 064057 (2008). arXiv:0806.2593
45. Bojowald, M., Reyes, R., Tibrewala, J.D.: Phys. Rev. D **80**, 084002 (2009). arXiv:0906.4767
46. Bojowald, M., Reyes, J.D.: Class.Quantum Grav. **26**, 035018 (2009). arXiv:0810.5119
47. Reyes, J.D.: Spherically symmetric loop quantum gravity: Connections to 2-dimensional models and applications to gravitational collapse. Ph.D. thesis, The Pennsylvania State University (2009)
48. Chiou, D.W., Vandersloot, K.: Phys. Rev. D **76**, 084015 (2007). arXiv:0707.2548
49. Ashtekar, A., Pawlowski, T., Singh, P.: Phys. Rev. D **74**, 084003 (2006). gr-qc/0607039
50. Cartin, D., Khanna, G.: Phys. Rev. D **73**, 104009 (2006). gr-qc/0602025

Chapter 9
Midisuperspace Models: Black Hole Collapse

Inhomogeneous models of various kinds provide important steps toward the full theory, adding the issue of infinitely many degrees of freedom. The simplest inhomogenous model is that of spherical symmetry, which provides interesting ways to test the formalism as well as applications in the context of black holes. In the vacuum case the model has a finite number of dynamical degrees of freedom, but the infinitely many kinematical ones already allow one to test field theoretic aspects.

9.1 Spherical Symmetry

Spherically symmetric models are constructed along the lines seen already for homogeneous ones. We start by taking the classical form of spherically symmetric connections and densitized triads [1]:

$$A = A_x(x)\tau_3 dx + (A_1(x)\tau_1 + A_2(x)\tau_2)d\vartheta + (A_1(x)\tau_2 - A_2(x)\tau_1)\sin\vartheta\, d\varphi + \tau_3\cos\vartheta\, d\varphi \tag{9.1}$$

and

$$E = E^x(x)\tau_3\sin\vartheta\frac{\partial}{\partial x} + (E^1(x)\tau_1 + E^2(x)\tau_2)\sin\vartheta\frac{\partial}{\partial\vartheta} + (E^1(x)\tau_2 - E^2(x)\tau_1)\frac{\partial}{\partial\varphi} \tag{9.2}$$

with real functions A_x, A_1, A_2, E^x, E^1 and E^2 on the radial manifold B coordinatized by x. (Again, we denote the radial coordinate by x to indicate that it is in general not identical to the area radius.) The functions E^x, E^1 and E^2 on B are canonically conjugate to A_x, A_1 and A_2:

$$\Omega_B = \frac{1}{2\gamma G}\int_B dx(dA_x \wedge dE^x + 2dA_1 \wedge dE^1 + 2dA_2 \wedge dE^2). \tag{9.3}$$

These expressions are spherically symmetric in the following sense: infinitesimal rotations act by Lie derivatives with respect to superpositions of vector fields

M. Bojowald, *Quantum Cosmology*, Lecture Notes in Physics 835,
DOI: 10.1007/978-1-4419-8276-6_9, © Springer Science+Business Media, LLC 2011

$X = \sin\varphi\partial_\vartheta + \cot\vartheta\cos\varphi\partial_\varphi$, $Y = -\cos\varphi\partial_\vartheta + \cot\vartheta\sin\varphi\partial_\varphi$, and $Z = \partial_\varphi$. The first term of A is clearly rotationally invariant, while for the others we have

$$\mathscr{L}_X A = (A_1\tau_1 + A_2\tau_2)\cos\varphi d\varphi - (-A_2\tau_1 + A_1\tau_2)\frac{\cos\varphi}{\sin\vartheta}d\vartheta$$

$$- \tau_3\left(\frac{\sin\varphi}{\sin\vartheta}d\varphi + \frac{\cos\vartheta\cos\varphi}{\sin^2\vartheta}d\vartheta\right)$$

$$= \left[A, \frac{\cos\varphi}{\sin\vartheta}\tau_3\right] + d\left(\frac{\cos\varphi}{\sin\vartheta}\tau_3\right)$$

$$\mathscr{L}_Y A = (A_1\tau_1 + A_2\tau_2)\sin\varphi d\varphi - (-A_2\tau_1 + A_1\tau_2)\frac{\sin\varphi}{\sin\vartheta}d\vartheta$$

$$- \tau_3\left(\frac{\cos\varphi}{\sin\vartheta}d\varphi - \frac{\cos\vartheta\sin\varphi}{\sin^2\vartheta}d\vartheta\right)$$

$$= \left[A, \frac{\sin\varphi}{\sin\vartheta}\tau_3\right] + d\left(\frac{\sin\varphi}{\sin\vartheta}\tau_3\right)$$

$$\mathscr{L}_Z A = 0.$$

Thus, any rotation of the connection amounts to just a gauge transformation, which is also true for the densitized triad (9.2). Expressions of this form were first used in the context of Yang–Mills theories [2, 3]. A general mathematical theory is based on the classification of invariant connections on symmetric principal fiber bundles [4, 5], which shows all possible classes of invariant connections. See also [6] for a summary.

These variables are subject to constraints, obtained by inserting the invariant forms into the full expressions. We have the Gauss constraint

$$G[\lambda] = \int_B dx\lambda(E^{x\prime} + 2A_1E^2 - 2A_2E^1) \approx 0 \tag{9.4}$$

generating U(1)-gauge transformations (the prime denoting a derivative by x), the diffeomorphism constraint

$$D_{grav}[N_x] = \int_B dx N_x(2A_1'E^1 + 2A_2'E^2 - A_xE^{x\prime}) \tag{9.5}$$

and the Hamiltonian constraint

$$C_{grav}[N] = (2G)^{-1}\int_B dx N\left(|E^x|((E^1)^2 + (E^2)^2)\right)^{-1/2} \tag{9.6}$$

$$\times\left(2E^x(E^1A_2' - E^2A_1') + 2A_xE^x(A_1E^1 + A_2E^2)\right.$$

$$+ (A_1^2 + A_2^2 - 1)\left((E^1)^2 + (E^2)^2\right) - (1+\gamma^2)$$

$$\left.\times\left(2K_xE^x(K_1E^1+K_2E^2)+(K_1^2+K_2^2)((E^1)^2+(E^2)^2)\right)\right) \tag{9.7}$$

$$=: -H^E[N] + P[N] \tag{9.8}$$

where H^E is the first (so-called Euclidean) part depending explicitly on connection components, and P the second part depending on extrinsic-curvature components (which are themselves functions of A_I and E^I). Spherically symmetric extrinsic curvature is written in a form analogous to (9.1), without the last term which is necessary only for connections but not for 1-forms, defining the components K_1 and K_2. The classical equations have been analyzed and solved in complex Ashtekar variables in [7–9], where Wheeler–DeWitt-type equations were provided as well.

A Wheeler–DeWitt quantization of midisuperspace models results in functional derivative equations where, for instance in the connection representation, the E^I become functional derivatives. A loop quantization, on the other hand, replaces all connection components by holonomies. To that end, it is useful to introduce variables

$$A_\varphi(x) := \sqrt{A_1(x)^2 + A_2(x)^2}, \qquad (9.9)$$

$$E^\varphi(x) := \sqrt{E^1(x)^2 + E^2(x)^2} \qquad (9.10)$$

and $\alpha(x)$, $\beta(x)$ defined by

$$\Lambda_\varphi^A(x) =: \tau_1 \cos \beta(x) + \tau_2 \sin \beta(x), \qquad (9.11)$$

$$\Lambda_E^\varphi(x) =: \tau_1 \cos(\alpha(x) + \beta(x)) + \tau_2 \sin(\alpha(x) + \beta(x)) \qquad (9.12)$$

for the internal directions

$$\Lambda_\varphi^A(x) := (A_1(x)\tau_2 - A_2(x)\tau_1)/A_\varphi(x), \qquad (9.13)$$

$$\Lambda_E^\varphi(x) := (E^1(x)\tau_2 - E^2(x)\tau_1)/E^\varphi(x). \qquad (9.14)$$

Similarly, we have $\Lambda_\vartheta^A(x) = -\tau_1 \sin \beta(x) + \tau_2 \cos \beta(x)$ and an analogous expression for Λ_E^ϑ. These variables are adapted to a loop quantization in that holonomies along integral curves of generators of the symmetry group are of the form $\exp(\delta A_\varphi \Lambda_\varphi^A)$.

However, E^φ is *not* the momentum conjugate to A_φ, which instead is given by

$$P^\varphi(x) := 2E^\varphi(x) \cos \alpha(x). \qquad (9.15)$$

Canonical coordinates are thus the conjugate pairs A_x, E^x; A_φ, P^φ; β, P^β with

$$P^\beta(x) := 2A_\varphi(x)E^\varphi(x) \sin \alpha(x) = A_\varphi(x)P^\varphi(x)\tan \alpha(x) \qquad (9.16)$$

The momenta as basic variables will directly be quantized (some of them smeared), resulting in flux operators with equidistant discrete spectra. But unlike in the full theory and (torsion-free) homogeneous models, the resulting quantum representation has a volume operator that does not commute with flux operators: classically we have $\{V, P^\varphi\} = 2E^\varphi \{\cos \alpha, P^\varphi\} \neq 0$ since $\cos \alpha = \Lambda_\varphi^A \cdot \Lambda_E^\varphi$ depends on the connection. This property follows since volume is determined by triad components, in particular E^φ which is related to P^φ, in a rather complicated way involving the connection component A_φ. Thus, the volume operator has eigenstates different from flux eigenstates, which makes the computation of commutators with holonomies more complicated [10]. An alternative way to derive volume eigenvalues in a quantization based on the variables A_x, A_1 and A_2 is pursued in [11, 12].

This issue is a consequence of inhomogeneity: The Gauss constraint now reads $\partial_x E^x + P^\beta = 0$, which would require $\sin\alpha = 0$ in a homogeneous model where $\partial_x E^x = 0$. With spatially varying E^x, however, $\cos\alpha$ is free to change and constitutes one of the degrees of freedom. The non-commutativity of volume and fluxes can thus be seen as a consequence of inhomogeneity, which is responsible for further technical complications; see also the discussion in [13]. Nevertheless, this issue can be rather straightforwardly dealt with in spherically symmetric models, which thus provide an ideal class of models between homogeneous ones and the full theory.

9.1.1 Canonical Transformation

By a canonical transformation which ensures that now E^φ plays the role of a basic momentum variable one can considerably simplify the formalism. Such a step will, of course, change the configuration variables, no longer being purely connection components. At first sight, this seems to render the procedure unsuitable for a loop quantization where basic operators make use of holonomies of the connection. After a transformation of the canonical variables, holonomies in general will be complicated functions of the new variables such that the new quantum representation would not be suitable for a loop quantization. It turns out, however, that the special form of a spherically symmetric spin connection and extrinsic curvature for a given triad leads to new variables which are ideally suited to a loop representation even from the dynamical point of view. As we will see, we just trade connection components for extrinsic-curvature components.

The co-triad corresponding to a densitized triad (9.2) is given by

$$e = e_x \tau_3 \mathrm{d}x + e_\varphi \Lambda_E^\vartheta \mathrm{d}\vartheta + e_\varphi \Lambda_E^\varphi \sin\vartheta \mathrm{d}\varphi \tag{9.17}$$

with

$$e_\varphi = \sqrt{|E^x|} \quad \text{and} \quad e_x = \mathrm{sgn}(E^x)\frac{E^\varphi}{\sqrt{|E^x|}}. \tag{9.18}$$

From this form, using general equations, one computes the spin connection

$$\Gamma = -(\alpha + \beta)' \tau^3 \mathrm{d}x + \frac{e_\varphi'}{e_x} \Lambda_E^\varphi \mathrm{d}\vartheta - \frac{e_\varphi'}{e_x} \Lambda_E^\vartheta \sin\vartheta \mathrm{d}\varphi + \tau_3 \cos\vartheta \mathrm{d}\varphi \tag{9.19}$$

and the extrinsic curvature for lapse function N and shift N^x,

$$K = N^{-1}(\dot{e}_x - (N^x e_x)')\tau_3 dx + N^{-1}(\dot{e}_\varphi - N^x e'_\varphi)\Lambda_E^\vartheta d\vartheta + N^{-1}(\dot{e}_\varphi - N^x e'_\varphi)\Lambda_E^\varphi \sin\vartheta d\varphi. \tag{9.20}$$

We define the φ-components of Γ and K as

$$\Gamma_\varphi := -\frac{e'_\varphi}{e_x} = -\frac{E^{x'}}{2E^\varphi}, \quad K_\varphi := N^{-1}(\dot{e}_\varphi - N^x e'_\varphi) \tag{9.21}$$

which combines to $A_\varphi = \sqrt{\Gamma_\varphi^2 + \gamma^2 K_\varphi^2}$ since the two internal φ-directions, Λ_E^ϑ and Λ_E^φ in the $d\varphi$-terms in (9.19) and (9.20), respectively, are internally orthogonal.

Many steps in the usual constructions of a loop quantization become more complicated when triad components are not among the basic canonical variables. On the other hand, applying a canonical transformation such that P^φ is replaced by E^φ may lead to more complicated configuration variables which are no longer related to holonomies in a simple way. Fortunately, some properties of the explicit form (9.19) of the spin connection and (9.20) of extrinsic curvature in spherical symmetry allow one to perform a suitable canonical transformation.

Since the momentum of A_x is already given by a triad component E^x, it will be left unchanged by our canonical transformation and we can focus on the variables A_φ, P^φ; β, P^β. Using the definitions (9.15) and (9.16) of P^φ and P^β in the canonical Liouville form and trading in E^φ for P^φ results in

$$\begin{aligned} P^\varphi dA_\varphi + P^\beta d\beta &= 2E^\varphi \cos\alpha \, dA_\varphi + P^\beta d\beta \\ &= E^\varphi d(2A_\varphi \cos\alpha) - 2E^\varphi A_\varphi d\cos\alpha + P^\beta d\beta \\ &= E^\varphi d(2A_\varphi \cos\alpha) + P^\eta d\eta. \end{aligned} \tag{9.22}$$

In the last line we now have E^φ as the momentum of the configuration variable $2A_\varphi \cos\alpha$, and the old $P^\eta := P^\beta$ as the momentum of the angle $\eta := \alpha + \beta$ determining the internal triad direction.

As a function of the original variables, $A_\varphi \cos\alpha$ looks complicated and does not seem to be related to holonomies. In fact, in terms of canonical variables α is a function of both A_φ and the momenta P^φ and P^β such that it cannot be expressed as a function of holonomies in the original variables alone. However, the structure of (9.19) and (9.20) shows that there is a simple geometrical meaning to the new configuration variable conjugate to E^φ. Here it is important to notice that the internal directions along a given angular direction of a spherically symmetric Γ in (9.19) are always internally perpendicular to those of E (note that Λ_E^ϑ and Λ_E^φ are exchanged in (9.19) compared to (9.2)), while the corresponding extrinsic-curvature components are parallel to those of E. Since A is obtained by summing Γ and K, we write

$$A_\varphi \Lambda_\varphi^A = \Gamma_\varphi \bar{\Lambda} + \gamma K_\varphi \Lambda$$

with $\Lambda := \Lambda_E^\varphi$ and $\bar{\Lambda} := \Lambda_E^\vartheta$. This implies

$$A_\varphi \cos\alpha = A_\varphi \Lambda_\varphi^A \cdot \Lambda = \gamma K_\varphi \tag{9.23}$$

where the left equality uses just the definition of α in (9.11) and (9.12). Thus, the new configuration variable is simply proportional to the extrinsic-curvature component K_φ.

Note that it is well known in the full theory that extrinsic-curvature components are conjugate to densitized-triad components. But as we have seen for Ashtekar–Barbero variables, this does not imply that the φ-components as defined here are conjugate, while E^1, E^2 would obviously be conjugate to A_1, A_2 as well as K_1, K_2. The nontrivial fact is that in contrast to the angular Ashtekar–Barbero connection components, the angular extrinsic-curvature component as configuration variable allows one to use triad components as momenta. As the derivation shows, this consequence depends crucially on properties of the spherically symmetric spin connection and extrinsic curvature. That E^φ is conjugate to K_φ follows from the fact that E and K have the same internal directions, while the orthogonality of internal directions in Γ to those of E is relevant for details of the canonical transformation.

With the new canonical triad variables, using

$$A'_\varphi \sin \alpha = (A_\varphi \sin \alpha)' - A_\varphi \alpha' \cos \alpha = \Gamma'_\varphi - \gamma K_\varphi \alpha'$$

we have the Euclidean part

$$H^E[N] = -(2G)^{-1} \int_B dx N(x) |E^x|^{-1/2} \left((\Gamma_\varphi^2 + \gamma^2 K_\varphi^2 - 1) E^\varphi + 2\gamma K_\varphi E^x (A_x + \eta') - 2 E^x \Gamma'_\varphi \right)$$

(9.24)

of the Hamiltonian constraint and the full constraint

$$H[N] = -(2G)^{-1} \int_B dx N(x) |E^x|^{-1/2} \left((1 - \Gamma_\varphi^2 + K_\varphi^2) E^\varphi \right.$$

$$\left. + 2\gamma^{-1} K_\varphi E^x (A_x + \eta') + 2 E^x \Gamma'_\varphi \right)$$

(9.25)

with $\Gamma_\varphi = -(E^x)'/2E^\varphi$ as a function of triad components as per (9.21). From (9.19) we have $\gamma^{-1}(A_x + \eta') = K_x$, and the constraint can directly be written as a function of the densitized triad and extrinsic curvature.

9.1.2 States

With the new spherically symmetric configuration variables $A_x, \gamma K_\varphi, \eta$ the construction of the quantum theory proceeds along the general lines of a loop representation. The sole connection component in the one-dimensional setting is A_x, which is represented by U(1)-holonomies $h_e(A) = \exp\left(\frac{1}{2}i \int_e A_x dx\right)$ along edges e in the one-dimensional radial manifold B. The other configuration fields are scalars, taking values in $\mathbb{R}$ (for γK_φ) and U(1) (for $e^{i\eta}$), respectively. They are thus represented by point holonomies $h_v(K_\varphi) = \exp(i\gamma K_\varphi(v))$ with γK_φ taking values in the

Bohr compactification of the real line, and $h_v(\eta) = \exp(i\eta(v))$ taking values in U(1). Both point holonomies are associated with vertices v: points in the radial manifold B. At this place, we will follow the general constructions outlined in Sect. 3.2.2.3.

A spin network state is obtained by evaluating holonomies in irreducible representations of their respective groups and multiplying them to a complex-valued function on the configuration space. Taking U(1)-representations $\rho_k : e^{i\alpha} \mapsto e^{ik\alpha}$ for integer k and $\bar{\mathbb{R}}_{\text{Bohr}}$-representations $\rho_\mu : z \mapsto e^{i\mu z}$ for real μ, we have

$$T_{g,k,\mu}(A) = \prod_{e \in g} \exp\left(\frac{1}{2}ik_e \int_e A_x(x)dx\right) \prod_{v \in V(g)} \exp(i\mu_v \gamma K_\varphi(v)) \exp(ik_v \eta(v)).$$

$$(9.26)$$

If we allow all edge labels $k_e \in \mathbb{Z}$ and vertex labels $\mu_v \in \mathbb{R}$ and $k_v \in \mathbb{Z}$ for arbitrary finite graphs g in the one-dimensional radial manifold B, we obtain a basis which is orthonormal in the product of Haar measures on the groups involved.

By definition in (9.9), A_φ is always non-negative, which is sufficient for the full space of connections because a sign change in both components A_1 and A_2 can always be compensated for by a gauge rotation. The extrinsic-curvature component K_φ, on the other hand, is measured relatively to the internal direction Λ_E^φ in (9.20) and thus both signs are possible: $K_\varphi \in \mathbb{R}$. This is a further advantage of using extrinsic curvature rather than connection components for the angular components: no boundary effects arise for instance in the discussion of self-adjointness of flux operators (see Sect. 2.2).

Flux operators are obtained from functional derivatives by configuration variables. For $\hat{E}^x$, we obtain the action

$$\hat{E}^x(x)f(h) = -2i\gamma\ell_P^2 \sum_e \frac{\partial f}{\partial h_e}\frac{\delta h_e}{\delta A_x(x)} = \frac{1}{2}\gamma\ell_P^2 \sum_{e \ni x} h_e \frac{\partial f}{\partial h_e} \tag{9.27}$$

where f can be any cylindrical function depending on the holonomies $h_e(A) = \exp\left(\frac{1}{2}i\int_e A_x dx\right)$. To simplify the notation we assumed that x lies only at boundary points of edges, which can always be achieved by subdivision, and which contributes the additional $\frac{1}{2}$. (Similar operators for flux or orbital area have been considered in [14, 15], with applications to horizon states.) The other flux components, P^φ and P^β, are density-valued scalars and thus will be turned into well-defined operators after integrating over regions $\mathscr{I} \subset B$. We obtain

$$\int_{\mathscr{I}} \hat{E}^\varphi f(h) = -2i\ell_P^2 \int_{\mathscr{I}} \frac{\delta}{\delta K_\varphi(x)}dx f(h) = -2i\ell_P^2 \int_{\mathscr{I}} dx \sum_v \frac{\partial f}{\partial h_v}\frac{\delta h_v}{\delta K_\varphi(x)}$$

$$= -2i\ell_P^2 \sum_{v \in \mathscr{I}} \frac{\partial}{\partial K_\varphi(v)} f(h) \tag{9.28}$$

and similarly

$$\int_{\mathscr{I}} \hat{P}^{\eta} f(h) = -2i\gamma \ell_{\mathrm{P}}^2 \sum_{v \in \mathscr{I}} \frac{\partial}{\partial \eta(v)} f(h). \tag{9.29}$$

On spin networks, this gives

$$\hat{E}^x(x) T_{g,k,\mu} = \gamma \ell_{\mathrm{P}}^2 \frac{k_{e^+(x)} + k_{e^-(x)}}{2} T_{g,k,\mu} \tag{9.30}$$

$$\int_{\mathscr{I}} \hat{E}^{\varphi} T_{g,k,\mu} = \gamma \ell_{\mathrm{P}}^2 \sum_{v \in \mathscr{I}} \mu_v T_{g,k,\mu} \tag{9.31}$$

$$\int_{\mathscr{I}} \hat{P}^{\eta} T_{g,k,\mu} = 2\gamma \ell_{\mathrm{P}}^2 \sum_{v \in \mathscr{I}} k_v T_{g,k,\mu} \tag{9.32}$$

where $e^{\pm}(x)$ are the two edges (or two parts of a single edge) meeting at x.

These operators allow us to quantize the volume $V(\mathscr{I}) = 4\pi \int_{\mathscr{I}} dx \sqrt{|E^x|} E^{\varphi}$ of a region $\mathscr{I} \times S^2 \subset B \times S^2$, resulting in the volume operator

$$\hat{V}(\mathscr{I}) = 4\pi \int_{\mathscr{I}} dx |\hat{E}^{\varphi}(x)| \sqrt{|\hat{E}^x(x)|} \tag{9.33}$$

where $\hat{E}^{\varphi}(x)$ is the distribution-valued operator

$$\hat{E}^{\varphi}(x) T_{g,k,\mu} = \gamma \ell_{\mathrm{P}}^2 \sum_{v \in B} \delta(v,x) \mu_v T_{g,k,\mu}.$$

Note that, just as A_{φ} in (9.9), also E^{φ} is defined to be non-negative in (9.10). Thus, only labels $\mu_v \geq 0$ should be allowed. Again, as in Bianchi models or the Kantowski–Sachs model, it is technically easier first to allow all values $\mu_v \in \mathbb{R}$ and in the end require physical states to be symmetric under $\mu_v \mapsto -\mu_v$ (solving the residual gauge transformation). We thus write explicit absolute values around $\hat{E}^{\varphi}(x)$ and μ_v. The volume operator then has eigenstates (9.26) with eigenvalues

$$V_{k,\mu} = 4\pi \gamma^{3/2} \ell_{\mathrm{P}}^3 \sum_v |\mu_v| \sqrt{\frac{1}{2} |k_{e^+(v)} + k_{e^-(v)}|}. \tag{9.34}$$

Here, the eigenvalues follow immediately with eigenstates identical to flux eigenstates.

The Gauss constraint

$$G[\lambda] = \int_B dx \lambda (E^{x\prime} + P^{\eta}) \approx 0 \tag{9.35}$$

is easily quantized to

$$\hat{G}[\lambda]T_{g,k,\mu} = \gamma \ell_P^2 \sum_v \lambda(v)(k_{e^+(v)} - k_{e^-(v)} + 2k_v)T_{g,k,\mu} = 0. \qquad (9.36)$$

One can formally derive this expression by assuming piecewise constant λ, for which $\int_{x_-}^{x_+} \lambda E^{x\prime}dx = \lambda(E^x(x_+) - E^x(x_-))$ can be used for each segment of constant λ, and eventually taking the limit in which smooth λ are obtained. Directly solving the Gauss constraint imposes the condition

$$k_v = -\frac{1}{2}(k_{e^+(v)} - k_{e^-(v)}) \qquad (9.37)$$

to be satisfied for gauge-invariant spin network states. In (9.26) we thus eliminate the integer-valued vertex labels k_v and obtain the general form of gauge-invariant spherically symmetric spin networks

$$T_{g,k,\mu} = \prod_e \exp\left(\frac{1}{2}ik_e \int_e (A_x + \eta')dx\right) \prod_v \exp(i\mu_v \gamma K_\varphi(v)). \qquad (9.38)$$

They depend only on the gauge invariant configuration variables $A_x + \eta' = \gamma K_x$ and γK_φ. At this stage, all configuration variables which enter are extrinsic-curvature components. Having used K_φ for point holonomies thus turns out to be a natural step at the level of gauge-invariant states. (Note that the presence of point holonomies depending on K_φ without integrations allows non-trivial gauge-invariant spin-network states with bivalent vertices in a one-dimensional manifold).

9.1.3 Basic Representation from the Full Theory

As in the example of reducing an anisotropic model to isotropy in Sect. 8.2.5, we can induce basic holonomy and flux operators from the full theory on spherically symmetric states. This demonstration will support the construction given so far by showing how basic operators of the model are related to those used in the full theory. Spherically symmetric states can again be interpreted as distributions in the full Hilbert space, but the dual action of arbitrary full operators applied to distributional symmetric states does in general not lead to another symmetric state. In fact, if we use symmetric distributions without taking further steps, we manage to embed symmetric states as distributions in the connection representation of the full theory, but implement symmetry conditions only for connections and not for triads. A consistent reduced model, already at the classical level, requires analogous conditions to be imposed for connections as well as the densitized triad. In a connection representation, triad conditions can only be imposed at the operator level. Even classically the Hamiltonian flow generated by a phase-space function would in general depart from the subspace of invariant connections if arbitrary triads were allowed (while the flow always stays inside the subspace of invariant connections and invariant triads if the symmetric model is well-defined).

There are, however, notable exceptions of objects whose flow does fix the space of invariant connections, and which allow us to obtain all operators for the basic variables directly from the full theory. This is the case for holonomies of A_x and A_φ, which commute with connections anyway. But we can also find special fluxes whose classical expressions generate a flow that stays in the subspace of invariant connections. For instance, for the τ_3-component of a full flux for a symmetry orbit S^2, $F_{S^2}^3(x) := -\int_{S^2} 2\mathrm{tr}(\tau_3(E(x)\lrcorner dx))d^2y$, we have

$$\{A_a^i(x), F_{S^2}^3\}|_{\mathcal{A}_{\mathrm{inv}}\times\mathcal{E}} = 8\pi\gamma G\delta_3^i\delta_a^x \int_{S^2} \delta(x, y)d^2y$$

which defines a distributional vector field on the phase space parallel to the subspace $\mathcal{A}_{\mathrm{inv}} \times \mathcal{E}$ of invariant connections (parallel to A_x). If we look at any other internal component, such as $F_{S^2}^2$ using τ_2, on the other hand, the Poisson bracket is proportional to $\delta_2^i\delta_a^x$, which is not parallel to the subspace of invariant connections where only the τ_3-component is non-vanishing for $\partial_x\lrcorner A$. Similarly, one can see that the flux

$$F_{\mathcal{I}\times S^1}^{\Lambda^A} := -2 \int_{\mathcal{I}\times S^1} \left(\mathrm{tr}(\Lambda_\varphi^A(x)(E(x)\lrcorner d\varphi))dxd\vartheta + \mathrm{tr}(\Lambda_\vartheta^A(x)(E(x)\lrcorner d\vartheta))dxd\varphi\right)$$

for a cylindrical surface along an interval $\mathcal{I} \subset B$ generates a flow parallel to A_φ leaving the space of invariant connections invariant. (We must use Λ^A rather than Λ_E: the resulting functions on phase space produce a flow along A_a^i-components, parallel to the space of invariant connections. Moreover, $\{A_a^i, \Lambda^A\} = 0$ while $\{A_a^i, \Lambda_E\} \neq 0$. The Λ_E, had we used them, would thus contribute to the flow.)

These two fluxes are sufficient for the basic momenta since

$$-2 \int_{S^2} \mathrm{tr}(\tau_3(E(x)\lrcorner dx))d^2y = 4\pi E^x(x)$$

and

$$-2 \int_{\mathcal{I}\times S^1} \left(\mathrm{tr}(\Lambda_\varphi^A(x)(E(x)\lrcorner d\varphi))dxd\vartheta + \mathrm{tr}(\Lambda_\vartheta^A(x)(E(x)\lrcorner d\vartheta))dxd\varphi\right)$$
$$= 2\pi \int_{\mathcal{I}} P^\varphi(x)dx$$

whose quantizations can thus be obtained directly from the full theory. The canonical pair is automatically produced in this way, with Λ^A-projections providing P^φ rather than E^φ as the momentum.

For the flow $F_{S^2}^3(x)$ we obtain the τ_3-component of a vector field along the x-direction. The pull-back to invariant connections, contained in the definition of a distributional symmetric state, ensures that the dual action of the flux operator for $F^3_{S^2}(x)$ on the distribution can be expressed by an SU(2)-invariant vector field on the reduced representation where only radial holonomies $h^{(e)}_x$ appear. (Here we use SU(2)-holonomies $h^{(e)}_x = \exp(\int_e A_x \tau_3 dx)$ along the radial direction and $h^{(x)}_\varphi = \exp(A_\varphi(x)\Lambda^A_\varphi)$ at vertices.). For the explicit expression we again assume that x is an endpoint of two edges, $e^+(x)$ and $e^-(x)$ which can be achieved by appropriate subdivision, and obtain

$$\hat{F}^3_{S^2}(x) = 4\pi i\gamma \ell^2_P \left(\text{tr}\left((\tau_3 h^{(e^+(x))}_x)^T \frac{\partial}{\partial h^{(e^+(x))}_x} \right) + \text{tr}\left((\tau_3 h^{(e^-(x))}_x)^T \frac{\partial}{\partial h^{(e^-(x))}_x} \right) \right).$$

(9.39)

Since τ_3 commutes with radial holonomies $h^{(e)}_x$, we do not need to distinguish between left and right-invariant vector field operators. The action of a derivative operator $\text{tr}\left((\tau_3 h^{(e)}_x)^T \partial/\partial h^{(e)}_x \right) = (\tau_3 h^{(e)}_x)^A{}_B \partial/\partial (h^{(e)}_x)^A{}_B$ with respect to $h^{(e)}_x$ then amounts to replacing $(h^{(e)}_x)^k$ by $k\tau_3(h^{(e)}_x)^k$ and $\tau_3(h^{(e)}_x)^k$ by $-\frac{1}{4}k(h^{(e)}_x)^k$. In this way, $\hat{E}^x(x) = (4\pi)^{-1}\hat{F}^3_{S^2}(x)$ results, in agreement with the reduced flux operator and its spectrum (9.30).

The operator $\hat{E}^x$ also appears in the Gauss constraint. A gauge-invariant state in the full theory satisfies $J_L(h^{(e^+(x))}_x) - J_R(h^{(e^-(x))}_x) + J_L(h^{(x)}_\varphi) - J_R(h^{(x)}_\varphi)$ if we interpret the reduced set of x- and φ-holonomies as a 4-vertex; so in particular $-2\text{tr}\left(\tau_3 \left(J_L(h^{(e^+(x))}_x) - J_R(h^{(e^-(x))}_x) + J_L(h^{(x)}_\varphi) - J_R(h^{(x)}_\varphi) \right) \right) = 0$. The operators $-2\text{tr}\left(\tau_3 J(h^{(e^\pm(x))}_x) \right)$ simply give operators $\hat{E}^x$ without a difference between right- and left-invariant ones, while for derivative operators with respect to φ-holonomies we have

$$-2\text{tr}\left(\tau_3 \left(J_L(h^{(x)}_\varphi) - J_R(h^{(x)}_\varphi) \right) \right) = -i\left(\text{tr}\left((h^{(x)}_\varphi \tau_3)^T \frac{\partial}{\partial h^{(x)}_\varphi} \right) \right.$$

$$\left. -\text{tr}\left((\tau_3 h^{(x)}_\varphi)^T \frac{\partial}{\partial h^{(x)}_\varphi} \right) \right)$$

$$= i\text{tr}\left([\tau_3, h^{(x)}_\varphi]^T \frac{\partial}{\partial h^{(x)}_\varphi} \right)$$

$$= i\text{tr}\left(\frac{\partial h^{(x)}_\varphi}{\partial \beta(x)} \frac{\partial}{\partial h^{(x)}_\varphi} \right) = i\frac{\partial}{\partial \beta(x)}$$

using $[\tau_3, h^{(x)}_\varphi] = \partial h^{(x)}_\varphi/\partial \beta(x)$ for the angle β in $\Lambda^A_\varphi(x) = \cos \beta(x)\tau_1 + \sin \beta(x)\tau_2$. The right-hand side is then simply proportional to $\hat{P}^\beta(x)$: the τ_3-component of the full quantum Gauss constraint when applied to distributional states is identical to the reduced Gauss constraint.

It remains to look at the full quantization of

$$\int_{\mathscr{I}} P^\varphi \mathrm{d}x = \frac{1}{2\pi} \int_{\mathscr{I}\times S^1} \left(\Lambda_\varphi^A \cdot (E \lrcorner \mathrm{d}\varphi)\mathrm{d}x\mathrm{d}\vartheta + \Lambda_\vartheta^A \cdot (E \lrcorner \mathrm{d}\vartheta)\mathrm{d}x\mathrm{d}\varphi \right)$$

acting on symmetric states. Since they are now Λ_φ^A-components of derivative operators with respect to h_φ, the end result is again a simple derivative operator acting on powers h_φ^μ with the same properties as in the reduced setting.

Alternatively, we may start by considering the subspace of the full phase space with invariant K_a^i, a subspace different from the one of invariant A_a^i as long as E_i^a remains unrestricted. (Classically, we obtain the same reduction as before only when E_i^a is invariant as well, providing the same sector of spherically symmetric models.) We would then use fluxes $F_{\mathscr{I}\times S^1}^{\Lambda_E}$ defined as before, except that Λ_E is used in place of Λ^A. The resulting flow leaves K_φ invariant since Λ_E are the internal directions of K_φ (and $\{K_\varphi, \Lambda_E\} = 0$). This flux provides the densitized-triad component E^φ, canonically conjugate to K_φ. Both representations used before can thus be reduced from the full setting; differences occur only in the intermediate subspaces used. Another difference arises at the quantum level: There are no holonomies associated with K_a^i in the full theory; thus, loop variables for spherically symmetric K-components cannot directly be induced. In other words, the canonical transformation employed in the simplification of geometrical operators, relating densitized-triad components directly to fluxes, cannot be implemented before the reduction at the quantum level.

9.1.4 Hamiltonian Constraint

Having derived the basic representation of the holonomy-flux algebra, we now follow the general steps of constructing Hamiltonian constraint operators. The end result is not unique, as usual, but has the general form [16]

$$\hat{C}[N] = \frac{\mathrm{i}}{2\pi G \gamma^3 \delta^2 \ell_{\mathrm{P}}^2} \sum_{v,\sigma=\pm 1} \sigma N(v)\mathrm{tr} \left(\left(h_\vartheta h_\varphi h_\vartheta^{-1} h_\varphi^{-1} - h_\varphi h_\vartheta h_\varphi^{-1} h_\vartheta^{-1} \right. \right.$$
$$+ 2\gamma^2 \delta^2 (1 - \hat{\Gamma}_\varphi^2)\tau_3 \Big) h_{x,\sigma}[h_{x,\sigma}^{-1}, \hat{V}]$$
$$+ \left(h_{x,\sigma} h_\vartheta (v + e^\sigma(v)) h_{x,\sigma}^{-1} h_\vartheta(v)^{-1} - h_\vartheta(v) h_{x,\sigma} h_\vartheta(v + e^\sigma(v))^{-1} h_{x,\sigma}^{-1} \right.$$
$$\left. + 2\gamma^2 \delta \int_{e^\sigma(v)} \hat{\Gamma}'_\varphi \Lambda(v) \right) h_\varphi[h_\varphi^{-1}, \hat{V}]$$
$$+ \left(h_\varphi(v) h_{x,\sigma} h_\varphi(v + e^\sigma(v))^{-1} h_{x,\sigma}^{-1} - h_{x,\sigma} h_\varphi(v + e^\sigma(v)) h_{x,\sigma}^{-1} h_\varphi(v)^{-1} \right.$$
$$\left. \left. + 2\gamma^2 \delta \int_{e^\sigma(v)} \hat{\Gamma}'_\varphi \bar{\Lambda}(v) \right) h_\vartheta[h_\vartheta^{-1}, \hat{V}] \right).$$

$$(9.40)$$

Here, δ is again a parameter that appears in the exponent of angular holonomies $h_\vartheta = \exp(\delta K_\varphi \bar{\Lambda})$ and $h_\varphi = \exp(\delta K_\varphi \Lambda)$, and will be related to lattice refinement schemes if it is taken as depending on the orbit area E^x. (Lattice refinement in inhomogeneous models, however, has not yet been fully formulated.) As matrix elements, these SU(2)-holonomies contain our basic holonomies $h_v[K_\varphi]$ from h_ϑ and h_φ, as well as $h_{e_\sigma}(v)[A_x]$ from $h_{x,\sigma}(v) := \exp(\int_{e_\sigma(v)} A_x \tau_3 dx)$, $e_\sigma(v)$ indicating the edge leaving v to the right ($\sigma = 1$) or the left ($\sigma = -1$). Moreover, matrix elements of $\Lambda(v) := \tau_1 \cos \eta(v) + \tau_2 \sin \eta(v)$ and $\bar{\Lambda}(v) := -\tau_1 \sin \eta(v) + \tau_2 \cos \eta(v)$ act by multiplication with holonomies $h_v(\eta)$.

In this inhomogeneous model we are dealing with many Hamiltonian constraint equations since the lapse function N can be varied freely at all vertices v, and $\hat{C}[N]$ must vanish for all choices. The constraint equation $\hat{C}[N]\psi = 0$ for all N can then be formulated as a set of coupled difference equations for states labeled by the triad quantum numbers k_e and μ_v, which have the form

$$
\hat{C}_{R+}(k_-, k_+ - 2)^\dagger \psi(\ldots, k_-, k_+ - 2, \ldots)
$$
$$
+ \hat{C}_{R-}(k_-, k_+ + 2)^\dagger \psi(\ldots, k_-, k_+ + 2, \ldots)
$$
$$
+ \hat{C}_{L+}(k_- - 2, k_+)^\dagger \psi(\ldots, k_- - 2, k_+, \ldots)
$$
$$
+ \hat{C}_{L-}(k_- + 2, k_+)^\dagger \psi(\ldots, k_- + 2, k_+, \ldots)
$$
$$
+ \hat{C}_0(k_-, k_+)^\dagger \psi(\ldots, k_-, k_+, \ldots) = 0, \tag{9.41}
$$

one for each vertex. Only the edge labels k_e are written explicitly in this difference expression, but states also depend on vertex labels μ_v on which the coefficient operators $\hat{C}_I$ act. The central coefficient is

$$
\hat{C}_0|\mu, \mathbf{k}\rangle = \frac{\ell_P}{2\sqrt{2}G\gamma^{3/2}\delta^2} \left(|\mu| \left(\sqrt{|k_+ + k_- + 1|} - \sqrt{|k_+ + k_- - 1|} \right) \right.
$$
$$
\times (|\mu_-, k_-, \mu + 2\delta, k_+, \mu_+\rangle + |\mu_-, k_-, \mu - 2\delta, k_+, \mu_+\rangle
$$
$$
- 2(1 + 2\gamma^2\delta^2(1 - \Gamma_\varphi^2(\mu, \mathbf{k})))|\mu_-, k_-, \mu, k_+, \mu_+\rangle)
$$
$$
\left. - 4\gamma^2\delta^2 \mathrm{sgn}_{\delta/2}(\mu)\sqrt{|k_+ + k_-|}\Gamma_\varphi'(\mu, \mathbf{k})|\mu_-, k_-, \mu, k_+, \mu_+\rangle \right)
$$
$$
+ \hat{H}_{\mathrm{matter}, v}|\mu_-, k_-, \mu, k_+, \mu_+\rangle \tag{9.42}
$$

and

$$
\hat{C}_{R\pm}(\mathbf{k})|\mu_-, \mu, \mu_+\rangle := \pm \frac{\ell_P}{4\sqrt{2}G\gamma^{3/2}\delta^2} \mathrm{sgn}_{\delta/2}(\mu)\sqrt{|k_+ + k_-|}
$$
$$
\times \left(|\mu_-, \mu + \frac{1}{2}\delta, \mu_+ + \frac{1}{2}\delta\rangle \right.
$$
$$
- |\mu_-, \mu + \frac{1}{2}\delta, \mu_+ - \frac{1}{2}\delta\rangle + |\mu_-, \mu - \frac{1}{2}\delta, \mu_+ + \frac{1}{2}\delta\rangle
$$
$$
\left. - |\mu_-, \mu - \frac{1}{2}\delta, \mu_+ - \frac{1}{2}\delta\rangle \right) \tag{9.43}
$$

$$\hat{C}_{L\pm}(\mathbf{k})|\mu_-,\mu,\mu_+\rangle := \pm \frac{\ell_P}{4\sqrt{2}G\gamma^{3/2}\delta^2} \mathrm{sgn}_{\delta/2}(\mu)\sqrt{|k_+ + k_-|}$$

$$\times \left(|\mu_- + \tfrac{1}{2}\delta, \mu + \tfrac{1}{2}\delta, \mu_+\rangle \right.$$

$$- |\mu_- - \tfrac{1}{2}\delta, \mu + \tfrac{1}{2}\delta, \mu_+\rangle + |\mu_- + \tfrac{1}{2}\delta, \mu - \tfrac{1}{2}\delta, \mu_+\rangle$$

$$\left. - |\mu_- - \tfrac{1}{2}\delta, \mu - \tfrac{1}{2}\delta, \mu_+\rangle \right)$$

$$\tag{9.44}$$

with

$$\mathrm{sgn}_{\delta/2}(\mu) := \frac{1}{\delta}(|\mu + \delta/2| - |\mu - \delta/2|) = \begin{cases} 1 & \text{for } \mu \geq \delta/2 \\ 2\mu/\delta & \text{for } -\delta/2 < \mu < \delta/2 \\ -1 & \text{for } \mu \leq -\delta/2 \end{cases}$$

We will discuss properties of this class of difference equations, in particular those related to singularity removal, in Sect. 9.3.

9.1.5 Lemaître–Tolman–Bondi Models and Gravitational Collapse

The difference equations of spherically symmetric models are difficult to solve. General properties of fundamental singularity resolution can be analyzed, as seen below, but specific properties of the approach to a singularity in gravitational collapse do not easily show up. As in isotropic models, effective equations would be useful. Here, however, in addition to complicated quantum back-reaction we have to face the anomaly-issue: the algebra of all constraints must still form a first-class set after quantum corrections are included. Anomaly-freedom is trivially realized in homogeneous models which have just a single constraint, but becomes a severe consistency condition in inhomogeneous situations. Not many examples are known in which at least some of the characteristic corrections of loop quantum gravity would provide consistent sets of effective constraints in explicit ways. Spherical symmetry is one example in which anomaly issues can be probed, linking to the physical application of gravitational collapse.

Gravitational collapse of dust clouds can conveniently be discussed with Lemaître–Tolman–Bondi (LTB) models, whose metrics have the general form

$$ds^2 = -N(x,t)^2 dt^2 + (R(x,t)')^2(dx + N^x(x,t)dt)^2 + R(x,t)^2(d\vartheta^2 + \sin^2\vartheta \, d\varphi^2).$$

$$\tag{9.45}$$

This class contains FLRW space-times as well as the Schwarzschild solution, but it can also describe space-times with local matter degrees of freedom in the form of dust. To describe the general class of models in terms of loop variables, we have to implement the condition $(E^x)' = 2E^\varphi$, implicitly imposed by (9.45), together with its dual condition $K'_\varphi = 2K_x \mathrm{sgn} E^x$ (which makes the diffeomorphism constraint identically satisfied) in general spherically symmetric models. This reduction can easily be done at the kinematical state level since the conditions refer directly to basic expressions of the quantization (provided one just exponentiates the curvature relation to result in holonomies) and can thus easily be formulated as conditions for kinematical states.

From the triad relation we derive a condition for fluxes simply by integrating over arbitrary radial intervals $\mathscr{I}$:

$$\int_{\mathscr{I}} E^\varphi = \frac{1}{2}|E^x|_{\partial\mathscr{I}} \qquad (9.46)$$

where $\partial\mathscr{I}$ is the boundary of $\mathscr{I}$ at which E^x is evaluated, taking into account orientation to have the correct signs. This relation can be imposed on triad eigenstates (9.38), where (9.30) and (9.31) imply

$$\mu_v = \frac{1}{2}(|k_{e^+(v)}| - |k_{e^-(v)}|) \qquad (9.47)$$

for any vertex v. This condition eliminates all vertex labels in favor of the edge labels which remain free, analogously to the function $|E^x| = R^2$ which classically determines the spatial part of an LTB metric completely.

On these reduced states, it turns out, the LTB condition for holonomy operators is already implemented. Upon integration and exponentiation, we have

$$\exp\left(\frac{1}{2}\mathrm{isgn}(E^x)\int_{v_1}^{v_2}(A_x + \eta')dx\right) = \exp\left(\frac{1}{2}i\gamma K_\varphi(v_1)\right)\exp\left(-\frac{1}{2}i\gamma K_\varphi(v_2)\right)$$

$$(9.48)$$

expressed solely in terms of elementary holonomy operators. This condition is realized in the sense that the left and right-hand sides, as multiplication operators, have the same action on solutions to the LTB condition satisfying (9.47). Indeed, the left-hand side simply increases the label of the edge between v_1 and v_2 (which we assume to be two adjacent vertices) by one. Thus, it changes both $k_{e^+(v_1)}$ and $k_{e^-(v_2)}$ by ± 1 depending on their sign. The two operators on the right-hand side, on the other hand, change the vertex label μ_{v_1} by $\frac{1}{2}$ and μ_{v_2} by $-\frac{1}{2}$ in the right way to respect the condition (9.47) if it was realized for the original state. (If there are vertices v between v_1 and v_2, $k_{e^+(v)}$ and $k_{e^-(v)}$ change by the same value such that (9.47) remains implemented without changing μ_v).

Notice that, unlike conditions for a symmetry reduction, the two LTB conditions for densitized triads and extrinsic curvature have vanishing Poisson brackets with each other (but not with the constraints). Thus, the curvature condition can indeed be implemented on the solution space of the triad condition. The implementation does not add further conditions for states because they are written in a specific polarization. LTB states are then simply represented by a chain of integer labels k_I for $I = 0, 1, \ldots$ which represents spatial discreteness (a one-dimensional lattice of independent sites) as well as the discreteness of quantum geometry (integer k_n as eigenvalues of the area radius squared). In a connection representation, they can be written as $T_k(z_0, z_1, \ldots) = \prod_I z_I^{k_I}$ where the assignment $I \mapsto z_I := \exp(\frac{1}{2}i \int_{e_I} \gamma K_x dx)$ is a generalized LTB connection.

While states can be reduced immediately to implement the LTB conditions, further conditions do result for composite operators because (9.48) must be used if the action of any operator is to be written on the LTB states where (9.47) has eliminated vertex labels. This provides reductions, for instance of constraint operators, such that characteristic quantum-gravity effects in loop operators can be carried over to constraints for an LTB model. Several consistent versions of different types of quantum corrections have been implemented and studied in [17] at the level of modified classical equations. So far, the situation regarding an effective picture of space-time around singularities, including classically naked ones, remains inconclusive.

LTB models have extensively been studied in Wheeler–DeWitt quantizations [18, 19], sometimes also suggesting discrete models. Instead of addressing the singularity, of prime interest in these investigations is the form of Hawking radiation and possible quantum-gravity corrections to it [20–24].

9.1.6 Further Applications of Spherical Symmetry

Spherically symmetric models, the simplest inhomogeneous reductions, allow us to explore several of the consistency issues of canonical quantum gravity. Also properties of refinement can be seen.

9.1.6.1 Consistent Deformation in the Presence of Dust

In order to probe the consistency of inverse-triad corrections in inhomogeneous models, we modify the Hamiltonian constraint

$$H^Q_{\mathrm{grav}}[N] = -\frac{1}{2G} \int dx N \left(\alpha |E^x|^{-\frac{1}{2}} K_\varphi^2 E^\varphi + 2\bar{\alpha} K_\varphi K_x |E^x|^{\frac{1}{2}} \right.$$
$$\left. + \alpha |E^x|^{-\frac{1}{2}} E^\varphi - \alpha_\Gamma |E^x|^{-\frac{1}{2}} \Gamma_\varphi^2 E^\varphi + 2\bar{\alpha}_\Gamma \Gamma'_\varphi |E^x|^{\frac{1}{2}} \right) \qquad (9.49)$$

by initially independent corrections α, $\bar{\alpha}$, α_Γ, $\bar{\alpha}_\Gamma$, and use

$$H^Q_{\mathrm{dust}}[N] = 4\pi \int dx N \sqrt{P_T^2 + \beta \frac{|E^x|}{(E^\varphi)^2}(P_T T' + P_\Phi \Phi')^2}.$$

as the matter source from dust in the canonical form of [25] with a correction function β. (See for instance [26] for the canonical formulation of spherically symmetric gravity in the presence of dust).

All correction functions depend on the triad, but one can see that a first-class algebra with the diffeomorphism constraint results only if there is no dependence on E^φ. Since the correction functions must be scalar in order to preserve the density weight of Hamiltonian densities, but E^φ in one spatial dimension transforms as a densitized scalar, it is reasonable that it cannot appear in correction functions. The radial component E^x remains a densitized vector field after symmetry reduction, and a densitized vector field in one spatial dimension transforms like a scalar without density weight. Thus, it can easily appear in correction functions. More precisely, we expect inverse-triad corrections to depend on $E^x(x)/\mathcal{N}(x)$ for refinement, corresponding to the size of a discrete patch on the spherical orbit at x.

We leave the diffeomorphism constraint $D_{\text{grav}} + D_{\text{dust}}$ uncorrected because it does not contain inverse-triad components and is quantized via its action on graphs, not giving rise to strong deviations from classical behavior. One obtains a consistent deformation of the whole constrained system provided that $\alpha_\Gamma = \alpha$, $\bar{\alpha}_\Gamma = \bar{\alpha}$. If this is the case, the Poisson bracket of two Hamiltonian constraints is [27, 28]

$$\{H^Q_{\text{grav}}[N] + H^Q_{\text{dust}}[N], H^Q_{\text{grav}}[M] + H^Q_{\text{dust}}[M]\}$$
$$= D_{\text{grav}}[\bar{\alpha}^2 |E^x|(E^\varphi)^{-2}(NM' - MN')]$$
$$+ D_{\text{dust}}[\beta|E^x|(E^\varphi)^{-2}(NM' - MN')]. \tag{9.50}$$

The right-hand side vanishes on the constraint surface only if $\beta = \bar{\alpha}^2$, providing in this case a consistent deformation of the classical spherically symmetric hypersurface-deformation algebra. For $\bar{\alpha} = 1$ the algebra is uncorrected even though there may still be corrections in the Hamiltonian constraint if $\alpha \neq 1$.

9.1.6.2 Poisson Sigma Models

Poisson sigma models [29–31] constitute an elegant formulation of a large class of gravitational models including the reduction to spherical symmetry in arbitrary dimensions. Their dynamics is provided by a two-dimensional field theory with action

$$S_{\text{PSM}} = -\frac{1}{2G} \int_\Sigma \left(A_i \wedge dX^i + \frac{1}{2}\mathscr{P}^{ij} A_i \wedge A_j \right) \tag{9.51}$$

for a Poisson tensor $\mathscr{P}^{ij}$ on a manifold M (antisymmetric and satisfying the Jacobi identity $\varepsilon_{ijk}\partial^i \mathscr{P}^{jk} = 0$ such that $\{f, g\} := \mathscr{P}^{ij}(\partial_i f)(\partial_j g)$ is a Poisson bracket), and fields $X^i \colon \Sigma \to M$, $A_i \colon T\Sigma \to T^*M$. For a 3-dimensional M, their relationship with geometrical variables is such that A_1 and A_2 form a co-dyad on the two-dimensional manifold Σ, A_3 is the spin connection on Σ, X^3 is the dilaton field (arising for instance as one of the scalar components in symmetry reduction) and X_1 and X_2 are auxiliary fields implementing the condition of torsion freedom.

In dilaton-gravity models [32] as a subclass of Poisson sigma models, the Poisson tensor has the form

$$\mathscr{P}^{ij} = \begin{pmatrix} 0 & -V(X^3)/2 & -X^1 + X^2 \\ V(X^3)/2 & 0 & X^1 + X^2 \\ X^1 - X^2 & -X^1 - X^2 & 0 \end{pmatrix} \tag{9.52}$$

with a free function $V(X^3)$, the dilaton potential. For instance, spherically symmetric gravity in D space-time dimensions, with metrics of the form

$$ds^2 = g_{\mu\nu}dx^\mu dx^\nu + \Phi^2(x^\mu)d\Omega^2_{S^{D-1}}$$

for two-dimensional reduced coordinates x^μ can be brought to dilaton-gravity form with $X^3 = \Phi^{D-2}$ and $V(X^3) = -(D-2)(D-3)(X^3)^{-1/(D-2)}$.

An interesting feature of Poisson sigma models is that this class of field theories is stable under consistent deformations [33]. Thus, a consistent deformation, for instance by some form of quantum corrections, must again be a Poisson sigma model. Since the Poisson tensor represents the only freedom in Poisson sigma models, consistent quantum corrections can be classified in terms of Poisson structures.

To make use of this feature in the context of corrections from loop quantum gravity, we must first reformulate spherically symmetric gravity in Ashtekar–Barbero variables as Poisson sigma models. We can do so by using the (rather involved) canonical transformation between the models worked out by [34]. In particular inverse-triad corrections can then be related to modifications of the dilaton potential, with conditions arising for consistency. Incidentally, the relationship between the models and their constraints can be used to formulate a loop quantization of the general class of dilaton models, including spherical reductions of higher-dimensional gravity for which no full loop quantization is known.

9.1.6.3 FLRW in LTB and Possible Derivations of Refinement Models

Classically, the LTB class contains FLRW models. If the relationship can also be established at the quantum level, using the states of Sect. 9.1.5, the link between inhomogeneous and homogeneous models can be used to derive possible lattice-refinement behaviors. At least some information in this direction can already be gained.

The LTB condition relates the densitized-triad components by $E^\varphi - \frac{1}{2}(E^x)' = 0$, and flat FLRW models within this class further satisfy $x(E^x)' - 2E^x = 0$.

Thus, $E^x \propto x^2$. In flat FLRW models, we have indeed $E^x = a(t)^2 x^2$ with the scale factor $a(t)$. The derivative condition provides an equation for this form of E^x with time-independent coefficients; $a(t)$ arises as an integration constant.

The FLRW condition can be used to write $x = 2E^x/(E^x)'$, and then allows us to see how the homogeneous coordinate x changes due to quantum effects that arise in the densitized-triad components. The relationship between a coordinate and phase-space function arises thanks to a gauge-fixing condition implementing additional symmetries. Using an analogous relation at the quantum level allows us to see how symmetry conditions affect the quantum representation.

In the present context, we are primarily interested in lattice refinement, which refers to lattice states embedded in a homogeneous space. If we relate vertex positions to the coordinate x and then to E^x via the gauge-fixing condition, we can see how the vertex distribution changes dynamically. As an "initial" configuration of maximally homogeneous form we assume the vertices of a one-dimensional lattice in an LTB model placed at positions $k_I = In$ with integers I and n, I labelling the vertices and n representing the flux eigenvalues of a discrete geometry of the corresponding FLRW model. The reduced dynamics, in elementary form and without lattice refinement, would merely change n by an integer, preserving the homogeneity of the distribution k_I. Homogeneity is, however, not preserved by the inhomogeneous LTB quantum dynamics, whose elementary steps map k_J to $k_J + 1$ for single J.

We read off changes in vertex positions from the inhomogeneous dynamics by using $x = 2E^x/(E^x)'$, with E^x at a place I after quantization having eigenvalues k_I, to define $x_I = 2k_I/\Delta k_I$ with $\Delta k_I = k_{I+1} - k_I$. For the initial configuration, $x_I = 2I$ is uniform as expected. A single elementary move of the inhomogeneous dynamics then maps k_J to $k_J + 1 = Jn + 1$ for a fixed J. The configuration can no longer be homogeneous, and the reconstructed vertex positions move. Near J, we have after the change $x_{J-1} = 2k_{J-1}/\Delta k_{J-1} = 2(J-1)n/(n+1)$ and $x_J = 2k_J/\Delta k_J = 2(Jn+1)/(n-1)$ while all other k_I do not move. In this way, the inhomogeneous dynamics implies changes in the lattice underlying the homogeneous reduction. Generically, the vertex $J-1$ moves to the left ($n/(n+1) < 1$ if $n \geq 0$) while the vertex J moves to the right ($(n+J^{-1})/(n-1) > 1$ if $n > 1$) by the creation of new flux. If we restrict attention to a finite region, which we always do in homogeneous models, some vertices may move in or out of this region due to the action of the Hamiltonian constraint. Even if the constraint does not generate new vertices and preserves the graph of states it acts on, reconstructed lattices in general do have changing vertex densities to be captured in lattice refinement. As realized in this example, gauge-fixing diffeomorphisms in the process of dynamically representing symmetries is expected to play an important role in deriving lattice refinement.

9.1.6.4 Symmetry After Quantization: Effective FLRW in LTB

The classical relationship between FLRW and LTB models can also be used to find explicit realizations of homogeneous solutions in quantum-corrected inhomogeneous space-times. Interestingly, obstructions to naive realizations of symmetry, or pure minisuperspace quantizations, arise [27].

From the Hamiltonian constraint (9.49) for $\bar{\alpha} = 0$, to be specific, we derive the equations of motion

$$R\dot{R}^2 = \alpha^2 F + R\alpha^2(\alpha^2\mathcal{E}^2 - 1)$$

and

$$2R\ddot{R} + \dot{R}^2 = 2(\dot{R}^2 + \alpha^4\mathcal{E}^2)\frac{\mathrm{d}\log\alpha}{\mathrm{d}\log R} - \alpha^2(1 - \alpha^2\mathcal{E}^2)$$

for $R = \sqrt{|E^x|}$, with a free function $\mathcal{E} = \sqrt{1 + \kappa}$. A classical line element

$$\mathrm{d}s^2 = -\mathrm{d}t^2 + \frac{R'^2}{1 + \kappa(x)}\mathrm{d}x^2 + R^2\mathrm{d}\Omega^2$$

results, which as used before is isotropic if $R = a(t)x$ and $\kappa = -kx^2$.

Assuming $R(t, x) = a(t)x$ and $\kappa(x) = -kx^2$ for the FLRW line element, the equation of motion

$$R\dot{R}^2 = \alpha^2 F + R\alpha^2(\alpha^2\mathcal{E}^2 - 1)$$

gives

$$a\dot{a}^2 + \alpha^2 a\frac{1 - \alpha^2 + kx^2\alpha^2}{x^2} = \frac{8\pi G}{3}\alpha^2 a^3\rho \tag{9.53}$$

for the scale factor, with a spatially constant dust density ρ in $F = (8\pi G/3)R^3\rho$. Here, the x-dependence cancels only for $\alpha = 1$, the classical case; otherwise no solution for $a(t)$ as a function just of time is possible. This result may suggest that dynamical embeddings of minisuperspace models in inhomogeneous ones are not possible at an exact level. Indeed, we have already seen that additional ingredients, such as lattice refinement which cannot be formulated fully in a minisuperspace context, are necessary.

Note that the considerations in this subsection are for $\bar{\alpha} = 1$, a case which modifies the constraints leaving their algebra unchanged. Thus, effective line elements are meaningful in this model, and classical manifold structures are still realized. One would not expect the notion of symmetry or homogeneous spaces to change; yet, the modified dynamics does not allow homogeneous solutions. For $\bar{\alpha} \neq 1$, quantum corrections change the algebra of constraints and thus the space-time structure. In this case, the form of symmetric spaces will have to be modified.

9.2 Models with Local Degrees of Freedom

In general $1 + 1$-dimensional midisuperspace models, local dynamical degrees of freedom exist. The form of an invariant connection in those cases is

$$A = A_x(x)\Lambda_x(x)\mathrm{d}x + A_y(x)\Lambda_y(x)\mathrm{d}y + A_z(x)\Lambda_z(x)\mathrm{d}z + \text{field-independent terms} \tag{9.54}$$

where the internal direction $\Lambda_I(x) \in su(2)$, $\mathrm{tr}(\Lambda_I(x)^2) = -\frac{1}{2}$, can be restricted further depending on the symmetry action. In general, however, they do not satisfy $\mathrm{tr}(\Lambda_I \Lambda_J) = -\frac{1}{2}\delta_{IJ}$, as it was the case in the spherically symmetric model with its non-trivial isotropy group, a condition that was responsible for the simplified structure of states and basic operators. Just as general Bianchi models require a diagonalization before a loop quantization can be evaluated explicitly, extra conditions would be helpful for midisuperspace models.

9.2.1 Models

In cylindrically symmetric models with a space manifold $\Sigma = \mathbb{R} \times (S^1 \times \mathbb{R})$, for instance, the symmetry group $S = S^1 \times \mathbb{R}$ acts freely, and invariant connections and triads have the form

$$A = A_x(x)\tau_3 dx + (A_1(x)\tau_1 + A_2(x)\tau_2)dz + (A_3(x)\tau_1 + A_4(x)\tau_2)d\varphi \tag{9.55}$$

$$E = E^x(x)\tau_3 \frac{\partial}{\partial x} + (E^1(x)\tau_1 + E^2(x)\tau_2)\frac{\partial}{\partial z} + (E^3(x)\tau_1 + E^4(x)\tau_2)\frac{\partial}{\partial \varphi} \tag{9.56}$$

such that, compared with (9.54), $\mathrm{tr}(\tau_3 \Lambda_z) = 0 = \mathrm{tr}(\tau_3 \Lambda_\varphi)$, but in general $\mathrm{tr}(\Lambda_z \Lambda_\varphi) \neq 0$.

The corresponding metric is

$$\begin{aligned}
ds^2 = (E^x)^{-1}(E^1 E^4 - E^2 E^3)dx^2 + E^x(E^1 E^4 - E^2 E^3)^{-1} \\
\times \left(((E^3)^2 + (E^4)^2)dz^2 - (E^2 E^4 + E^1 E^3)dz d\varphi + ((E^1)^2 + (E^2)^2)d\varphi^2 \right)
\end{aligned} \tag{9.57}$$

which is not diagonal. To simplify the model further one may require the metric to be diagonal, which physically corresponds to selecting a particular polarization of Einstein–Rosen waves. This is achieved by imposing the additional condition $E^2 E^4 + E^1 E^3 = 0$ which, in order to yield a non-degenerate symplectic structure, has to be accompanied by a suitable condition for the connection components. With a quadratic condition for the triad components, the condition for connection components takes a different form. It can be derived, for instance, by using the classical Hamiltonian and ensuring that the triad condition is preserved in time. In terms of extrinsic curvature, its off-diagonal components will be required to vanish, imposing a restriction on connection components. We will see this explicitly in the section on Gowdy models, following [35].

Polarized cylindrical waves of this form have perpendicular internal directions since now $\mathrm{tr}(\Lambda_z \Lambda_\varphi) = 0$ for both A and E, and similar simplifications as in the spherically symmetric case can be expected. The form of the metric now is

$$ds^2 = (E^x)^{-1}E^z E^\varphi dx^2 + E^x \left(E^\varphi / E^z dz^2 + E^z / E^\varphi d\varphi^2 \right) \tag{9.58}$$

with

$$E^z := \sqrt{(E^1)^2 + (E^2)^2}, \quad E^\varphi := \sqrt{(E^3)^2 + (E^4)^2}. \tag{9.59}$$

Einstein–Rosen waves are usually represented in the form

$$ds^2 = e^{2(\gamma - \psi)}dr^2 + e^{2\psi}dz^2 + e^{-2\psi}r^2d\varphi^2 \tag{9.60}$$

with only two free functions γ and ψ. Compared with (9.58) one function has been eliminated by gauge-fixing the diffeomorphism constraint.

In fact, this form can be obtained from the more general (9.58) by a field-dependent coordinate change [36]: The symmetry reduction leads to a space-time metric $ds^2 = e^\Lambda dU dV + W(e^{-\Psi}dX^2 + e^\Psi dY^2)$ which indeed has a spatial part as in (9.58) with three independent functions Λ, $W = E^x$ and $\Psi = \log(E^\varphi/E^z)$. One then introduces $t := \frac{1}{2}(V - U)$ and $\rho := \frac{1}{2}(V + U)$, and fixes the diffeomorphism gauge by $W = \rho$ such that

$$ds^2 = e^\Lambda(-dt^2 + d\rho^2) + \rho(e^{-\Psi}dX^2 + e^\Psi dY^2). \tag{9.61}$$

Finally, defining $\Lambda = 2(\gamma - \psi)$, $e^{-2\psi}\rho := e^{-\Psi}$ and renaming $X =: \varphi$, $Y =: z$ leads to the metric (9.60).

A Gowdy model of type T^3 [37] has the line element

$$ds^2 = e^{2a}(-dT^2 + d\vartheta^2) + T(e^{2W}dX^2 + e^{-2W}dY^2)$$

with two free functions a and W depending only on T and ϑ but not on X and Y. Einstein's equation can be solved exactly in this case, resulting in [38]

$$W(T, \vartheta) = \alpha + \beta \log T + \sum_{n=1}^{\infty} (a_n J_0(nT) \sin(n\vartheta + \gamma_n) + b_n N_0(nT) \sin(n\vartheta + \delta_n))$$

in terms of Bessel functions J_0 and N_0, and with constant parameters α, β, a_n, b_n, γ_n and δ_n. Homogeneous Kasner solutions are obtained for the special case $\beta = 1/2$, $a_n = b_n = 0$. Unless $b_n = 0$ and $\beta = 1/2$, there is a curvature singularity at $T = 0$ with diverging $R_{abcd}R^{abcd}$.

9.2.2 Connection Formulation

Initial steps for a loop quantization of midisuperspace models with local degrees of freedom have been provided by [9, 11, 12, 39], and most completely by [35, 40]. Invariant connections and triads for the symmetry type considered here are

$$A_a^i \tau_i dx^a = A_\vartheta \tau_3 d\vartheta + (A_x^1 \tau_1 + A_x^2 \tau_2)dx + (A_y^1 \tau_1 + A_y^2 \tau_2)dy \tag{9.62}$$

$$E_i^a \tau^i \frac{\partial}{\partial x^a} = E^\vartheta \tau_3 \frac{\partial}{\partial \vartheta} + (E_1^x \tau^1 + E_2^x \tau^2) \frac{\partial}{\partial x} + (E_1^y \tau^1 + E_2^y \tau^2) \frac{\partial}{\partial y}. \tag{9.63}$$

As in Sect. 9.1, we define

$$A_x^1 \tau_1 + A_x^2 \tau_2 =: A_x \Lambda_A^x, \quad A_y^1 \tau_1 + A_y^2 \tau_2 =: A_y \Lambda_A^y \tag{9.64}$$

$$E_1^x \tau^1 + E_2^x \tau^2 =: E^x \Lambda_x^E, \quad E_1^y \tau^1 + E_2^y \tau^2 =: E^y \Lambda_y^E \tag{9.65}$$

with gauge invariant A_x, A_y, E^x and E^y (which will no longer be canonically conjugate to each other). The internal directions are parameterized by four angles:

$$\Lambda_x^E = \tau_1 \cos \beta + \tau_2 \sin \beta, \quad \Lambda_A^x = \tau_1 \cos(\alpha + \beta) + \tau_2 \sin(\alpha + \beta) \tag{9.66}$$

$$\Lambda_y^E = -\tau_1 \sin \bar{\beta} + \tau_2 \cos \bar{\beta}, \quad \Lambda_A^y = -\tau_1 \sin(\bar{\alpha} + \bar{\beta}) + \tau_2 \cos(\bar{\alpha} + \bar{\beta}). \tag{9.67}$$

The Gauss constraint

$$G = \frac{1}{8\pi \gamma G} (\partial_\vartheta E^\vartheta + A_x^1 E_2^x - A_x^2 E_1^x + A_y^1 E_2^y - A_y^2 E_1^y) \tag{9.68}$$

removes one of the four angles as gauge and allows us to solve for the momentum of another one. The solution procedure is conveniently expressed in the canonical pairs $(A_\vartheta, E^\vartheta)$, $(A_x \cos \alpha, E^x)$, $(A_y \cos \bar{\alpha}, E^y)$, (β, P^β) with $P^\beta = -E^x A_x \sin \alpha$, and $(\bar{\beta}, P^{\bar{\beta}})$ with $P^{\bar{\beta}} = -E^y A_y \sin \bar{\alpha}$. The angles can also be arranged as (ξ, P^ξ) with $\xi = \beta - \bar{\beta}$ and $P^\xi = \frac{1}{2}(P^\beta - P^{\bar{\beta}})$, and (η, P^η) with $\eta = \beta + \bar{\beta}$ and $P^\eta = \frac{1}{2}(P^\beta + P^{\bar{\beta}})$. In these variables, the Gauss constraint simplifies to

$$G = \frac{1}{8\pi \gamma G} (\partial_\vartheta E^\vartheta + 2P^\eta).$$

From the triad in this parameterization we now determine the corresponding line element and compare with the form usually used:

$$ds^2 = \cos \xi \frac{E^x E^y}{E^\vartheta} d\vartheta^2 + \frac{E^\vartheta}{\cos \xi} \left(\frac{E^y}{E^x} dx^2 + \frac{E^x}{E^y} dy^2 - 2 \sin \xi \, dx dy \right). \tag{9.69}$$

Also here, a loop quantization will be simplest in the case of a diagonal metric, allowing an Abelianization. For the off-diagonal components to vanish we have to impose the polarization condition $\xi = 0$ or $\beta = \bar{\beta}$. Then indeed the internal triad components in all three spatial directions will form an orthogonal triple as in the case of spherical symmetry (but with independent components in the x-and y-directions). The corresponding condition to be imposed for connection components can be derived from the requirement that $\xi = 0$ be preserved in time [35]: $\{\xi, H\} = 0$ using the Hamiltonian constraint. As a result, $2P^\xi + E^\vartheta \partial_\vartheta \log(E^y/E^x) = 0$, which is then automatically preserved in time. This equation can also be seen to be equivalent to $K_{xy} = 0$, such that $\dot{q}_{xy} = 0$ and the diagonal behavior is preserved.

Using the polarization condition and solving the Gauss constraint removes the variables (ξ, P^{ξ}) and (η, P^{η}). We are then left with three pairs which turn out to be $(A_{\vartheta}, E^{\vartheta})$, (K_x, E^x) and $K_y, E^y)$ with extrinsic-curvature components appearing for connection components, in a way similar to the spherically symmetric case. The only difference is that there are three rather than two independent canonical pairs per point, which will leave one local degree of freedom once all constraints are considered. After having solved the Gauss constraint, the remaining ones are the diffeomorphism constraint

$$D = \frac{1}{8\pi G} \left(E^x \partial_{\vartheta} K_x + E^y \partial_{\vartheta} K^y - \gamma^{-1} A_{\vartheta} \partial_{\vartheta} E^{\vartheta} \right) \tag{9.70}$$

and the Hamiltonian constraint

$$H = -\frac{1}{8\pi G} \frac{1}{\sqrt{|E^{\vartheta}| E^x E^y}} \left(K_x E^x K_y E^y + \gamma^{-1}(K_x E^x + K_y E^y) A_{\vartheta} E^{\vartheta} \right.$$
$$\left. + \frac{1}{4} \left((\partial_{\vartheta} E^{\vartheta})^2 - (E^{\vartheta} \partial_{\vartheta} \log(E^y/E^x))^2 \right) \right) + \frac{1}{8\pi G} \partial_{\vartheta} \frac{E^{\vartheta} \partial_{\vartheta} E^{\vartheta}}{\sqrt{|E^{\vartheta}| E^x E^y}}. \tag{9.71}$$

Also for this model and related ones, a loop representation can directly be formulated and constraint operators constructed [40]. Adapting the construction of states and operators as we have seen them in spherical symmetry does not lead to additional difficulties, the only difference being that we have one additional degree of freedom per point on B, given by K_z in Einstein–Rosen models or K_y in Gowdy models. For the new degree of freedom, kinematically, we have additional holonomies $\exp(i\mu_z \gamma K_z)$ in vertices of spin network states.

9.2.3 Hybrid Quantization

As an intermediate stage between loop quantized mini- and midisuperspace models, one can begin by writing a Gowdy model as a Bianchi model plus inhomogeneous field perturbations whose dynamics is governed by a self-interacting Hamiltonian. The Bianchi background can be quantized by standard loop techniques, and for the inhomogeneities a Fock-space representation is simpler and better understood. In the resulting hybrid model [41] the inhomogeneities appear somewhat similar to anisotropies in the perturbative treatment of Sect. 8.2.5, at least kinematically. The dynamics of hybrid models is not induced from a loop Hamiltonian for the Gowdy degrees of freedom; it is constructed by separate loop and Fock quantizations for the background variables and inhomogeneities, respectively. (Fock representations within loop quantization have also been discussed in [42, 43]).

This procedure provides several interesting aspects. First, one can attempt to generalize statements about singularities from homogeneous models to inhomogeneous systems, and in fact bounce solutions have been obtained in some cases [44] based on tree-level equations. This result, by itself, is not very surprising because the inhomogenous contributions serve as a matter term in the Bianchi background very similar to what is obtained with kinetic domination. The model also provides a time-dependent Hamiltonian with the usual choice of an internal time. In this context, interesting questions about the unitarity of evolution arise, which have been studied extensively at the level of non-hybrid Fock quantizations of Gowdy models [45–49]

9.3 Properties

Various $1 + 1$-dimensional models, including those with local physical degrees of freedom, can be loop quantized in an Abelianized manner with explicit expressions for their Hamiltonian constraint. In all cases, a triad representation exists thanks to the Abelian behavior, which allows one to write the constraint equation as a difference equation for wave functions. It can be solved with suitable initial and boundary values, providing crucial insights into whether or not quantum hyperbolicity, the fundamental singularity removal mechanism seen in homogeneous models, can be extended to inhomogeneous situations.

9.3.1 Non-Singular Behavior

Difference equations encountered here, such as (41) constitute large sets of coupled partial difference equations for a wave function $\psi(\mu_v, k_e)$ defined at all values of vertex labels μ_v and edge labels k_e. To define a solution scheme [50], amounting to a well-posed initial/boundary-value problem giving rise to a unique solution once sufficient data are specified, we proceed iteratively from vertex to vertex. To have a starting point in this procedure, we consider a finite region only, requiring spatial boundary conditions. We then start at one side ∂ of the spatial lattice, which could be put into an asymptotically flat regime where a highly semiclassical uncorrelated state can be chosen. We will also have to specify initial values in local internal time, for which a good choice is the edge labels k_e, quantizing E^x. As we evolve to small k_e, a spatial slice approaches the singularity when one or more of the edge labels vanish. Initial values thus are posed by specifying the wave function at a fixed (and non-vanishing) value for all k_e.

Doing so, we have given the boundary values for all μ_∂ and $k_+(\partial) =: k_-$ of the wave function as well as values for large positive $k_e = k_0$ and $k_0 - 1$ at all edges e; we have specified the initial situation, for instance again by a semiclassical state on the initial slice far away from the singularity. The set of difference equations is then to be solved for $\hat{C}_{R+}\psi(k_-, k_+ - 2)$ in terms of values of the wave function following from the initial conditions. We thus move one step further in the recurrence because we now have information about the wave function at $k_+ - 2$ for a smaller edge label (our local internal time) evolving toward the classical singularity.

Next, we have to know how to find ψ from its image under $\hat{C}_{R+}$. An inversion of $\hat{C}_{R+}$ can be done by specifying conditions for the wave function at small μ. These conditions take the form of boundary values for the recurrence, not in space (the radial line) but in midisuperspace since μ quantizes the field space variable E^φ. (This boundary is not in the singular part of minisuperspace but represents an ordinary boundary. Boundary conditions can thus safely be chosen there without precluding one to address the singularity problem. This happens in exactly the same way as in homogeneous models.) Whether or not there is a physical singularity will be determined by trying to evolve through the classical singularity at $k_e = 0$ in the quantized midisuperspace. One crucial difference to cosmological models is that the coefficients $\hat{C}_I(k)$ are not only functions of the local internal time, k_+, studied in the iteration but also of neighboring labels such as k_-. These labels do not take part in the difference equation under consideration for the k_+-evolution; the dependence on them has been determined in iteration steps for previous vertices. This new feature coming from the inhomogeneous context has a bearing on the singularity issue. Any extension of the homogeneous non-singularity results will thus be non-trivial.

Singularities are removed in the sense of quantum hyperbolicity if the difference equation determines the wave function everywhere on midisuperspace once initial and boundary conditions have been chosen away from classical singularities, in a connected component of the classical superspace of non-degenerate metrics. The simplest realization is by a difference equation with non-zero coefficients everywhere. However, since this property is not automatically realized with an equation coming from a general construction of the Hamiltonian constraint, it has to be checked explicitly. With the difference equation derived before for spherically symmetric or other midisuperspace models, it turns out that a symmetric constraint indeed leads to non-zero functions $C_I(k)$ which consequently will not pose a problem to the evolution. All values of the wave function, at positive as well as negative k, are determined uniquely by the difference equations and the initial and boundary values chosen. Evolution thus continues through the classical singularity at zero k: there is no quantum singularity even in midisuperspace models. Other quantization choices can lead to quantum singularities, providing selection criteria to formulate the quantum theory with implications also for the full framework.

Thus, the same mechanism as in homogeneous models contributes to the removal of spherically symmetric classical singularities. Key features are that densitized triads as basic variables in quantum geometry provide us with a local internal time taking values at two sides of the classical singularity, distinguished from each other by orientation, combined with a quantum evolution that connects both sides. No conceptually new ingredients are necessary for inhomogeneous singularities, only an application of the general scheme to the new and more complicated situation.

As in cosmological models the argument applies only to space-like singularities such as the one encountered in the Schwarzschild solution: we have to evolve a spatial slice toward the classical singularity in internal time and test whether it will stop. A time-like or null singularity would require a different mechanism for being resolved, which is not known at present. In fact, investigations based on quantum-corrected classical equations for LTB models [17] have not resulted in a clear mechanism by which non-spacelike singularities could be removed effectively. Cases like negative-mass solutions seem to remain singular, a welcome property helping to rule out unwanted solutions that would lead to instability [51].

The scenario described here based on the general form of difference equations for Abelianizable midisuperspace models does not only apply to vacuum black holes but also to spherically symmetric matter systems as well as Gowdy models. In such cases, compared to spherical symmetry, there would be new labels for matter fields or local gravitational degrees of freedom, and a new contribution to the constraint from the matter Hamiltonian. Such contributions do not change the structure of the difference equation, and the same conclusions apply. The fundamental mechanism of singularity removal, based on quantum hyperbolicity, is general and applies well beyond homogeneous models; the absence of singularities can be demonstrated even in situations with local gravitational degrees of freedom.

9.3.2 Lattice Refinement and Anomaly Freedom

Comparing midisuperspace constructions with the full theory, there is a difference which is visible only in inhomogeneous models: the issue whether or not the constraint creates new edges and vertices, or just changes labels of existing ones. As already discussed in general terms, such a question has a strong bearing on lattice refinement, and thus indirectly affects also homogeneous models. An advantage of midisuperspace models is that they can show refinement behaviors explicitly while still remaining tractable. In our treatment so far we did not include explicit refinement along the radial line, which is not part of the initial constructions of constraint operators in the full theory but has already been considered as potential modifications of the procedures [52, 53]. In those cases, not creating new edges and vertices but instead linking existing vertices by holonomy operators might better explain the presence of correlations at an intuitive level [52], or be of advantage for consistent constructions of the whole set of quantum constraints. However, it makes checking and ensuring anomaly-freedom much more complicated.

The main problem of an anomalous quantization would be that too many states could be removed when imposing inconsistent constraints, leaving insufficiently many physical solutions. At least qualitatively, we can check this issue with the constraint we used. If there is no matter field present we expect just one classical physical degree of freedom, the Schwarzschild mass M. In our solution scheme we started with a boundary state ψ_{∂} corresponding to a wave function for this degree of freedom. Since the state can be specified freely, it is already clear that we do not lose too many solutions. A more complicated question is whether the number of independent physical solutions is correct, that is not too large either. The main difficulty here is the role of semiclassical states to be compared with the number of classical solutions. New quantum solutions may always arise, but for the correct classical limit a quantum theory should not produce more semiclassical solutions than could have a correspondence with classical ones. In the iteration for midisuperspace models we solve one difference equation for ψ at each vertex, such that any freedom here would provide new quantum degrees of freedom. Since the difference equation for ψ has the same form as that in homogeneous loop quantum cosmology, the number of quantum degrees of freedom turns out to be related to the initial value problem of quantum cosmology. Dynamical initial conditions [54, 55] thus appear to be important also to verify the correct classical limit of loop quantum gravity.

If dynamical initial conditions are strong enough, as they are in some isotropic models, solutions for ψ at each vertex are unique and the mass, quantized by the boundary state, is the only quantum degree of freedom. Currently, however, the question of how strong available conditions are is still open. Ultimately, an answer will also depend on issues of imposing the physical inner product or of precisely defining semiclassical states. But at least a simple counting of free variables supports the connection to initial conditions: The vacuum spherically symmetric model has difference equations in three independent variables, an edge label k and two neighboring vertex labels μ. Homogeneous loop quantum cosmology gives rise to an equation of similar structure, also with three variables. If we assume that there is a mechanism for a unique solution in homogeneous models, it will as well apply to black holes of a given mass. Adding matter fields (or more gravitational degrees of freedom as in Einstein–Rosen waves) increases the number of independent variables to five in inhomogeneous models (two new vertex labels) as opposed to four in homogeneous matter models. The type of difference equations thus agrees in homogeneous and inhomogeneous models in vacuum without local degrees of freedom, suggesting an agreement also in the number of solutions. But when local classical degrees of freedom are present, there is additional room for free variables also in quantum states.

As indicated by several examples and applications, the structure of the Hamiltonian-constraint equation arising from models of loop quantum gravity can potentially provide explanations for issues as diverse as the singularity problem in cosmology and black hole physics, initial conditions in quantum cosmology, the semiclassical limit, issues of quantum degrees of freedom, and the anomaly problem which is related to covariance and space-time structure. Many specific details and realizations remain to be checked in generality. Still, such connections between seemingly unrelated issues in quantum gravity can be seen as support for the overall internal consistency of the whole theory and, hopefully, provide guidance for future developments.

References

1. Bojowald, M.: Class. Quantum Grav. **21**, 3733 (2004). gr-qc/0407017
2. Cordero, P.: Ann. Phys. **108**, 79 (1977)
3. Cordero, P., Teitelboim, C.: Ann. Phys. **100**, 607 (1976)
4. Kobayashi, S., Nomizu, K.: Foundations of Differential Geometry, vol 1. Wiley, New York (1963)
5. Brodbeck, O.: Helv. Phys. Acta. **69**, 321 (1996). gr-qc/9610024
6. Bojowald, M.: Canonical Gravity and Applications: Cosmology, Black Holes, and Quantum Gravity. Cambridge University Press, Cambridge (2010)
7. Thiemann, T., Kastrup, H.A.: Nucl. Phys. B **399**, 211 (1993). gr-qc/9310012
8. Kastrup, H.A., Thiemann, T.: Nucl. Phys. B **425**, 665 (1994). gr-qc/9401032

9. Husain, V., Smolin, L.: Nucl. Phys. B **327**, 205 (1989)
10. Bojowald, M., Swiderski, R.: Class. Quantum Grav. **21**, 4881 (2004). gr-qc/0407018
11. Neville D.E. gr-qc/0511005
12. Neville D.E. gr-qc/0511006
13. Bojowald, M.: Class. Quantum Grav. **23**, 987 (2006). gr-qc/0508118
14. Dasgupta, A.: JCAP **0308**, 004 (2003). hep-th/0305131
15. Dasgupta, A.: Class. Quantum Grav. **23**, 635 (2006). gr-qc/0505017
16. Bojowald, M., Swiderski, R.: Class. Quantum Grav. **23**, 2129 (2006). gr-qc/0511108
17. Bojowald, M., Harada, T., Tibrewala, R.: Phys. Rev. D **78**,064057 (2008). arXiv:0806.2593
18. Vaz, C., Witten, L.: Phys. Rev. D **60**, 024009 (1999). arXiv:gr-qc/9811062
19. Vaz, C., Witten, L., Singh, T.P.: Phys. Rev. D **63**, 104020 (2001). arXiv:gr-qc/0012053
20. Barve, S., Singh, T.P., Vaz, C., Witten, L.: Nucl. Phys. B **532**, 361 (1998). gr-qc/9802035
21. Kiefer, C., Müller-Hill, J., Singh, T.P., Vaz, C.: Phys. Rev. D **75**, 124010 (2007). gr-qc/0703008
22. Vaz, C., Gutti, S., Kiefer, C., Singh, T.P.: Phys. Rev. D **76**, 124021 (2007). arXiv:0710.2164
23. Tibrewala, R., Gutti, S., Singh, T.P., Vaz, C.: Phys. Rev. D **77**, 064012 (2008). arXiv:0712.1413
24. Vaz C., Tibrewala R., Singh T.P. arXiv:0805.0519
25. Brown, J.D., Kuchař, K.V.: Phys. Rev. D **51**, 5600 (1995)
26. Kiefer, C., Müller-Hill, J., Vaz, C.: Phys. Rev. D **73**, 044025 (2006). gr-qc/0512047
27. Bojowald, M., Reyes, J.D., Tibrewala, R.: Phys. Rev. D **80**, 084002 (2009). arXiv:0906.4767
28. Reyes J.D.: Spherically symmetric loop quantum gravity: connections to two-dimensional models and applications to gravitational collapse. Ph.D. thesis, The Pennsylvania State University (2009)
29. Ikeda, N., Izawa, K.I.: Prog. Theor. Phys. **90**, 237 (1993). hep-th/9304012
30. Ikeda, N.: Ann. Phys. **235**, 435 (1994). hep-th/9312059
31. Schaller, P., Strobl, T.: Mod. Phys. Lett. A **9**, 3129 (1994). hep-th/9405110
32. Grumiller, D., Kummer, W., Vassilevich, D.V.: Phys. Rept.**369**, 327 (2002). hep-th/0204253
33. Izawa, K.I.: Prog. Theor. Phys. **103**, 225 (2000). hep-th/9910133
34. Bojowald, M., Reyes, J.D.: Class. Quantum Grav. **26**, 035018 (2009). arXiv:0810.5119
35. Banerjee, K., Date, G.: Class. Quantum Grav. **25**, 105014 (2008). arXiv:0712.0683
36. Bičák, J., Schmidt, B.: Phys. Rev. D **40**, 1827 (1989)
37. Gowdy, R.H.: Ann. Phys. **83**, 203 (1974)
38. Isenberg, J., Moncrief, V.: Ann. Phys. **199**, 84 (1990)
39. Hinterleitner F., Major S.: arXiv:1006.4146
40. Banerjee, K., Date, G.: Class. Quantum Grav. **25**, 145004 (2008). arXiv:0712.0687
41. Martín-Benito, M., Garay, L.J., Mena Marugán, G.A.: Phys. Rev. D **78**, 083516 (2008). arXiv:0804.1098
42. Varadarajan, M.: Phys. Rev. D **64**, 104003 (2001). gr-qc/0104051
43. Varadarajan, M.: Phys. Rev. D **66**, 024017 (2002). gr-qc/0204067
44. Brizuela, D., Mena Marugán, G.A., Pawlowski, T.: Class. Quantum Grav. **27**, 052001 (2010). arXiv:0902.0697
45. Mena Marugán, G.: Phys. Rev. D **56**, 908 (1997). gr-qc/9704041
46. Corichi, A., Cortez, J., Quevedo, H.: Phys. Rev. D **67**, 087502 (2003)
47. Cortez, J., Mena Marugan, G.A.: Phys. Rev. D **72**, 064020 (2005)
48. Corichi, A., Cortez, J., Mena Marugan, G.A., Velhinho, J.M.: Phys. Rev. D **76**, 124031 (2007). arXiv:0710.0277
49. Cortez, J., Mena Marugan, G.A., Olmedo, J., Velhinho, J.M.: Phys. Rev. D **83**, 025002 (2011). arXiv:1101.2397
50. Bojowald, M.: Phys. Rev. Lett .**95**, 061301 (2005). gr-qc/0506128
51. Horowitz, G.T., Myers, R.C.: Gen. Rel. Grav. **27**, 915 (1995). gr-qc/9503062
52. Smolin L. gr-qc/9609034
53. Giesel, K., Thiemann, T.: Class. Quantum Grav. **24**, 2465 (2007). gr-qc/0607099
54. Bojowald, M.: Phys. Rev. Lett. **87**, 121301 (2001). gr-qc/0104072
55. Bojowald, M.: Gen. Rel. Grav. **35**, 1877 (2003). gr-qc/0305069

Chapter 10
Perturbative Inhomogeneities

An issue of considerable interest for cosmology, and one major testing ground for quantum gravity, is the inclusion of perturbative inhomogeneities around homogeneous models. A successful implementation will allow one to check the stability and robustness of effects such as singularity resolution seen in isotropic settings, but also lead the way to cosmological applications of structure formation. As always in inhomogeneous settings, the daunting anomaly issue has to be faced when corrections from a canonical quantization are included in general relativity. Also the derivation of all quantum back-reaction terms is quite involved, and so no complete formulation of perturbative inhomogeneities yet exists. Nevertheless, several effects can already be seen, shedding light on the quantum space–time structure, and by an analysis of the anomaly issue one is making contact with the full theory when specific forms of consistent corrections are derived.

10.1 Formalism

Explicit calculations in loop quantum gravity are most easily done when configuration variables, in particular the triad, can be assumed to be diagonal. Examples seen so far for mechanisms to allow diagonalization are isotropy subgroups of symmetry groups, polarization conditions, or an explicit diagonalization in Bianchi models. For cosmological perturbations, diagonalizations of the triad can only be achieved by gauge choices of different modes. While gauge-fixed treatments of quantization must be treated with considerable care, at least some information on quantum corrections can be gained in this way. This information, in turn, can be used as a starting point to be completed in gauge-invariant treatments.

10.1.1 Linear Modes in Cosmological Perturbations

Before setting up a quantization of cosmological perturbations by loop techniques it will be useful to review the standard decomposition of a linearized metric as used in

M. Bojowald, *Quantum Cosmology*, Lecture Notes in Physics 835,
DOI: 10.1007/978-1-4419-8276-6_10, © Springer Science+Business Media, LLC 2011

cosmology; see [1] for more details about the corresponding canonical framework. For perturbations around a flat FLRW model, the general form in conformal time, with a background lapse function $N(t) = a(t)$, is

$$ds^2 = a(t)^2 \left(-(1 + 2\phi(\mathbf{x}, t))dt^2 + 2(\partial_a B(\mathbf{x}, t) + F_a(\mathbf{x}, t))dt dx^a \right.$$
$$\left. + \left((1 - 2\psi(\mathbf{x}, t))\delta_{ab} + \partial_a \partial_b E(\mathbf{x}, t) + 2\partial_a f_b + h_{ab}^{TT}(\mathbf{x}, t) \right) dx^a dx^b \right)$$
$$(10.1)$$

where $\partial^a F_a = 0$, $\partial^a f_a = 0$, $\partial^a h_{ab}^{TT} = 0$ and $h_{ab}^{TT} \delta^{ab} = 0$. The four fields ϕ, ψ, B and E constitute the scalar mode, the two transverse vector fields F_a and f_a the vector mode, and the transverse-traceless h_{ab}^{TT} the tensor mode.

By applying a Lie derivative to the background metric one determines which of the free fields in the linearized metric can be removed by changing coordinates. It turns out that one can always choose a space–time gauge such that ϕ and ψ are the only scalar modes and F_a is the only vector mode. The transverse-traceless field h_{ab}^{TT} for tensor modes is already gauge invariant and cannot be reduced by coordinate choices. Disentangling gauge from evolution is a standard procedure in the classical setting; see also Sect. 10.3.1.2. However, if quantum effects correct the constraints, not only the equations of motion will be modified but also gauge transformations and gauge-invariant objects. In such a context of quantum space–time structures, independent coordinate transformations to determine gauge transformations kinematically may not be available. At this stage, the full canonical formalism, providing gauge and evolution at the same time, becomes crucial. (Sometimes one uses classical information about gauge-invariant quantities before one quantizes, as for instance in [2, 3]. While the required equations may simplify considerably, not all possible quantum effects can be included in this way. We will see examples for effects from quantum modifications of gauge structures later in this chapter).

Selecting modes and choosing a gauge can justify the use of diagonal densitized triads to simplify calculations in a loop quantization. For instance, for scalar modes we can assume an inhomogeneous densitized triad to be of the form $E_i^a = a^2(1 - 2\psi(x))\delta_i^a$, which is diagonal. It turns out that this restriction is not general enough for a loop quantization: a discrete cubic lattice, for instance, which does not have the same spin label at each edge and in this sense is inhomogeneous, cannot have all its vertices isotropic. We thus use an anisotropic (but diagonal) form $E_i^a = \tilde{p}^{(i)}(x)\delta_i^a$, of a linearized densitized triad even though the classical metric we aim to describe does have all its diagonal elements at a fixed point identical. This choice still corresponds to longitudinal gauge with $E = 0$.

If we then choose a vanishing shift vector $N^a = 0$, which is partially a gauge choice ($B = 0$) and partially an assumption about the absence of vector modes, extrinsic curvature $K_a^i = \tilde{k}_{(i)}(x)\delta_a^i$ is diagonal, too. (The Ashtekar–Barbero connection, on the other hand, will not be diagonal because it has non-diagonal contributions from the spin connection. It is of the form $A_a^i = \tilde{k}_{(i)}(x)\delta_a^i + \psi_I(x)\varepsilon_a^{iI}$ whose off-diagonal part ψ_I arising from the spin connection may, if A_a^i is to be included in holonomies, be dealt with perturbatively.)

For diagonal E_i^a and K_a^i one can perform many derivations as explicitly as seen in several of the symmetric models. Calculations of specific quantum correction terms thus simplify in some gauges, but would be more complicated in others. For the anomaly issue, on the other hand, too early a gauge fixing would be damaging since not all the freedom required in the interrelations of different gauge flows could be seen. After having computed candidate quantum corrections in a given gauge, one has to go back to the non-gauge fixed situation and see which extra terms—called counterterms in what follows—are required for an anomaly-free system of constraints in the presence of quantum corrections. Counterterms, or a whole effective set of constraints, should ultimately follow directly from a quantum Hamiltonian, not just by consistency arguments implemented at an effective level. By checking how specific counterterms can arise in expectation values of consistent constraint operators one has means to test the consistency of the full theory of loop quantum gravity. At this level, different gauges can be implemented by using suitable states peaked on classical geometries when computing expectation values.

In this chapter we will organize the constructions as follows:

- We first compute candidate quantum corrections to show their general forms and types. This part also serves as further introduction to the general methods of loop quantum gravity.
- We then use the results at a partial effective level by inserting the corrections, suitably parameterized, into the classical expressions. This step by itself, owing to its incomplete realization of correction terms, is unlikely to provide consistent equations, but by evaluating the consistency conditions of anomaly-freedom, secondary corrections (the counterterms) will be derived.
- With consistent equations at our disposal, we will then analyze the cosmological effects and applications they provide.

10.1.2 Basic Operators

We now begin explicit computations of quantum-geometry correction terms [4]. In addition to the diagonal gauge for a linearized geometry, we assume states to be based on regular cubic lattices, where each vertex v has a unique edge $e_{v,I}$ pointing in one of the three spatial directions I, $I = 1, 2, 3$, as in Fig. 10.1 (and one pointing in the opposite direction which we will denote as $e_{v,-I}$.) Similarly, each edge $e_{v,I}$ has a unique transversal surface $S_{v,I}$. The assumption of a lattice state allows systematic derivations; qualitative features of generic states can be captured in this way. As in spherically symmetric models, we directly use the components of extrinsic curvature in holonomies: $h_{v,I} = \exp(\gamma \tau_I \int_{e_{v,I}} dt \tilde{k}_I(e_{v,I}(t)))$ to exploit their diagonality in longitudinal gauge. Similarly, fluxes will be of the form $F_{v,I} = \int_{S_{v,I}} d^2y\, \tilde{p}^I(y)$. In a regular lattice, each edge has the same coordinate length ℓ_0, and transversal surfaces covering the whole space without overlap have coordinate areas ℓ_0^2. This parameter ℓ_0 is the same that we encountered in the discussion of lattice refinement in homogeneous systems.

Fig. 10.1 A regular lattice
with three edges $e_{v,I}$ (*solid*)
emerging from a vertex in
the three Cartesian directions
and one transversal surface
$S_{v,I}$ (*centered at the dot*)

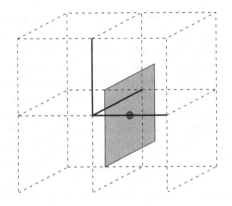

10.1.2.1 Lattice States

For a nearly isotropic geometry we assume $\tilde{k}_I$ to be approximately constant along
every edge. Thus,

$$h_{v,I} = \exp\left(\int_{e_{v,I}} d\lambda\gamma\tilde{k}_I\tau^I\right) \approx \cos\left(\ell_0\gamma\tilde{k}_I(v+\tfrac{1}{2}I)/2\right) + 2\tau_I \sin\left(\ell_0\gamma\tilde{k}_I(v+\tfrac{1}{2}I)/2\right)$$

$$(10.2)$$

where $v + \tfrac{1}{2}I$ denotes, in a slight abuse of notation, the midpoint of the edge. We use
(10.2) as the most symmetric relation between holonomies and continuous fields.
Similarly,

$$F_{v,I} = \int_{S_{v,I}} \tilde{p}^I(y)\mathrm{d}^2y \approx \ell_0^2\tilde{p}^I\left(v + \frac{1}{2}I\right).$$

$$(10.3)$$

(Note that the surface $S_{v,I}$ is defined to be centered at the midpoint of the edge
$e_{v,I}$.) Assuming almost-constant fields along edges and surfaces requires the lattice to be fine enough, which will be true in regimes where fields are not strongly
varying. For more general regimes this assumption has to be dropped and non-local
objects appear even in effective approximations because the local function $\tilde{k}_I$ cannot
fully be reconstructed in terms of $h_{v,I}$ if edges of fixed lengths are used. Since the
recovered classical fields must be continuous, they can arise only if quantizations of
$h_{v,I}$ and $F_{v,I}$, respectively, for nearby lattice links do not have too widely differing
expectation values in a semiclassical state. Otherwise, continuous classical fields can
only be recovered after a process of coarse-graining.

In addition to the assumption of slowly-varying fields on the lattice scale, we have
made use of the diagonality of extrinsic curvature which allows us to evaluate the
holonomy in a simple way without taking care of the factor ordering of su(2)-values
along the path. We can thus re-formulate the theory in terms of U(1)-holonomies, or
functions

$$\eta_{v,I} = \exp\left(i\int_{e_{v,I}} d\lambda\gamma\tilde{k}_I/2\right) \approx \exp\left(i\ell_0\gamma\tilde{k}_I(v + \tfrac{1}{2}I)/2\right) \qquad (10.4)$$

along all lattice links $e_{v,I}$ as they appear in (10.2). On the lattice, a basis of all possible states is then given by specifying an integer label $\mu_{v,I}$ for each edge starting at a vertex v in direction I and defining

$$\langle\tilde{k}(x)| \ldots, \mu_{v,I}, \ldots\rangle := \prod_{v,I}\exp\left(i\mu_{v,I}\int_{e_{v,I}} d\lambda\gamma\tilde{k}_I/2\right) \qquad (10.5)$$

as the functional form of the state $|\ldots, \mu_{v,I}, \ldots\rangle$ in the k-representation. This construction extends the previous representations of midisuperspace models to the fully inhomogeneous lattice case.

Such a form of the states is a consequence of the representation of holonomies. States are functions of U(1)-holonomies, and any such function can be expanded in terms of irreducible representations which for U(1) are just integer powers. The analogous procedure would be more complicated if we allowed all possible, including non-diagonal, curvature components as one is doing in the full theory. In such a case, one would not be able to reduce the original SU(2)-holonomies to simple phase factors and more complicated multiplication rules would have to be considered, not to speak of a devastatingly complicated volume operator. In particular, one would have to make sure that matrix elements of holonomies are multiplied with one another in such a way that functions invariant under SU(2)-gauge rotations result [5]; see Sect. 3.2.2.3. Additional vertex labels are then required which we do not need in the perturbative situation. Nevertheless, we are able to capture crucial holonomy issues while avoiding technical difficulties of explicitly doing calculations with SU(2).

For the same reason we have simple multiplication operators given by holonomies associated with lattice links,

$$\hat{\eta}_{v,I}|\ldots, \mu_{v',J}, \ldots\rangle = |\ldots, \mu_{v,I} + 1, \ldots\rangle. \qquad (10.6)$$

(Only the label associated with the vertex appearing in the holonomy changes.) Furthermore, there are derivative operators with respect to $\tilde{k}_I$, quantizing the conjugate triad components. Just as holonomies are obtained by integrating the connection or extrinsic curvature, densitized-triad components are integrated on surfaces as in (10.3), before they can be quantized. For a surface S of lattice-plaquette size intersecting a single edge $e_{v,I}$ outside a vertex, we have the flux

$$\hat{F}_{v,I}|\ldots, \mu_{v',J}, \ldots\rangle = 4\pi\gamma\ell_P^2\mu_{v,I}|\ldots, \mu_{v',J}, \ldots\rangle \qquad (10.7)$$

or

$$\hat{\mathscr{F}}_{v,I}|\ldots, \mu_{v',J}, \ldots\rangle = 2\pi\gamma\ell_P^2(\mu_{v,-I} + \mu_{v,I})|\ldots, \mu_{v',J}, \ldots\rangle. \qquad (10.8)$$

if the intersection happens to be at the vertex.

As already noted, even for scalar perturbations which classically have triads proportional to the identity, distinct $\tilde{p}^I(v)$-components have to be treated as independent at the quantum level. One cannot assume all edge labels around any given vertex to be identical while still allowing inhomogeneity. Moreover, operators require local edge holonomies that change one edge label $\mu_{v,I}$ independently of the others. Similarly, corresponding operators $\hat{F}_{v,I}$ and $\hat{F}_{v,J}$ ($I \neq J$) act on different links emerging from a vertex v and in general have independent eigenvalues. To pick a regime of scalar modes, one will choose a state whose edge fluxes are peaked close to the same triad value in all directions and whose holonomies are peaked close to the same exponentiated extrinsic curvature values, thus giving effective equations for a single scalar-mode function. This restriction, requiring fluxes to be identical in different directions around a vertex, cannot be done at the level of operators.

10.1.2.2 Reduction to Isotropy

Before we continue with the discussion of quantum corrections, we now use the constructions presented so far to shed more light on the role of inhomogeneities in the reduction of basic holonomy and flux operators from the full theory to models. A triad eigenstate (10.5) is of the form

$$\psi_{\{\mu_{v,I}\}}[h_{v,I}] = \prod_{v,I} \eta_{v,I}^{\mu_{v,I}} = \langle k_J(x)| \dots, \mu_{v,I}, \dots \rangle$$

with $k_I(x) = \ell_0 \tilde{k}_I(x)$ and $\eta_{v,I} \approx \exp(i\gamma k_I(v + \frac{1}{2}I)/2)$. This classical field corresponds to an isotropic connection $\tilde{c}$ if $\gamma \tilde{k}_I(x) = \tilde{c}$, or $\gamma k_I(x) = \ell_0 \mathcal{V}^{-1/3} c = \mathcal{N}^{-1/3} c$, for all I and x. For an exactly isotropic connection, the restriction of the state then becomes the isotropic one

$$\psi_\mu(c) = \exp(i\mu c/2) =: \langle c|\mu \rangle \quad \text{with} \quad \mu = \mathcal{N}^{-1/3} \sum_{v,I} \mu_{v,I}. \tag{10.9}$$

Following the general procedure as in Sect. 8.2.5, there is a map σ taking an isotropic state of the reduced model to a distributional state $(\mu| = \sigma(|\mu\rangle)$ in the inhomogeneous setting such that

$$(\mu| \dots, \mu_{v,I}, \dots\rangle = \langle \mu| \dots, \mu_{v,I}, \dots\rangle|_{\gamma \tilde{k}_I(x) = \tilde{c}} \quad \text{for all} \quad | \dots, \mu_{v,I}, \dots\rangle \tag{10.10}$$

with the inner product on the right-hand side taken in the isotropic Hilbert space, using the restricted state $| \dots, \mu_{v,I}, \dots\rangle|_{\gamma \tilde{k}_I(x) = \tilde{c}}$ as an isotropic state (10.9).

Link holonomies as multiplication operators simply reduce to multiplication operators on isotropic states. Fluxes for lattice sites, however, do not map isotropic states to other isotropic ones. This can easily be seen using $(\mu|\hat{F}_{v,J}| \dots, \mu_{v,I}, \dots\rangle = 4\pi\gamma\ell_{\rm P}^2 \mu_{v,J}| \dots, \mu_{v,I}, \dots\rangle$ on states

$$|\psi_{v,I}\rangle := |0, \ldots, 0, 1, 0, \ldots, 0\rangle$$

which have non-zero labels only on one lattice link $e_{v,I}$. We then have

$$(\mu|\hat{F}_{v,I}|\psi_{v,I}\rangle = 4\pi\gamma\ell_{\mathrm{P}}^2\mu_{v,I}\delta_{\mu,1} \quad \text{and} \quad (\mu|\hat{F}_{v,I}|\psi_{v+I,I}\rangle = 0$$

since the flux surface and the non-trivial link do not intersect in the second case, and thus $(1|\hat{F}_{v,I}|\psi_{v,I}\rangle \neq (1|\hat{F}_{v,I}|\psi_{v+I,I}\rangle$. However, $(v|\psi_{v,I}\rangle = (v|\psi_{v+I,I}\rangle$ for any isotropic state $(v|$. Thus, $(\mu|\hat{F}_{v,I}$ cannot be a superposition of isotropic distributional states, and flux operators associated with a single link do not map the space of isotropic states to itself. (The above formulas show that $(1|$ cannot be contained in a decomposition of $(\mu|\hat{F}_{v,I}$ in basis states, but we can repeat the arguments with arbitrary values for the non-zero label in $|\psi_{v,I}\rangle$ to show that no isotropic state $(\mu|$ can be contained in the decomposition.)

Instead, we should use extended fluxes which are closer to homogeneous expressions. First, we extend a lattice-site flux $\hat{F}_{v,I}$ to span through a whole plane in the lattice, leading to $\sum_{v':v'_I=v_I}\hat{F}_{v',I}$ (with v_I the Ith Cartesian component of the position vector for v). This sum corresponds to a homogeneous flux in the whole box of size $\mathcal{V}$ but is still not translationally invariant because the plane $\{v' : v'_I = v_I\}$ is distinguished. We make it homogeneous on the lattice by averaging along the direction I transversal to the plane. We obtain a sum over all lattice vertices within the box:

$$\hat{p}^I := \mathcal{N}^{-1/3} \sum_v \hat{F}_{v,I}$$

including a factor $\mathcal{N}^{-1/3}$ from averaging in one direction. (There are $\mathcal{N}^{1/3}$ parallel vertex-intersecting planes in the box.) Finally, we take the directional average

$$\hat{p} = \frac{1}{3}\sum_I \hat{p}^I = \frac{1}{3\mathcal{N}^{1/3}}\sum_{v,I}\hat{F}_{v,I}$$

to define the isotropic flux operator.

Now, to find the action of this operator on a distributional state $(v|$ we compute

$$(v|\hat{p}|\ldots, \mu_{v,I}, \ldots\rangle = \frac{4\pi\gamma\ell_{\mathrm{P}}^2}{3\mathcal{N}^{1/3}}\sum_{v',J}\mu_{v',J}(v|\ldots, \mu_{v,I}, \ldots\rangle = \frac{4}{3}\pi\gamma\ell_{\mathrm{P}}^2\mu\delta_{v,\mu}$$

where μ is defined in terms of $\mu_{v,I}$ as in (10.9). This result agrees with the isotropic flux operator defined in isotropic models,

$$\hat{p}\sigma(|\mu\rangle) = \sigma(\hat{p}|\mu\rangle), \tag{10.11}$$

and in particular shows that $\hat{p}$, unlike $\hat{p}_{v,I}$, maps an isotropic distribution to another such state. Thus, the isotropic representation in loop quantum cosmology follows from the inhomogeneous lattice representation along the lines of a symmetry reduction at the quantum level. Notice that this procedure leads directly to an operator for p

rather than $\tilde{p}$ without explicitly introducing the box size $\mathcal{V}$, which would correspond here to the averaging volume. By being forced to extend fluxes over the whole lattice in order to make them homogeneous, the combination $\mathcal{V}^{2/3}\tilde{p} = p$ automatically arises.

10.1.2.3 General Behavior of Quantum Corrections

Having the basic operators $h_{v,I}$ and $F_{v,I}$ in a lattice setting suitable for cosmological perturbations, more complicated ones and in particular the constraints can be constructed by methods of the full theory. As we have seen, local approximations of the basic operators do not depend directly on the classical fields $\tilde{p}^I(x)$ and $\tilde{k}_I(x)$ as components of the densitized triad and extrinsic curvature but on quantities $p^I(x) := \ell_0^2 \tilde{p}^I(x)$ and $k_I(x) := \ell_0 \tilde{k}_I(x)$ rescaled by factors of the lattice link size ℓ_0. This re-scaling, which occurs automatically by the general definitions of basic variables used in a loop quantization, has several advantages. As a minor one, it makes the basic variables independent of coordinates and provides them unambiguously with dimensions of length squared for p^I while k_I becomes dimensionless. (Otherwise, one could choose to put dimensions in coordinates or in metric components which sometimes makes arguments for the expected relevance of quantum corrections confusing.)

> This scaling also happens in homogeneous models as already seen, but in that case, especially in spatially flat models, it is initially the artificial volume $\mathcal{V}$ of a box chosen at will which enters. Since this size is not observable, one has to be careful with interpretations of basic homogeneous variables. Moreover, in this context the scale factor, for instance, as the isotropic analog of $\tilde{p}^I$ could be multiplied by an arbitrary constant and thus the total scale would have no meaning even when multiplied by the analog of ℓ_0^2. Thus, correction functions depending on this quantity in an isotropic model require an additional assumption on how the total scale is fixed. One has to go beyond what can be constructed in a pure minisuperspace model by using lattice refinement. Then, the normalization of p^I relative to ℓ_0 is provided.

Although the magnitude of the p^I is coordinate independent, unlike the value of the scale factor, say, its relation to the (coordinate-dependent) classical field depends on ℓ_0 and thus on the lattice size. It may thus appear that p^I is dependent on artificial structures, but this is clearly not the case because it derives directly from a coordinate-independent flux. The lattice variables are defined independently of coordinates, just by attaching labels $\mu_{v,I}$ to lattice links. Once they have been specified and the lattice has been embedded in a spatial manifold, their relation to classical metric fields can be determined. To be sure, the classical fields such as metric components, do depend on the coordinate choice when they are tensorial. But also the relation between p^I and the classical metric depends on the lattice spacing measured in the same coordinates that have been chosen for the representation of the classical metric. Thus, our basic quantities are coordinate independent and coordinates enter only when classical descriptions are recovered in a semiclassical limit.

In inhomogeneous situations the quantities p^I appear in quantum corrections, their values determining unambiguously when corrections become important; see also Sect. 5.1. Classically, the values of p^I depend on the plaquette size and the geometry, but the quantum theory has these fluxes as elementary quantum excitations. Values of p^I thus directly characterize a state, just like the particle number does for the Fock space of quantum field theory. By reference to a quantum state the fluxes have unambiguous meaning, and their values enter quantum correction terms. When flux sizes are close to the Planck scale, quantum corrections from inverse-triad operators will become large. Or, if the p^I become too large, approaching the Hubble length squared or a typical wave length squared, discreteness effects would become noticeable even in usual regimes of scales which have already been tested. In this way, quantum effects can be probed by several independent conditions in different regimes. These scaling properties are important for an interpretation of corrections and for comparisons with other approaches.

Now, recall the usual expectation that quantum gravity gives rise to low-energy effective actions with higher-curvature terms such as $(16\pi G)^{-1} \int d^4 x \sqrt{|\det g|} \ell_P^2 R^2$ or $(16\pi G)^{-1} \int d^4 x \sqrt{|\det g|} \ell_P^2 R_{\mu\nu\rho\sigma} R^{\mu\nu\rho\sigma}$ added on to the Einstein–Hilbert action $(16\pi G)^{-1} \int d^4 x \sqrt{|\det g|} R$. Irrespective of details of numerical coefficients, there are two key aspects: The Planck length $\ell_P = \sqrt{G\hbar}$ must appear for dimensional reasons in the absence of any other length scale, and higher spatial as well as time derivatives arise with higher powers of $R_{\mu\nu\rho\sigma}$. In canonical variables, one expects higher powers and higher spatial derivatives of extrinsic curvature and the triad, together with components of the inverse metric necessary to define scalar quantities from higher curvature powers (which forces one to raise indices on the Riemann tensor, for instance). Higher time derivatives, on the other hand, are more difficult to see in a canonical treatment and correspond to the presence of independent quantum variables without classical analog [6], as in Chaps. 5 and 13.

Any quantization starts from the purely classical action (or Hamiltonian) where $\hbar$ and thus ℓ_P vanishes. In effective equations of the resulting quantum theory, quantum corrections depending on $\hbar$ will nevertheless emerge. As a first step in deriving such effective equations, we have non-local holonomy terms in a Hamiltonian operator which through its expectation values in semiclassical states will give rise to similar contributions of the same functional form of $k_I(v)$. At first sight, however, these expressions do not agree with expectations from higher-curvature actions. One can easily see that the use of holonomies (10.4) in a Hamiltonian constraint implies higher powers of extrinsic curvature by expanding the trigonometric functions, and higher spatial derivatives of extrinsic curvature by Taylor expanding discrete displacements. Moreover, higher spatial derivatives of the triad may arise from similar non-local terms in the spin-connection contribution. But there are no factors of the Planck length in such higher powers (all factors of G and $\hbar$ are written out explicitly and not "set equal to one"). In fact, by definition $k_I(v)$ is dimensionless since it is obtained by multiplying the curvature component $\tilde{k}_I(v)$ with ℓ_0 in which all possible dimensions cancel. Higher-power terms here do not need any dimensionful prefactor.

Moreover, there are no components of the inverse metric (which would be $1/\tilde{p}^I(v)$ for our diagonal triads) in contrast to what is required in higher-curvature terms.

These seemingly contradictory properties are reconciled if we make use of the relationship (10.3) to write $\ell_0^2 = p^I/\tilde{p}^I$. Inverse-metric components $1/\tilde{p}^I$ then directly occur in combination with $\tilde{k}_J$ factors as required for higher-curvature terms. After eliminating ℓ_0, there are now factors of p^I multiplying the corrections. These are basic variables of the quantum theory, the fluxes whose eigenvalues proportional to $\mu_{v,I}$ determine the fundamental discreteness. Thus, factors of the Planck length occurring in low-energy effective actions are replaced by the state-specific quantities p^I. While the Planck length $\ell_P = \sqrt{G\hbar}$ is expected to appear for dimensional reasons without bringing in information about quantum gravity (it can just be computed using classical gravity for G and quantum mechanics for $\hbar$), the p^I are determined by a state of quantum gravity. If expressed through labels $\mu_{v,I}$, the Planck length also appears, but the factor it provides may be enlarged when $\mu_{v,I} > 1$. Moreover, the lattice labels are dynamical (and in general inhomogeneous) and may thus change in time, in contrast to ℓ_P.

10.1.3 Composite Operators

For explicit estimates of quantum effects we have to look at candidate constraint operators. An important ingredient in the construction of constraints is the volume operator. Using the classical expression $V = \int d^3x\sqrt{|\tilde{p}^1\tilde{p}^2\tilde{p}^3|}$ we introduce the lattice volume operator $\hat{V} = \sum_v \prod_{I=1}^3 \sqrt{|\hat{\mathscr{F}}_{v,I}|}$ which, using (10.8), has eigenvalues

$$V(\{\mu_{v,I}\}) = \left(2\pi\gamma\ell_P^2\right)^{3/2} \sum_v \prod_{I=1}^3 \sqrt{|\mu_{v,I} + \mu_{v,-I}|}. \tag{10.12}$$

As already used in homogeneous and midisuperspace models, we construct operators for co-triad components based on classical identities such as

$$\left\{A_a^i, \int \sqrt{|\det E|}d^3x\right\} = 2\pi\gamma G\varepsilon^{ijk}\varepsilon_{abc}\frac{E_j^b E_k^c}{\sqrt{|\det E|}}\text{sgn}\det(E_l^d) = 4\pi\gamma G e_a^i. \tag{10.13}$$

The resulting operators are of the form $h_e[h_e^{-1}, \hat{V}]$ for SU(2)-holonomies along suitable edges e, for instance

$$\text{tr}(\tau^i h_{v,I}[h_{v,I}^{-1}, \hat{V}_v]) \sim -\frac{1}{2}i\hbar\ell_0\widehat{\{A_a^i, V_v\}} \tag{10.14}$$

for $h_{v,I}$ as in (10.2). Factors of the link size ℓ_0 are needed in reformulating Poisson brackets through commutators with holonomies, which are provided by the discretized integration measure in spatial integrations as they occur in the Hamiltonian constraint.

As in the models encountered before, we have a Hamiltonian constraint operator with vertex contributions of the form

$$\hat{H}_v = \frac{1}{16\pi G} \frac{2i}{8\pi \gamma G \hbar} \frac{N(v)}{8} \sum_{IJK} \sum_{\sigma_I \in \{\pm 1\}} \sigma_1 \sigma_2 \sigma_3 \varepsilon^{ijk}$$

$$\times \operatorname{tr}(h_{v,\sigma_I I}(A) h_{v+\sigma_I I, \sigma_J J}(A) h_{v+\sigma_J J, \sigma_I I}(A)^{-1} h_{v,\sigma_J J}(A)^{-1}$$

$$\times h_{v,\sigma_K K}(A)[h_{v,\sigma_K K}(A)^{-1}, \hat{V}]) \tag{10.15}$$

summed over all non-planar triples of edges in all possible orientations. The combination

$$h_{v,\sigma_I I}(A) h_{v+\sigma_I I, \sigma_J J}(A) h_{v+\sigma_J J, \sigma_I I}(A)^{-1} h_{v,\sigma_J J}(A)^{-1}$$

gives a single plaquette holonomy with tangent vectors $e_{v,\sigma_I I}$ and $e_{v,\sigma_J J}$. Compared to homogeneous models, where we would multiply only two independent holonomies for the two directions of the loop, we are using loops made of four independent holonomies when edges are moved parallelly. Taking all possible vertices, the coupling of infinitely many degrees of freedom is realized in this way, although we usually continue to work with finite graphs contained in a compact region, probing only finitely many degrees of freedom.

The use of holonomies provides a non-polynomial realization of the Hamiltonian constraint, the precise form of which depends on the parameter ℓ_0. To check the correct continuum limit, assuming small and weakly-varying connection components compared to the edge scale, we expand in ℓ_0. The leading term of the integrand is of the order ℓ_0^3 which automatically results in a Riemann-sum representation of the first term in (10.15). Since one needs to assume that the lattice is sufficiently fine for classical values of the fields A_a^i in regimes and gauges of interest, there are certainly states corresponding to coarser lattices on which stronger quantum corrections may result. As usual, semiclassical behavior is not realized on all states but only for a select class. Ensuring the correct classical limit, our arguments show that for any low-curvature classical configuration a chosen lattice leads only to small quantum corrections such that sufficiently many semiclassical states exist.

Using extrinsic curvature as the basic quantity entering in holonomies, rather than the connection, allows simplifications of the whole constraint by easily combining the remaining quadratic terms in K_a^i with the first term of the constraint; we do not need to use squares of multiple commutators from quantizing a Poisson bracket expression for extrinsic curvature in terms of the Euclidean constraint and volume as in (4.4). Writing

$$F_{ab}^i = 2\partial_{[a}\Gamma_{b]}^i + 2\gamma\partial_{[a}K_{b]}^i - \varepsilon_{ijk}(\Gamma_a^j + \gamma K_a^j)(\Gamma_b^k + \gamma K_b^k)$$
$$= 2\partial_{[a}\Gamma_{b]}^i + 2\gamma\partial_{[a}K_{b]}^i - \gamma\varepsilon_{ijk}(\Gamma_a^j K_b^k + \Gamma_b^k K_a^j) - \varepsilon_{ijk}(\Gamma_a^j\Gamma_b^k + \gamma^2 K_a^j K_b^k) \tag{10.16}$$

we obtain a term $2\gamma \partial_{[a} K^i_{b]} - \gamma^2 \varepsilon_{ijk} K^j_a K^k_b$ resembling a "curvature" 2-form $F^i_{ab}(\gamma K)$ as computed from extrinsic curvature alone, a curvature term of the spin connection as well as cross-terms $\varepsilon_{ijk}(\Gamma^j_a K^k_b + \Gamma^k_b K^j_a)$. With the diagonality conditions used so far, the cross-terms disappear and only the "K-curvature" term and spin connection terms remain to be quantized: We write the constraint for scalar modes in longitudinal gauge as $H[N] = H_K[N] + H_\Gamma[N]$ with

$$H_K[N] := \frac{1}{8\pi G} \int_\Sigma d^3x\, N \,|\det E|^{-1/2} \left(\varepsilon_{ijk} \gamma \partial_a K^i_b + K^j_a K^k_b \right) E^{[a}_j E^{b]}_k \qquad (10.17)$$

(combined with the $(1 + \gamma^2)$-term in the full constraint) and

$$H_\Gamma[N] := \frac{1}{8\pi G} \int_\Sigma d^3x\, N \,|\det E|^{-1/2} \left(\varepsilon_{ijk} \partial_a \Gamma^i_b - \Gamma^j_a \Gamma^k_b \right) E^{[a}_j E^{b]}_k. \qquad (10.18)$$

(Also the term of H_K containing $\partial_a K^i_b$ drops out for diagonal variables, such that the constraint is explicitly γ-independent). Both constraint contributions can rather easily be dealt with by standard techniques, using K-holonomies around a loop for the first one and direct quantizations of Γ^i_a for the second. The split-off spin connection components are quantized separately, which is possible in the perturbative treatment on a background, see below, and then added on to the operator.

Following the general procedure, first for H_K, we obtain vertex contributions

$$\hat{H}_{K,v} = -\frac{1}{16\pi G} \frac{2i}{8\pi \gamma^3 G\hbar} \frac{N(v)}{8} \sum_{IJK} \sum_{\sigma_I \in \{\pm 1\}} \sigma_1 \sigma_2 \sigma_3 \varepsilon^{IJK}$$

$$\times \operatorname{tr}\left(h_{v,\sigma_I I} h_{v+\sigma_I I, \sigma_J J} h^{-1}_{v+\sigma_J J, \sigma_I I} h^{-1}_{v,\sigma_J J} h_{v,\sigma_K K} \left[h^{-1}_{v,\sigma_K K}, \hat{V}_v \right] \right). \qquad (10.19)$$

As before, $h_{v,I}$ denotes a K-holonomy along the edge oriented in the positive I-direction and starting at a vertex v, but we also include the opposite direction $h_{v,-I}$ in the sum to ensure rotational invariance. In some of the holonomies, $v + I$ is again used for the vertex adjacent to v in the positive I-direction. The $\{IJK\}$-summation is taken over all possible orientations of the IJ-loop and a transversal K-direction. For notational brevity, we introduce, as in [4],

$$c_{v,I} := \frac{1}{2}\operatorname{tr}(h_{v,I}), \qquad s_{v,I} := -\operatorname{tr}(\tau_{(I)} h_{v,I}) \qquad (10.20)$$

such that (10.2) becomes $h_{v,I} = c_{v,I} + 2\tau_I s_{v,I}$. In a continuum approximation, we have

$$c_{v,I} = \cos\left(\gamma k_I \left(v + \frac{1}{2} I \right)/2 \right), \qquad s_{v,I} = \sin\left(\gamma k_I \left(v + \frac{1}{2} I \right)/2 \right) \qquad (10.21)$$

where again $k_I(v) = \ell_0 \tilde{k}_I(v)$. The traces in (10.19) can then be seen to be of a form

$$\frac{i}{8\pi\gamma G\hbar}\text{tr}(h_{v,I}h_{v+I,J}h_{v+J,I}^{-1}h_{v,J}^{-1}h_{v,K}[h_{v,K}^{-1},\hat{V}_v])$$

$$= -\varepsilon_{IJK}\left\{\left[(c_{v,I}c_{v+J,I}+s_{v,I}s_{v+J,I})c_{v,J}c_{v+I,J}+(c_{v,I}c_{v+J,I}-s_{v,I}s_{v+J,I})\right.\right.$$

$$\left.\left.\times s_{v,J}s_{v+I,J}\right]\hat{A}_{v,K}\right\}$$

$$+ \varepsilon_{IJK}^2\left\{\left[(c_{v,I}s_{v+J,I}-s_{v,I}c_{v+J,I})s_{v,J}c_{v+I,J}+(s_{v,I}c_{v+J,I}+c_{v,I}s_{v+J,I})\right.\right.$$

$$\left.\left.\times c_{v,J}s_{v+I,J}\right]\hat{B}_{v,K}\right\},$$

$$(10.22)$$

where

$$\hat{A}_{v,K} := \frac{1}{4\pi i\gamma G\hbar}\left(\hat{V}_v - c_{v,K}\hat{V}_v c_{v,K} - s_{v,K}\hat{V}_v s_{v,K}\right),$$

$$\hat{B}_{v,K} := \frac{1}{4\pi i\gamma G\hbar}\left(s_{v,K}\hat{V}_v c_{v,K} - c_{v,K}\hat{V}_v s_{v,K}\right).$$

$$(10.23)$$

Only the second line contributes after contracting with ε_{IJK}, and results in

$$\hat{H}_{K,v} = \frac{-N(v)}{64\pi\gamma^2 G}\sum_{IJK}\sum_{\sigma_I\in\{\pm 1\}}\left((s_{v,\sigma_I I,\sigma_J J}^- s_{v,\sigma_J J}c_{v+\sigma_I I,\sigma_J J}\right.$$

$$\left.+s_{v,\sigma_I I,\sigma_J J}^+ c_{v,\sigma_J J}s_{v+\sigma_I I,\sigma_J J})\hat{B}_{v,\sigma_K K}\right), \quad (10.24)$$

with

$$s_{v,\sigma_I I,\sigma_J J}^\pm := \sin\left(\frac{1}{2}\gamma(k_{\sigma_I I}(v+\sigma_I I/2)\pm k_{\sigma_I I}(v+\sigma_J J+\sigma_I I/2))\right).$$

In the homogeneous case the first term in the sum (10.24) vanishes and the leading contribution is

$$4\sin(\gamma k_I/2)\cos(\gamma k_I/2)\sin(\gamma k_J/2)\cos(\gamma k_J/2)\hat{B}_{v,K}, \quad (10.25)$$

in agreement with our earlier constructions directly in homogeneous models, where $\gamma k_I = \gamma\ell_0\tilde{k}_I = \ell_0\mathcal{V}^{-1/3}c = c/\mathcal{N}^{1/3}$.

10.1.4 Quantum Corrections

Our two types of quantum-geometry corrections are visible from expression (10.24): Using commutators to quantize inverse densitized-triad components implies eigenvalues of $\hat{B}_{v,I}$ which differ from the classical expectation at small labels $\mu_{v,I}$. Moreover, using holonomies contributes higher-order terms in extrinsic curvature together with higher spatial derivatives when sines and cosines are expanded in regimes of slowly-varying fields. To see the forms of correction terms, we will now show the next-leading order of higher powers and spatial derivatives of $\tilde{k}_I(v)$, making the general considerations in Sect. 10.1.2.3 explicit, before dealing with inverse-triad corrections. Calculations here serve to illustrate the general form; no unique derivations of curvature corrections exist at the present stage of developments.

10.1.4.1 Curvature

We expand the Hamiltonian explicitly in ℓ_0 after writing $k_I = \ell_0 \tilde{k}_I$. This implements a slowly-varying-field approximation with respect to the lattice spacing. For the (I, J)-plaquette, a single term in the sum (10.24) becomes

$$
\begin{aligned}
2(s^-_{v,I,J} & s_{v,J} c_{v+I,J} + s^+_{v,I,J} c_{v,J} s_{v+I,J}) \\
= {} & \gamma^2 \ell_0^2 \tilde{k}_I \tilde{k}_J + \frac{1}{2} \gamma^2 \ell_0^3 \left(\tilde{k}_I \partial_J \tilde{k}_J + \tilde{k}_J \partial_I \tilde{k}_I + 2\tilde{k}_J \partial_I \tilde{k}_I \right) \\
& + \frac{1}{8} \gamma^2 \ell_0^4 \left(\tilde{k}_I \partial_J^2 \tilde{k}_J + \tilde{k}_J \partial_I^2 \tilde{k}_I + 4(\tilde{k}_I \partial_I^2 \tilde{k}_J + \tilde{k}_I \partial_I \partial_J \tilde{k}_J + \partial_I \tilde{k}_I \partial_I \tilde{k}_J \right. \\
& \left. + \partial_J \tilde{k}_I \partial_I \tilde{k}_J) + 2\partial_I \tilde{k}_I \partial_J \tilde{k}_J - \frac{4}{3} \gamma^2 \tilde{k}_I \tilde{k}_J (\tilde{k}_I^2 + \tilde{k}_J^2) \right) + O(\ell_0^5). \quad (10.26)
\end{aligned}
$$

Link labels $\tilde{k}_I$ were initially introduced as values of the extrinsic-curvature components evaluated at midpoints of edges in the continuum approximation (10.21) of our basic non-local variables. The expression above is written in terms of just two components $\tilde{k}_I(v)$ and $\tilde{k}_J(v)$ obtained by Taylor expanding the midpoint evaluations around the vertex v. For a fixed direction K there are in total eight terms to be included in the sum (10.24). They are obtained from (10.26) by taking into account the four plaquettes in the (I, J)-plane meeting at vertex v and considering both orientations in which each plaquette can be traversed.

Summing over all four plaquettes (each traversed in both directions), the cubic terms drop out and we are left with

$$
\begin{aligned}
& \gamma^2 \ell_0^2 \tilde{k}_I \tilde{k}_J - \frac{\gamma^4 \ell_0^4}{6} \tilde{k}_I \tilde{k}_J (\tilde{k}_I^2 + \tilde{k}_J^2) \\
& + \frac{\gamma^2 \ell_0^4}{8} (\tilde{k}_I \partial_J^2 \tilde{k}_J + \tilde{k}_J \partial_I^2 \tilde{k}_I + 2(\tilde{k}_I \partial_I^2 \tilde{k}_J + \tilde{k}_J \partial_J^2 \tilde{k}_I + \partial_I \tilde{k}_I \partial_I \tilde{k}_J \\
& + \partial_J \tilde{k}_I \partial_J \tilde{k}_J)) + O(\ell_0^5). \quad (10.27)
\end{aligned}
$$

The first term, when combined with $\hat{B}_{v,K}$ and summed over all triples IJK, reproduces the correct classical limit of the constraint H_K.

In the final expression, the factor ℓ_0^2 in the leading term together with a factor ℓ_0 from $\hat{B}_{v,K}$ through (10.14) combine to give the Riemann measure of the classical integral. Higher-order terms, however, come with additional (coordinate dependent) factors of ℓ_0 in (10.27) which are not absorbed in the measure. The result is independent of coordinates since the whole construction (10.24) in terms of k_I is coordinate independent. But for a comparison with higher-curvature terms we have to formulate corrections in terms of $\tilde{k}_I$ and $\tilde{p}^I$ as these are the components of classical extrinsic curvature and densitized triad tensors. Higher-order terms in the expansion are already formulated with $\tilde{k}_I$ in coordinate independent combinations with ℓ_0-factors. For a comparison with low-energy effective actions, it remains to interpret the additional ℓ_0 factors, as explained in Sect. 10.1.2.3, by eliminating them in favor of components of the inverse metric. At this stage, the corrections obtained from loop quantum gravity have been reconciled with general expectations of effective actions. Also, the fact that the cubic term in ℓ_0 in (10.27) drops out is in agreement with higher-curvature corrections since in that case only even powers of the length scale ℓ_P can occur.

This limit, in summary, was obtained in two steps: we first performed the continuum approximation by replacing holonomies with mid-point evaluations of extrinsic-curvature components. After this step we are still dealing with a non-local Hamiltonian since each vertex contribution now refers to evaluations of the classical field at different points. In a second step, we Taylor expanded these evaluations around the central vertex v, which gives a local result and corresponds to a further, slowly-varying-field approximation.

10.1.4.2 Inverse Triad

A direct calculation using (10.5) and (10.12), whose details are very similar to calculations in homogeneous models, shows that $\hat{B}_{v,K}$, as defined in (10.23), commutes with all flux operators and thus has the flux eigenstates as eigenbasis. The action

$$\hat{B}_{v,K}|\ldots,\mu_{v,K},\ldots\rangle := \left(2\pi\gamma\ell_P^2\right)^{1/2}\sqrt{|\mu_{v,I}+\mu_{v,-I}||\mu_{v,I}+\mu_{v,-J}|}$$
$$\times\left(\sqrt{|\mu_{v,K}+\mu_{v,-K}+1|}-\sqrt{|\mu_{v,K}+\mu_{v,-K}-1|}\right)$$
$$\times|\ldots,\mu_{v,K},\ldots\rangle$$
$$(10.28)$$

directly shows the eigenvalues, which do not agree exactly with the classical expectation $e_K(v) = \sqrt{|p^I(v)p^J(v)/p^K(v)|} \sim \sqrt{|\mu_{v,I}\mu_{v,J}/\mu_{v,K}|}$ (indices such that $\varepsilon^{IJK} = 1$) for the co-triad (10.13). But for large values $\mu_{v,I} \gg 1$ the classical expectation is approached as an expansion of the eigenvalues shows.

More generally, we compute eigenvalues of the operators

$$\hat{B}_{v,K}^{(r)} := \left(2\pi\gamma\ell_P^2\right)^{-1}\frac{\hat{V}^r|_{\mu_{v,K}+1}-\hat{V}^r|_{\mu_{v,K}-1}}{r}$$
$$(10.29)$$

where the subscript of the volume operator indicates that its eigenvalue in a lattice state is computed according to (10.12) with a shifted label of the link $e_{v,K}$. The eigenvalues are

Fig. 10.2 Behavior of the
correction function $\alpha^{(r)}$ in
(10.31) [4]. It approaches
one from above for large
arguments. For small
arguments, the function is
increasing from zero and
reaches a peak value larger
than one. Also shown is the
limiting case $r = 2$ which
does not show a peak but a
constant correction function
for $\mu > 1$

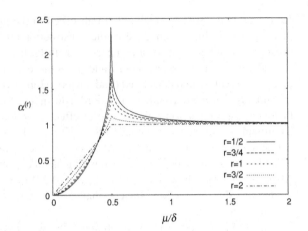

$$B_{v,K}^{(r)} := \frac{1}{r} \left(2\pi \gamma \ell_{\mathrm{P}}^2\right)^{3r/2-1} |\mu_{v,I} + \mu_{v,-I}|^{r/2} |\mu_{v,J} + \mu_{v,-J}|^{r/2}$$
$$\times \left(|\mu_{v,K} + \mu_{v,-K} + 1|^{r/2} - |\mu_{v,K} + \mu_{v,-K} - 1|^{r/2}\right). \qquad (10.30)$$

compared to the classical expectation

$$(2\pi \gamma \ell_{\mathrm{P}}^2)^{3r/2-1} |\mu_{v,I} + \mu_{v,-I}|^{r/2} |\mu_{v,J} + \mu_{v,-J}|^{r/2} |\mu_{v,K} + \mu_{v,-K}|^{r/2-1}$$

for $V^{r-1} e_K$.

Inverse-triad corrections are obtained by extracting the deviations from $e_K(v)$ which $B_{v,K}$ receives on smaller scales. We introduce the correction function as a multiplier $\alpha_{v,K}(\mu_{v,I})$ such that $B_{v,K} = \alpha_{v,K} e_K(v)$ and $\alpha_{v,K} \to 1$ for $\mu_{v,K} \gg 1$. Comparing the eigenvalues of $\hat{B}_{v,K}$ with those of flux operators in the combination $\sqrt{|\mathscr{F}_{v,I} \mathscr{F}_{v,J} / \mathscr{F}_{v,K}|}$ suitable for $e_K(v)$, we find

$$\alpha_{v,K}^{(r)} = \frac{1}{r} |\mu_{v,K} + \mu_{v,-K}|^{1-r/2} \left(|\mu_{v,K} + \mu_{v,-K} + 1|^{r/2} - |\mu_{v,K} + \mu_{v,-K} - 1|^{r/2}\right).$$
$$(10.31)$$

Some examples of r are seen in Fig. 10.2. For regular quantizations of inverse-triad components, $0 < r < 2$.

Having computed the operators and their eigenvalues for general diagonal configurations, we now specialize the correction function to perturbations of the scalar mode. We reduce the number of independent labels by imposing $\mu_{v,I} + \mu_{v,-I} = \mu_{v,J} + \mu_{v,-J}$ for arbitrary I and J. This corresponds to a metric proportional to the identity δ_{ab}, as it is realized for scalar perturbations; see Sect. 10.1.1. We then assign a new variable $p(v) = 2\pi \gamma \ell_{\mathrm{P}}^2 (\mu_{v,I} + \mu_{v,-I})$ to each vertex v, which is independent of the direction of the edge I and describes the diagonal part of the triad. Quantum numbers in eigenvalues of the lattice operators can then be replaced by $p(v)$, allowing us to compare the resulting functions with the classical ones. The remaining subscript v indicates that the physical quantities are vertex-dependent, inhomogeneous.

Averaging over the plaquette orientations in the constraint then becomes trivial and
the total correction reads

$$\alpha^{(r)}[p(v)] = \frac{1}{\pi r \gamma \ell_{\mathrm{P}}^2} |p(v)|^{1-r/2} \left(|p(v) + 2\pi\gamma\ell_{\mathrm{P}}^2|^{r/2} - |p(v) - 2\pi\gamma\ell_{\mathrm{P}}^2|^{r/2} \right).$$
(10.32)

Using $p(v) = \ell_0^2 \tilde{p}(v) = \ell_0^2 a^2$, an expression as in (3.61) results, but with δ in
(3.61) fixed. Explicitly using an underlying discrete state has provided a normalization of inverse-triad corrections, as anticipated in the lattice-refinement picture.

A parameter with a consequence similar to δ of the isotropic setting, but in discretized form, arises by considering a larger set of quantization ambiguities. Some
quantization ambiguities have already been included by using the same classical
expression of inverse-triad components in terms of $\hat{B}_{v,K}^{(r)}$ with different values of r.
In addition to ambiguities in the exponent r one could use different representations for
holonomies before taking the trace, rather than only the fundamental representation
understood in (10.14) and (10.19); see also [7]. In this case, we have

$$\hat{B}_{v,K}^{(r,j)} = \frac{3}{irj(j+1)(2j+1)} \left(2\pi\gamma\ell_{\mathrm{P}}^2 \right)^{-1} \mathrm{tr}_j \left(\tau^K h_{v,K} \hat{V}_v^r h_{v,K}^{-1} \right)$$
(10.33)

labeled by the two ambiguity parameters, r for the exponent of volume in commutators and j for the irreducible representation used for holonomies and traces.

Arguments against the use of $j \neq 1/2$ in Hamiltonian operators have been put forward based
on the presence of classically unexpected solutions for constraints modified by holonomy
corrections with higher spins [8, 9]. However, these arguments can be circumvented by
suitable sums of higher-spin contributions so as to remove unwanted solutions. Moreover,
quantum theories commonly produce solutions not expected classically, and new solutions
from higher spins usually have large non-classical values of curvature.

Eigenvalues of such operators can be expressed as

$$B_{v,K}^{(r,j)} = \frac{3}{rj(j+1)(2j+1)} \left(2\pi\gamma\ell_{\mathrm{P}}^2 \right)^{3r/2-1}$$
$$\times |\mu_{v,I} + \mu_{v,-I}|^{r/2} |\mu_{v,J} + \mu_{v,-J}|^{r/2} \sum_{m=-j}^{j} m|\mu_{v,K} + \mu_{v,-K} + 2m|^{r/2}$$
(10.34)

leading to the general class of correction functions

$$\alpha_{v,K}^{(r,j)} = \frac{3}{rj(j+1)(2j+1)} |\mu_{v,K} + \mu_{v,-K}|^{1-r/2} \sum_{m=-j}^{j} m \left| \mu_{v,K} + \mu_{v,-K} + 2m \right|^{r/2}.$$
(10.35)

After imposing isotropy the last expression becomes

$$\alpha^{(r,j)} = \frac{6}{rj(j+1)(2j+1)} |\mu|^{1-r/2} \sum_{m=-j}^{j} m |\mu + m|^{r/2}. \tag{10.36}$$

The computation of traces involving higher representations of SU(2) as they occur in inverse-triad operators can be simplified using the following ingredients [10]; see also [7]. First, for the normalization we note that

$$\text{tr}(\tau_i^{(j)} \tau_k^{(j)}) = -\frac{1}{3} j(j+1)(2j+1)\delta_{ik} \tag{10.37}$$

with generators $\tau_i^{(j)}$ of the j-representation. In inverse-triad operators such as (10.33), in which only a single generator is used as a factor and in holonomies, the matrix multiplications and the trace can be computed easily when the generator used is diagonalized, the diagonalization being possible thanks to the cyclic commutativity of the trace operation. Then, in the j-representation the diagonalized $h_{v,K}$ reads

$$h_{v,K}^{(j)} = \exp(c\tau_k^{(j)}) = \begin{pmatrix} e^{-ijc} & 0 & \cdots & 0 & 0 \\ 0 & e^{-i(j-1)c} & & & 0 \\ \vdots & & \ddots & & \vdots \\ 0 & & & e^{i(j-1)c} & 0 \\ 0 & 0 & \cdots & 0 & e^{ijc} \end{pmatrix} \tag{10.38}$$

or $(h_{v,K}^{(j)})_{\alpha\beta} = e^{i(\alpha-j)c}\delta_{(\alpha)\beta}$ for $0 \le \alpha, \beta \le 2j$. For instance,

$$(h_{v,K}^{(j)})_{\alpha\beta} \left[(h_{v,K}^{(j)})_{\beta\gamma}^{-1}, \hat{V} \right] = \delta_{(\alpha)\gamma}(\hat{V} - e^{i(\alpha-j)c}\hat{V}e^{-i(\alpha-j)c})$$

and traces reduce to summations such as the one seen in (10.34).

For large j, the sum in $\alpha^{(r,j)}$ can be approximated well as long as μ is not too close to j. (We present the resulting expressions for inverse-triad corrections as an illustration of their general form. By this we do not suggest that large j should necessarily be expected in the dynamical expressions of loop quantizations.) For $\mu > j$, absolute values can be omitted as all the expressions in the sum are positive. Approximating the summation by integration [4, 10] yields

$$\alpha^{(r,j)} = \frac{6\tilde{\mu}^{1-r/2}}{r} \left(\frac{1}{r+4} \left((\tilde{\mu} + 1)^{r/2+2} - (\tilde{\mu} - 1)^{r/2+2} \right) \right.$$
$$\left. - \frac{\tilde{\mu}}{r+2} \left((\tilde{\mu} + 1)^{r/2+1} - (\tilde{\mu} - 1)^{r/2+1} \right) \right) \quad \text{for } \tilde{\mu} > 1$$

where $\tilde{\mu} := \mu/j$. For $\mu < j$, the terms in the sum corresponding to $m < \mu$ and $m > \mu$, respectively, must be considered separately before the absolute value can be dropped. The end result, however, is very similar to the previous one,

$$\alpha^{(r,j)} = \frac{6\tilde{\mu}^{1-r/2}}{r} \left(\frac{1}{r+4} \left((\tilde{\mu} + 1)^{r/2+2} - (1 - \tilde{\mu})^{r/2+2} \right) \right.$$
$$\left. - \frac{\tilde{\mu}}{r+2} \left((\tilde{\mu} + 1)^{r/2+1} + (1 - \tilde{\mu})^{r/2+1} \right) \right) \quad \text{for } \tilde{\mu} < 1.$$

Fig. 10.3 Comparison
between the correction
function (10.35) and its
approximation (10.39). The
spikes are smeared out by the
approximation [4]

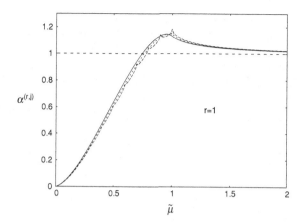

After some rearrangements, these two expressions can be combined into a single
one as

$$\alpha^{(r,j)} = \frac{6\tilde{\mu}^{1-r/2}}{r(r+2)(r+4)}\left((\tilde{\mu}+1)^{r/2+1}(r+2-2\tilde{\mu})\right.$$
$$\left.+\operatorname{sgn}(\tilde{\mu}-1)|\tilde{\mu}-1|^{r/2+1}(r+2+2\tilde{\mu})\right). \tag{10.39}$$

The approximation is compared to the exact expression of the correction function obtained through eigenvalues in Fig. 10.3. As one can see, spikes, which arise
from absolute values whenever μ passes an integer as long as it is less than j, are
smeared out by the approximation (except for the point $\tilde{\mu} = 1$ where the approximation remains non-differentiable at second order which is not visible from the
plot). The general trend, however, is reproduced well by the approximation also
to the left of the peak. For applications in effective equations, the approximation
might even be considered more realistic than the exact eigenvalue expression because
those equations would be based on semiclassical states. Since such states cannot be
eigenstates of the triad but must rather be peaked on a certain expectation value,
they will automatically give rise to a smearing-out of the spikes in the eigenvalues.
(The eigenvalue function will be convoluted with the profile of the semiclassical state
if we use a semiclassical state $\psi_{\bar{\mu}}(\mu) = \phi_{\mu-\bar{\mu}}$ peaked at some $\bar{\mu}$ with profile ϕ_{μ} to
write $_{\bar{\mu}}\langle\psi|\hat{B}|\psi\rangle_{\bar{\mu}} = \sum_{\mu}|\phi_{\mu-\bar{\mu}}|^2 B_{\mu} = \sum_{\mu}|\phi_{\mu-\bar{\mu}}|^2\alpha_{\mu}e_{\mu} = \alpha_{\bar{\mu}}e_{\bar{\mu}} +$ moments. In
the last steps, we have factored the inverse-triad operator into the correction function
and co-triad eigenvalues. Deviations from triad eigenstates can be captured either by
convolution with the state profile, or by adding moment terms as in general effective
equations; see also Sect. 10.2.)

This class of correction functions parameterized by two ambiguity parameters r
and j captures the most important general properties of such functions, including
the position of their maxima at $\tilde{\mu} \approx 1$ (or $\mu \approx j$) and the initial power-law increase
for small μ (determined by r). In its role for the peak, the spin label j is analogous

to the ambiguity parameter δ in isotropic models; see (3.61). The inhomogeneous treament has not only discretized the allowed values by relating them to spin labels (which in homogeneous treatments are mixed with the coordinate length of edges). It has also given indications that small values close to the fundamental representation of SU(2) may be preferred. Corrections thus arise for elementary lattice fluxes not too far from the Planck scale.

Asymptotically, all correction functions have the correct classical limit on large scales, such as

$$\alpha^{(r,j)}(\tilde{\mu}) \approx 1 + \frac{(r-2)(r-4)}{40}\tilde{\mu}^{-2} + O(\tilde{\mu}^{-4}) \to 1 \tag{10.40}$$

for (10.32). Moreover, for small μ the correction function goes to zero as

$$\alpha^{(r,j)}(\tilde{\mu}) \approx (2\tilde{\mu})^{2-r/2}, \tag{10.41}$$

which ensures boundedness of the quantized co-triad $e_{(K)} \propto \alpha\sqrt{\tilde{\mu}} \propto \tilde{\mu}^2$ (using $r = 1$ for this case as in (10.13)), when $\tilde{\mu} \to 0$. The same is true for higher j since the sum (10.36) of odd terms when evaluated at $\mu = 0$ gives zero.

So far, the holonomies we used only contributed the extrinsic-curvature terms H_K to the Hamiltonian but no spin connection terms of H_Γ at all. In the procedure followed here, we have to quantize $\Gamma^i_a[E]$ directly which is possible in the perturbative regime where line integrals of the spin connection have covariant meaning. This gives rise to one further correction function from the expression of the spin connection

$$\Gamma^i_I = -\varepsilon^{ijk}e^b_j\left(\partial_{[I}e^k_{b]} + \frac{1}{2}e^c_k e^l_a \partial_{[c}e^l_{b]}\right), \tag{10.42}$$

as it also contains a co-triad (10.13). We again make use of the fact that the triad and its inverse have a diagonal form $e^I_i = E^I_i/\sqrt{|\det E|} = e^{(I)}\delta^I_i$ and $e_I = e_{(I)}\delta^i_I$ with components given by $e^I = p^I/\sqrt{|\det E|} = (e_I)^{-1}$. The spin connection then simplifies to

$$\Gamma^i_I = \varepsilon^{ic}_I e^{(c)}\partial_c e_{(I)} \tag{10.43}$$

and in terms of components of the densitized triad reads

$$\Gamma^i_I = \frac{1}{2}\varepsilon^{ij}_I \frac{p^{(j)}}{p^{(I)}}\left(\sum_J \frac{\partial_j p^J}{p^J} - 2\frac{\partial_j p^I}{p^I}\right). \tag{10.44}$$

For a complete quantization of H_Γ we observe that we need terms of the form $\ell_0^2\Gamma^i_a\Gamma^j_a$ and $\ell_0^2\partial_a\Gamma^i_b$ in the constraint since one factor ℓ_0 of the Riemann measure will be absorbed in the commutator $\hat{B}_{v,I}$. To quantize $\ell_0\Gamma^i_a$, we combine ℓ_0 with the partial derivative ∂_I in (10.43) to approximate a lattice difference operator Δ_I defined by $(\Delta_I f)_v = f_{v+I} - f_v$ for any lattice function f. A well-defined lattice operator results once a prescription for quantizing the inverse triad has been chosen. Again one can make use of Poisson identities for the classical inverse which, however, allows more freedom than for the combination of triad components we saw in the Hamiltonian constraint. For any choice we obtain a well-defined operator, which would not be available without the perturbative treatment since the full spin connection is not a tensorial object.

An explicit example can most easily be derived by writing the spin connection integrated along a link $e_{v,I}$ as it might appear in a holonomy,

$$\int_{e_{v,I}} d\lambda \dot{e}_I^a \Gamma_a^i \approx \ell_0 \Gamma_I^i = \varepsilon_I^{ic} e^{(c)} \ell_0 \partial_c e_{(I)} \approx \varepsilon_I^{iK} \frac{p^{(K)}}{\sqrt{|\det E|}} \Delta_K e_{(I)}$$

using the lattice difference operator $\Delta_I \approx \ell_0 \partial_I$. We then have to deal with the inverse powers explicit in the fraction and implicit in the co-triad e_I. The latter is standard, replacing e_I by $\ell_0^{-1} h_I \{h_I^{-1}, V_v\}$ based on (10.13). The inverse determinant in the fraction cannot be absorbed in the resulting Poisson bracket because it does not commute with the derivative. Moreover, absorbing a single inverse in a single co-triad would lead to a logarithm of V_v in the Poisson bracket which does not allow a densely defined quantization. It can, however, be absorbed in the flux $\ell_0^2 p^K$ if we do not use the basic flux operator $\hat{F}_{v,K}$ but the classically equivalent expression

$$F_{v,K} \approx \ell_0^2 p^K = \frac{1}{2} \ell_0^2 \delta_{(K)}^k \varepsilon_{kij} \varepsilon^{KIJ} e_I^i e_J^j$$

$$= -\frac{1}{4} (4\pi \gamma G)^{-2} \sum_{IJ} \sum_{\sigma_I \in \{\pm 1\}} \sigma_I \sigma_J \varepsilon^{IJK} \text{tr}(\tau_{(K)} h_I \{h_I^{-1}, V_v\} h_J \{h_J^{-1}, V_v\}) \quad (10.45)$$

(which is analogous to expressions used in [11]). Since there are now two Poisson brackets, we can split the inverse V_v evenly among them, giving rise to square roots of V_v in the brackets:

$$\frac{p^K}{\sqrt{|\det E|}} \approx \ell_0 \frac{F_{v,K}}{V_v}$$

$$= -\frac{\ell_0}{16\pi^2 \gamma^2 G^2} \sum_{IJ} \sum_{\sigma_I \in \{\pm 1\}} \sigma_I \sigma_J \varepsilon^{IJK} \text{tr}\left(\tau_{(K)} h_I \left\{h_I^{-1}, \sqrt{V_v}\right\}\right.$$

$$\left. \times h_J \left\{h_J^{-1}, \sqrt{V_v}\right\}\right). \quad (10.46)$$

The remaining factor of ℓ_0 is absorbed in e_I inside the derivative which is quantized following the standard procedure. A well-defined quantization of spin-connection components follows, one that is not local in a vertex since the difference operator connects to the next vertex. Similarly, the derivative of the spin connection needed in the Hamiltonian constraint leads to further connections to next-to-next neighbors.

Explicitly, one can write an integrated spin connection operator quantizing $\Gamma_{v,I}^i := \int_{e_{v,I}} d\lambda \dot{e}_I^a \Gamma_a^i$ as

$$\hat{\Gamma}_{v,I}^i = \varepsilon_I^{iK} \left(\frac{1}{16\pi^2 \gamma^2 \ell_P^2} \sum_{J,L,\sigma_J,\sigma_L} \sigma_J \sigma_L \varepsilon^{JLK} \text{tr}\right.$$

$$\times \left(\tau_{(K)} h_J [h_J^{-1}, \hat{V}_v^{1/2}] h_L [h_L^{-1}, \hat{V}_v^{1/2}]\right)$$

$$\times \left. \Delta_K \left(\frac{i}{2\pi \gamma \ell_P^2} \text{tr}(\tau^{(I)} h_I [h_I^{-1}, \hat{V}_v])\right)\right). \quad (10.47)$$

Replacing the commutators by classical expressions times correction functions $\alpha^{(r)}$ (and $\alpha = \alpha^{(1)}$) defined as before leads to an expression

$$(\Gamma_I^i)_{\text{corr}} = \alpha^{(1/2)}(p^i)\alpha^{(1/2)}(p^I)\varepsilon_I{}^{ic}e^{(c)}\partial_c(\alpha(p^I)e_I)$$

$$= \alpha^{(1/2)}(p^i)\alpha^{(1/2)}(p^I)\left(\alpha(p^I)\Gamma_I^i + \sum_{K \neq I}\alpha'(p^I)p^K\Gamma_K^i\right)$$

for the corrected spin-connection components. For scalar modes, using that all p^I at a given point are equal, this can be written with a single correction function

$$\beta[p(v)] = \alpha^{(1/2)}[p(v)]^2(\alpha[p(v)] + 2p\alpha'[p(v)]) \qquad (10.48)$$

for Γ_I^i, where $\alpha' = d\alpha/dp$.

Having well-defined quantizations of the spin connection at one's disposal in perturbative settings, one could use them to implement the reality conditions of complex Ashtekar variables.

Matter Hamiltonian. Another important source of inverse-triad corrections is presented by matter Hamiltonians, which we demonstrate for a scalar field φ. The general quantization steps have been described in Sects. 3.2.2.3 and 4.2.2, which can directly be adapted to the lattice setting. We represent $\exp(i\nu_v\varphi(v))$ directly at vertices, and quantize the momentum via

$$P_v := \int_{R_v} d^3x\,\pi \approx \ell_0^3\pi(v)$$

where R_v is a cubic region around the vertex v of the size of a single lattice site.
The matter Hamiltonian

$$H_\varphi[N] = \int d^3x\,N(x)\left(\frac{1}{2\sqrt{\det h}}\pi(x)^2 + \frac{E_i^a E_i^b}{2\sqrt{\det h}}\partial_a\varphi(x)\partial_b\varphi(x) + \sqrt{\det h}\,W(\varphi)\right)$$

with a potential $W(\varphi)$ requires inverse-triad corrections at two places, based on two versions of the general identity

$$\{A_a^i, V_v^r\} = 4\pi\gamma G r V_v^{r-1}e_a^i \qquad (10.49)$$

so as to supply the correct inverse powers. Any such Poisson bracket will be quantized to

$$\dot{e}_K^a\{A_a^i, V_v^r\} \mapsto \frac{-2}{i\hbar\ell_0}\text{tr}(\tau^i h_{v,K}[h_{v,K}^{-1}, \hat{V}_v^r])$$

using holonomies $h_{v,I}$ in the direction I with tangent vector $\dot{e}_K^a$.
Since holonomies in our lattice states have internal directions τ_K for the direction K, we can compute the trace explicitly and obtain

$$\widehat{V_v^{r-1}e_K^i} = \frac{-2}{8\pi i r\gamma\ell_P^2\ell_0}\sum_{\sigma\in\{\pm1\}}\sigma\text{tr}(\tau^i h_{v,\sigma K}[h_{v,\sigma K}^{-1}, \hat{V}_v^r]) = \frac{1}{2\ell_0}(\hat{B}_{v,K}^{(r)} - \hat{B}_{v,-K}^{(r)})\delta_{(K)}^i$$

$$(10.50)$$

where, for symmetry, we use both edges touching the vertex v along the direction K and $\hat{B}_{v,K}^{(r)}$ is the generalized version of (10.23):

$$\hat{B}_{v,K}^{(r)} := \frac{1}{4\pi i \gamma G \hbar r} \left(s_{v,K} \hat{V}_v^r c_{v,K} - c_{v,K} \hat{V}_v^r s_{v,K} \right) \tag{10.51}$$

The exponent used for the gravitational part was $r = 1$, and $r = 1/2$ already occurred in the spin connection. The scalar Hamiltonian introduced in [12, 13], which we closely follow here, uses $r = 1/2$ for the kinetic term and $r = 3/4$ for the gradient term. With the relations in Sect. 4.2.2 one can replace the inverse powers in the scalar Hamiltonian as follows: For the kinetic term, we discretize

$$\int d^3x \frac{\pi^2}{\sqrt{\det h}} \approx \sum_v \ell_0^3 \frac{\pi(v)^2}{\sqrt{\det h(v)}} \approx \sum_v \frac{P_v^2}{V_v}.$$

Then, the classically singular

$$\frac{1}{V_v} = \left(\frac{\ell_0^3}{6} \varepsilon^{abc} \varepsilon_{ijk} \frac{e_a^i e_b^j e_c^k}{V_v^{3/2}} \right)^2 = \left(\frac{\ell_0^3}{6(2\pi\gamma G)^3} \varepsilon^{abc} \varepsilon_{ijk} \{A_a^i, V_v^{1/2}\}\{A_b^j, V_v^{1/2}\}\{A_c^k, V_v^{1/2}\} \right)^2 \tag{10.52}$$

is quantized to the square of

$$\frac{1}{48} \varepsilon^{ijk} \varepsilon_{ijk} (\hat{B}_{v,I}^{(1/2)} - \hat{B}_{v,-I}^{(1/2)}) \delta_{(I)}^i (\hat{B}_{v,J}^{(1/2)} - \hat{B}_{v,-J}^{(1/2)}) \delta_{(J)}^j (\hat{B}_{v,K}^{(1/2)} - \hat{B}_{v,-K}^{(1/2)}) \delta_{(K)}^k.$$

Similarly, we discretize the gradient term by

$$\int d^3x \frac{E_i^a E_i^b}{\sqrt{\det h}} \partial_a \varphi \partial_b \varphi \approx \sum_v \ell_0^3 \frac{E_i^a(v) E_i^b(v)}{\sqrt{\det h(v)}} (\partial_a \varphi)(v)(\partial_b \varphi)(v)$$

$$\approx \sum_v \frac{p^I(v) p^J(v)}{V_v} \Delta_I \varphi_v \Delta_J \varphi_v$$

where we replace spatial derivatives ∂_a by lattice differences Δ_I. Now, using

$$\delta_{(I)}^i \frac{p^I(v)}{V_v^{1/2}} = \ell_0^2 \frac{E_i^I(v)}{V_v^{1/2}} = \frac{\ell_0^2}{6} \frac{\varepsilon^{Ibc} \varepsilon_{ijk} e_b^j e_c^k \operatorname{sgn} \det(e_a^l)}{V_v^{1/2}}$$

$$= \frac{\ell_0^2}{6(3\pi\gamma G)^2} \operatorname{sgn} \det(e_a^l) \varepsilon^{Ibc} \varepsilon_{ijk} \{A_b^j, V_v^{3/4}\}\{A_c^k, V_v^{3/4}\}$$

we quantize the metric contributions to the gradient term by

$$\frac{1}{24^2} \varepsilon^{IKL} \varepsilon_{ijk} (\hat{B}_{v,K}^{(3/4)} - \hat{B}_{v,-K}^{(3/4)}) \delta_{(K)}^j (\hat{B}_{v,L}^{(3/4)} - \hat{B}_{v,-L}^{(3/4)}) \delta_{(L)}^k$$

$$\times \varepsilon^{JMN} \varepsilon_{imn} (\hat{B}_{v,M}^{(3/4)} - \hat{B}_{v,-M}^{(3/4)}) \delta_{(M)}^m (\hat{B}_{v,N}^{(3/4)} - \hat{B}_{v,-N}^{(3/4)}) \delta_{(N)}^n. \tag{10.53}$$

In addition to the fact that we are using different values for r in each term in the gravitational and matter parts, giving rise to different correction functions, the matter terms are less unique than the gravitational term and can be written with different parameters r. Instead of using $r = 1/2$ in the kinetic term, for instance, we could use the class of relations

$$\frac{1}{\sqrt{|\det E|}} = \frac{|\det e|^k}{|\det E|^{(k+1)/2}} = \left| \frac{1}{6} \varepsilon^{abc} \varepsilon_{ijk} (4\pi G\gamma)^3 \right. $$

$$\left. \times \{A_a^i, V^{(2k-1)/3k}\} \{A_b^j, V^{(2k-1)/3k}\} \{A_c^j, V^{(2k-1)/3k}\} \right|^k$$

for any positive integer k to write the inverse determinant through Poisson brackets not involving the inverse volume (see also the appendix of [14]). This ambiguity determines an integer family of quantizations with $r_k = (2k-1)/3k > \frac{1}{3}$. For $k = 2$ we obtain the previous expression, but other choices are possible. Moreover, using the same r in all terms arising in gravitational and matter Hamiltonians can only be done in highly contrived ways, if at all. There is thus no clearly distinguished value. On regular lattice states, all ingredients are composed to a Hamiltonian operator (4.8).

Implementing inverse-triad corrections, (10.32) can be used to write the corrected matter Hamiltonian on a conformally flat space $q_{ab} = |p(x)|\delta_{ab}$ as

$$H_\varphi = \int_\Sigma d^3x\, N(x) \left(\frac{\nu[p(x)]}{2|\tilde{p}(x)|^{3/2}} \pi(x)^2 + \frac{\sigma[p(x)])|\tilde{p}(x)|^{\frac{1}{2}} \delta^{ab}}{2} \right. $$

$$\left. \times \partial_a\varphi \partial_b\varphi + |\tilde{p}(x)|^{\frac{3}{2}} W(\varphi) \right), \tag{10.54}$$

where a comparison with (4.8) shows that we have correction functions

$$\nu[p(v)] = \alpha^{(1/2)}[p(v)]^6 \quad \text{and} \quad \sigma[p(v)] = \alpha^{(3/4)}[p(v)]^4. \tag{10.55}$$

10.2 Quantum Corrections in Effective Equations

In the preceding section, we have computed several examples for individual quantum-geometry corrections by separate constructions. In a fully quantum corrected Hamiltonian, they all arise at the same time, in combination with quantum back-reaction. The general effective Hamiltonian follows from a background-state expansion as in Chap. 5, or by the expansions of Chap. 13 using Poisson geometry. Correction terms are then automatically organized in powers of $\hbar$, possibly combined with powers of the lattice spacing ℓ_0 if a continuum approximation is performed as well.

From the explicit calculations of quantum-geometry corrections it may not be obvious to see how they enter effective equations obtained by expanding around a semiclassical state. Inverse-triad corrections, for instance, were computed via the eigenvalues of inverse-triad operators on triad eigenstates, which are not at all semiclassical. Nevertheless, as we will show in this section, such calculations capture the quantum-geometry contribution from inverse-triad corrections.

10.2.1 Holonomy Corrections

Holonomy corrections arise because of a classical modification of the constraints, inserting holonomies for connection components. As quantum-geometry corrections, they can simply be expanded at a classical level, without worrying about semiclassical states or physical Hilbert-space issues. Expansions of two different types are present in inhomogeneous situations, both organized by the lattice spacing ℓ_0: an expansion of exponentials in terms of higher orders of connection or extrinsic-curvature components, and a derivative expansion of the non-local integrations along curves involved in holonomies. In both cases, the specific routing of curves used for holonomies in the Hamiltonian is relevant, as well as their relation to curves in the graph of a discrete state acted on. State properties are thus important for details of the correction terms, but not so much their semiclassical nature.

More specifically, we can write the expansions for lattice states, with an edge starting at a vertex v, in the form

$$\sin\left(\frac{1}{2}\gamma \int_{e(v)} \tilde{k}_I d\lambda\right) \sim \sin\left(\frac{1}{2}\gamma \ell_0(\tilde{k}_I(v) + \ell_0 \partial_I \tilde{k}_I(v) + \cdots)\right)$$

$$\sim \sin\left(\frac{1}{2}\ell_0\tilde{c}\right) + \ell_0 \delta\tilde{c} \cos\left(\frac{1}{2}\ell_0\tilde{c}\right) + \cdots$$

The first expansion is the derivative expansion of the integrated field, the second one an expansion around the isotropic field value $\tilde{c}$, introducing inhomogeneities $\delta\tilde{c}$. The leading term of the expansion produces the isotropic constraint.

10.2.2 Inverse-Triad Corrections

Inverse-triad operators such as $\hat{B}_{v,K}$ in (10.23) arise from commutators of holonomies with positive powers of the volume operator, obtained by quantizing Poisson brackets used to rewrite inverse-triad components [12, 15]. In an inhomogeneous setting, contributions to inverse-triad corrections come from individual lattice sites, with elementary lattice operators as fluxes $\hat{F}$ determining the excitation level of links, and holonomies $\hat{h}$ changing the excitation levels. In the Abelian case, and dropping numerical factors, these operators satisfy the algebra $[\hat{F}, \hat{h}] = \ell_P^2 \hat{h}$. Inverse-triad operators then have the form

$$\hat{B} = \frac{1}{2qG\hbar}(\hat{h}^\dagger|\hat{F}|^q\hat{h} - \hat{h}|\hat{F}|^q\hat{h}^\dagger) = \frac{1}{iqG\hbar}(\hat{s}|\hat{F}|^q\hat{c} - \hat{c}|\hat{F}|^q\hat{s})$$

with some power q, and splitting the holonomy $\hat{h} = \hat{c} + i\hat{s}$ in self-adjoint and anti-self-adjoint contributions, as it occurs by writing the exponential as a combination of sine and cosine.

To leading order in $\hbar$, the Poisson bracket of the flux $|F|^q$ with a connection component is obtained, which is identical to the classical inverse. One might thus think that inverse-volume corrections in an effective Hamiltonian obtained by a background-state or moment expansion arise only in terms of higher moments, not by correcting expressions for expectation values. This expectation, however, is not fully correct, because it ignores an identity realized for holonomy operators: the basic operators $(\hat{F}, \hat{h}, \hat{h}^\dagger)$ are subject to the reality condition $\hat{h}\hat{h}^\dagger = 1$. If this identity is used, the order of some operator products in a background-state expansion

$$\hat{B} = B(\langle\hat{F}\rangle) + \sum_{a,b,c} B_{a,b,c}(\langle\hat{F}\rangle, \langle\hat{h}\rangle, \langle\hat{h}^\dagger\rangle)(\hat{F} - \langle\hat{F}\rangle)^a(\hat{h} - \langle\hat{h}\rangle)^b(\hat{h}^\dagger - \langle\hat{h}\rangle^\dagger)^c$$

with the classical inverse $B(\langle\hat{F}\rangle)$, or of some moments in a moment expansion, can be reduced compared to what it appears to be initially. In particular, the initial order $a+b+c$ of a term with exponents a, b, c above can be reduced to $a + |b - c|$. If this procedure is followed consistently, inverse-triad corrections do arise which take the form of correction functions depending on the expectation values of basic operators, in particular on $\langle\hat{F}\rangle$, but not on moments.

To see this explicitly, let us look at the products of operators $\hat{h}|\hat{F}|^q\hat{h}^\dagger$ and $\hat{h}^\dagger|\hat{F}|^q\hat{h}$ as they appear in inverse-triad operators. Using the basic commutators $[\hat{h}, \hat{F}] = -\ell_P^2\hat{h}$ and $[\hat{h}^\dagger, \hat{F}] = \ell_P^2\hat{h}^\dagger$, together with the reality condition $\hat{h}\hat{h}^\dagger = 1$, we reorder terms as $\hat{h}\hat{F} = (\hat{F} - \ell_P^2)\hat{h}$ and $\hat{h}^\dagger\hat{F} = (\hat{F} + \ell_P^2)\hat{h}^\dagger$ and rewrite

$$\hat{h}|\hat{F}|^q\hat{h}^\dagger = |\hat{F} - \ell_P^2|^q = |\langle\hat{F}\rangle - \ell_P^2|^q \sum_{k=0}^{\infty}\binom{r}{k}\frac{(\hat{F} - \langle\hat{F}\rangle)^k}{|\langle\hat{F}\rangle - \ell_P^2|^k} \tag{10.56}$$

$$\hat{h}^\dagger|\hat{F}|^q\hat{h} = |\hat{F} + \ell_P^2|^q = |\langle\hat{F}\rangle + \ell_P^2|^q \sum_{k=0}^{\infty}\binom{r}{k}\frac{(\hat{F} - \langle\hat{F}\rangle)^k}{|\langle\hat{F}\rangle + \ell_P^2|^k}. \tag{10.57}$$

In this way, cancelling holonomy operators as far as possible, we are reducing orders of operator products to the smallest possible values. Corrections from higher orders in the product expansion, or from higher moments of a state if an expectation value is taken, remain, but additional moment-independent corrections (the $\pm\ell_P^2$-terms for $k = 0$ in the expansion) have been transferred to the expectation value dependence of inverse-triad corrections.

The leading terms of background-state expansions are then combined to

$$\frac{1}{2qG\hbar}(\hat{h}^\dagger|\hat{F}|^q\hat{h} - \hat{h}|\hat{F}|^q\hat{h}^\dagger) = \frac{|\langle\hat{F}\rangle + \ell_P^2|^q - |\langle\hat{F}\rangle - \ell_P^2|^q}{2q\ell_P^2} + \cdots \tag{10.58}$$

whose leading term in an $\hbar$-expansion is $|\langle\hat{F}\rangle|^{q-1}$, an inverse of the flux eigenvalue if $q < 1$. If we include higher-order corrections in $\hbar$, a correction function as seen before for inverse-triad corrections results, even if no higher moments of the expansion are considered.

Unlike holonomy corrections, inverse-triad corrections have the form of an $\hbar$-expansion without modifying the classical expression at leading order. They incorporate quantum corrections resulting from quantum geometry, conceptually distinguished from quantum back-reaction as it results from the dynamics. Since all these corrections in semiclassical regimes are organized in an $\hbar$-expansion, one must compare the specific correction terms in order to see which ones might be more dominant. Corrections from quantum back-reaction are arranged in orders of moments of a state, in the present context obtained by taking an expectation value of background-state expansions (10.56). We are thus comparing terms of the order $\ell_P^2/\langle\hat{F}\rangle$ from the inverse-triad expansion (10.58) with relative moments in an expansion starting with $(\Delta F)^2/\langle\hat{F}\rangle^2 = (\hat{F} - \langle\hat{F}\rangle)^2/\langle\hat{F}\rangle^2$. Inverse-triad corrections are dominant if flux fluctuations are smaller than the Planck length squared. Since elementary flux values are quantized in multiples of the Planck length squared, this result is reasonable: we can notice the discreteness, and in particular inverse-triad corrections as a result of the discreteness, only if quantum fluctuations are not so large that they wash out the discreteness scale. For a semiclassical state we expect the relative fluctuations to be small, while flux expectation values in an elementary state are about Planckian. Inverse-triad corrections are thus larger than terms obtained from the moment expansion.

Quantum fluctuations of individual plaquette fluxes are less than Planckian. Fluctuations of larger regions, whose fluxes are sums of elementary fluxes, are correspondingly larger: For weakly correlated plaquette states, the macroscopic fluctuation squared is the sum of all microscopic fluctuations squared.

We conclude that the terms written in (10.58) contain the leading quantum corrections in their first few orders of $\hbar$. The result agrees with the formulas obtained earlier for inverse-triad eigenvalues. Correction functions as derived earlier can thus be used for any semiclassical state provided only that relative fluctuations are small, as realized in the context of semiclassical geometry. One need not assume that the geometry is described by triad eigenstates, which would not be realistic.

For these considerations, we should use only the first $\hbar$-terms in an expansion of (10.58), as higher orders would become comparable to the neglected moment terms. Inverse-triad corrections are thus reliable when they refer to the asymptotic deviations from the classical value obtained for flux values larger than the peak position in Fig. 10.2. Corrections for fluxes smaller than the peak position require an inclusion of at least some of the moment terms, which have not been fully derived yet.

In this regime, identifying $|\langle\hat{F}\rangle| = v^{2/3}$ with the elementary scale of lattice refinement, we expand correction functions as $\bar{\alpha} = 1 + \alpha_0(\ell_P/v^{1/3})^m$ where the coefficient α_0 and (to some degree) the integer exponent m depend on quantization ambiguities. Lattice refinement for some a-dependence of $v(a)$ then leads to $\bar{\alpha} = 1 + \alpha_0(a_*/a)^\sigma$ with a reference scale a_* and $\sigma = (2x + 1)m$: From the basic identities of lattice refinement, $\mathcal{N}v = \mathcal{V}a^3$ and $\mathcal{N} = \mathcal{N}_0 a^{-6x}$, we have $\ell_P/v^{1/3} = (\mathcal{N}/\mathcal{V})^{1/3}\ell_P/a = (\mathcal{N}_0/\mathcal{V})^{1/3}\ell_P/a^{2x+1}$. The exponent σ appearing in inverse-triad corrections that include lattice refinement may thus be close to zero, unlike the initial exponent m which, by an expansion of inverse-triad values, usually takes the value $m = 4$. (Inverse-triad expressions are even functions of flux components, so that corrections in terms of the inverse scale factor are at least of the power four.) The a-dependence of inverse-triad corrections is thus stronger if lattice refinement is ignored, which helps to rule out some parameterizations; see [16] which also provides further parameter estimates.

If moments can be ignored, correction functions for inverse-triad corrections take exactly the form derived from eigenvalues. In particular, they can be written as functions only of the flux expectation values, not of holonomies. In an effective Hamiltonian, correction functions then depend only on the densitized triad, not on the connection. As one can easily see, this restricted dependence on the phase-space variables leads to significant simplifications in the computation for instance of the constraint algebra, as seen below. Nevertheless, in general one should expect also a weak dependence on the connection if non-Abelian holonomies are used. In this case, the holonomy-flux algebra, which is essential for the derivation of inverse-triad corrections, changes. For instance, we have reality conditions more complicated than $\hat{h}\hat{h}^\dagger = 1$ as used above. Holonomies in the commutators for inverse-triad corrections no longer cancel completely, leaving a connection dependence [17]. Such calculations are more difficult to perform, but as we will see below, a connection dependence can be included at an effective level in the form of counterterms. Thus, such terms are not completely neglected. In any case, a U(1)-approximation or Abelianization is an important tool not just in loop quantum cosmology but also in the full theory (see for instance [18–20]). An analysis of inverse-triad corrections in cosmology, parameterized in a sufficiently general form, thus tests the full theory of loop quantum gravity.

10.2.3 Types of Corrections

To summarize, holonomy corrections can be computed by expanding lattice holonomies as polynomials in curvature components, in combination with a slowly-varying-field approximation that provides higher spatial derivatives. The form of inverse-triad corrections follows from a consideration of inverse-triad operators based on algebraic properties of the holonomy-flux algebra, not just on their eigenvalues. The resulting correction functions depend on the spatial metric and its derivatives via the elementary fluxes, whose values are typically near Planckian and can then give rise to sizable corrections. The set of quantum corrections is completed by quantum back-reaction terms, which are more complicated to compute in a field-theoretic setting and have not been treated reliably yet. They are nevertheless required from the perspective of higher-curvature terms, and can be expected to play an important role for anomaly-freedom and covariance.

Since there are several corrections of different types, one may ask whether any of them could be analyzed in separation, without requiring all corrections to be derived at once. This is in fact the case, especially for inverse-triad corrections. They are of a type which cannot be produced by the other corrections. For instance, holonomy corrections always produce higher-order terms in the connection, unlike inverse-triad corrections, and quantum back-reaction terms always include moments of a state, not just expectation values. If one is interested in anomaly-freedom which requires a test whether the corrected constraints can form a first-class algebra, one must see, among other things, whether anomalies obtained from one correction could be cancelled by terms from another correction. For anomalies that may result from inverse-triad corrections, this is not the case: the Poisson bracket of terms affected by holonomy corrections always produces a term with higher-order corrections (at least higher orders of the background connection if an expansion by inhomogeneities is used). Terms containing moments always have a Poisson bracket containing a moment, which follows from the Poisson structure of the quantum phase space discussed in detail in Chap. 13. Terms with inverse-triad corrections contain neither higher powers of the connection, nor moments. Inverse-triad corrections must thus provide an anomaly-free consistent deformation of the classical constraints on their own; otherwise the theory cannot be consistent no matter what the other corrections look like. And if inverse-triad corrections deform the classical hypersurface-deformation algebra non-trivially, the result shows properties of quantum space–time structure that cannot be eliminated by including other corrections. We will now turn to an analysis of these questions.

10.3 Anomaly-Freedom of Quantum Corrections

We have obtained candidate terms for quantum-geometry corrections from the lattice construction. Although they show us what types of corrections we should expect, they cannot be the final form because they assumed specific structures of states and used gauge fixings. Due to the gauge fixing we have lost access to the full behavior of gauge transformations, which makes it difficult, if not impossible, to address anomalies. In fact, if we were to derive cosmological perturbation equations from a linear expansion of the resulting equations of motion around FLRW space–times with the corrections obtained so far, we would get an inconsistent set lacking any non-trivial solutions.

10.3.1 Consistency Issues

To highlight the consistency issue, we take a look at the classical cosmological perturbation equations. In longitudinal gauge as used so far, we have

$$\partial_c \left(\dot{\psi} + \mathcal{H}_{\mathrm{conf}} \phi \right) = 4\pi G \dot{\bar{\varphi}} \partial_c \delta \varphi \tag{10.59}$$

$$\nabla^2 \psi - 3 \mathcal{H}_{\mathrm{conf}} \left(\dot{\psi} + \mathcal{H}_{\mathrm{conf}} \phi \right) = 4\pi G \left(\dot{\bar{\varphi}} \delta \dot{\varphi} - \dot{\bar{\varphi}}^2 \phi + a^2 V_{,\varphi}(\bar{\varphi}) \delta \varphi \right)$$

$$\ddot{\psi} + \mathscr{H}_{\text{conf}} \left(2\dot{\psi} + \dot{\phi}\right) + \left(2\dot{\mathscr{H}}_{\text{conf}} + \mathscr{H}_{\text{conf}}^2\right)\phi = 4\pi G \left(\dot{\bar{\varphi}}\dot{\delta\varphi} - a^2 V_{,\varphi}(\bar{\varphi})\delta\varphi\right) \quad (10.60)$$

$$\partial_a \partial^b (\phi - \psi) = 0 \qquad\qquad (10.61)$$

$$\dot{\delta\varphi} + 2\mathscr{H}_{\text{conf}}\dot{\delta\varphi} - \nabla^2 \delta\varphi + a^2 V_{,\varphi\varphi}(\bar{\varphi})\delta\varphi + 2a^2 V_{,\varphi}(\bar{\varphi})\phi - \dot{\bar{\varphi}}\left(\dot{\phi} + 3\dot{\psi}\right) = 0 \qquad\qquad (10.62)$$

for scalar metric perturbations $h_{ab} = a^2(1 - 2\psi)\delta_{ab}$ and a perturbed lapse function $N = a(1 + \phi)$ (in conformal time). As matter source we have assumed a scalar field $\varphi = \bar{\varphi} + \delta\varphi$, and $\mathscr{H} = \dot{a}/a$ is the conformal Hubble parameter, with a derivative by conformal time η.

This set of equations appears overdetermined: there are five differential equations for three unknown functions. But since the equations are not independent, which one can explicitly check by solving (10.61) by $\phi = \psi$ and deriving the Klein–Gordon equation (10.62) from the first three equations (also using the background equations), non-trivial solutions for perturbations exist. Quantization now introduces corrections by means of effective equations, which amend the classical ones by correction terms. For a consistent implementation, corrections must preserve the fact that the equations are not independent. Inconsistencies could, for instance, arise from anomalies in the constraint algebra, implying that the first two perturbation equations, which are free of second-order time derivatives and pose constraints on the initial values, would not be preserved by evolution. In particular, it is not consistent to use only the corrections for the background equations, which can rather easily be derived in isotropic models using the effective techniques discussed earlier. Since the scale factor and background scalar field appear in coefficients of the perturbation equations, they must satisfy specific equations of motion to be consistent. If quantum corrections are introduced in the background, they must also be introduced in the perturbation equations, and they are highly restricted by consistency.

The second important consistency issue is that of gauge invariance. When a constrained theory is quantized, the constraints receive quantum corrections. Corrections then enter evolution equations as just discussed, but they also affect gauge flows generated by the constraints. Thus, also expressions of gauge invariant variables must be corrected. This feature can be implemented correctly only if we do not fix the gauge in the first place before putting in corrections. In our derivation of candidate correction terms we did fix the gauge (to make explicit calculations possible by Abelianization), and so the derivations performed so far are incomplete. We will see in what follows how the condition of consistency implies what additional corrections must be present that have not been seen in gauge-fixed derivations of correction functions.

10.3.1.1 Metric Perturbations

Again for scalar modes but now in a general gauge, we have perturbations ϕ, ψ, E and B where $N = a(1+\phi)$ as before and $h_{ab} = a^2(1-2\psi)\delta_{ab}+\partial_a\partial_b E$, $N^a = \partial^a B$, as they appear in a line element

$$\mathrm{d}s^2 = a^2(\eta)\left(-(1+2\phi)\mathrm{d}\eta^2 + 2\partial_a B\mathrm{d}\eta\mathrm{d}x^a - ((1-2\psi)\delta_{ab} + 2\partial_a\partial_b E)\mathrm{d}x^a\mathrm{d}x^b\right). \tag{10.63}$$

Not all these fields are independent perturbations because they mix under changes of coordinates, which represent gauge transformations in the classical manifold picture. There are two independent gauge-invariant combinations, the Bardeen variables [21]

$$\Psi = \psi - \mathcal{H}_{\text{conf}}(B - \dot{E}) \tag{10.64}$$

$$\Phi = \phi + \left(B - \dot{E}\right)^{\bullet} + \mathcal{H}_{\text{conf}}(B - \dot{E}) \tag{10.65}$$

and a similar quantity for the gauge-invariant matter perturbation $\delta\varphi^{\text{GI}}$. Here, another consistency issue arises: if the constraints are not compatible with each other and with the evolution equations, gauge invariance will not be preserved in time. Gauge artefacts will then couple into the perturbation equations; evolution equations cannot be expressed solely in terms of gauge-invariant variables.

In quantum gravity, we do not necessarily expect a classical space–time picture with its coordinate transformations to hold. At this stage, a Hamiltonian formulation offers many advantages because the consistency issues can be addressed at a purely algebraic level: the constraint algebra must remain first class after including quantum corrections. Then, constraint equations are preserved in time, preventing over-determinedness, and gauge invariance of perturbations is preserved, preventing couplings to gauge artefacts.

To show the classical Hamiltonian derivation first [1, 22], we start with the Hamiltonian

$$H_{\text{grav}}[N] = \frac{1}{16\pi\gamma G}\int_{\Sigma} \mathrm{d}^3x\, N \left(\varepsilon_{ijk}F^i_{ab} - 2(1+\gamma^2)K^i_a K^j_b\right) \frac{E^{[a}_i E^{b]}_j}{\sqrt{|\det E|}}$$

and perturb it. For all fields X we write $X = \bar{X} + \delta X$ such that $\int \mathrm{d}^3x\delta X = 0$ is a pure perturbation and $\bar{X} = \mathcal{V}^{-1}\int \mathrm{d}^3xX$ is the average over our integration region of size $\mathcal{V}$, the same that we also use for the homogeneous background model. By these conditions, we split the fields in background quantities and perturbations without overcounting, keeping them clearly separate. Specifically, we have $E^a_i = \bar{p}\delta^a_i + \delta E^a_i$, $K^i_a = \bar{k}\delta^i_a + \delta K^i_a$, $N = \bar{N} + \delta N$ and $N^a = \delta N^a$ without a background shift when we perturb around an isotropic model. The perturbation of the densitized triad is related to the metric perturbations by

$$\delta E_i^a = -2a^2 \psi \delta_i^a + a^2 \left(\frac{1}{3} \delta_i^a \Delta - \partial^a \partial_i \right) E \qquad (10.66)$$

and δK_a^i is related to $\dot{\psi}$ at the level of equations of motion; for now we treat it as an independent phase-space variable.

With these preparations, we expand $(16\pi G)^{-1} \int d^3x \left(\bar{N} H^{(2)} + \delta N H^{(1)} \right)$ with

$$H^{(1)} = -4\mathcal{H}_{\text{conf}} a \delta_j^c \delta K_c^j - \frac{\mathcal{H}_{\text{conf}}^2}{a} \delta_c^j \delta E_j^c + \frac{2}{a} \partial_c \partial^j \delta E_j^c$$

and

$$H^{(2)} = a \delta K_c^j \delta K_d^k \delta_k^c \delta_j^d - a(\delta K_c^j \delta_j^c)^2 - \frac{2\mathcal{H}_{\text{conf}}}{a} \delta E_j^c \delta K_c^j$$

$$- \frac{\mathcal{H}_{\text{conf}}^2}{2a^3} \delta E_j^c \delta E_k^d \delta_c^k \delta_d^j + \frac{\mathcal{H}_{\text{conf}}^2}{4a^3} (\delta E_j^c \delta_c^j)^2 - \frac{\delta^{jk}}{2a^3} (\partial_c \delta E_j^c)(\partial_d \delta E_k^d).$$

Similarly, the diffeomorphism constraint $D_{\text{grav}}[N^a] = (8\pi G)^{-1} \int_\Sigma N^a F_{ab}^i E_i^b$ with $N^a = \delta N^a$ ($\bar{N}^a = 0$) can be expanded. This implies constraint equations $\delta H / \delta \phi = \mathcal{H}^{(1)} = 0$, which becomes $\nabla^2 \psi - 3\mathcal{H}(\dot{\psi} + \mathcal{H}_{\text{conf}} \phi)$, and (for a scalar M determining the shift perturbation via $\delta N^a = \partial^a M$) $\delta D / \delta M = 0$ which becomes $\partial_c (\dot{\psi} + \mathcal{H}_{\text{conf}} \phi)$. For $\bar{N} = a$, $\delta N = a\phi$ we generate equations of motion in conformal time such as $\delta \dot{E}_i^a = \{\delta E_i^a, H[a(1 + \phi)]\}$, using

$$\{\delta K_a^i(x), \delta E_j^b(y)\} = 8\pi G \delta_a^b \delta_j^i \delta(x, y).$$

(There is no shift perturbation for equations of motion of scalar modes.)

For gauge transformations, we consider lapse and shift linear in perturbations:

$$H[\bar{N}\xi^0] = -\frac{1}{16\pi G} \int d^3x \bar{N} \xi_0 \left(4\mathcal{H}_{\text{conf}} a \delta_c^j \delta K_c^j + \frac{\mathcal{H}_{\text{conf}}^2}{a} \delta_c^j \delta E_j^c - \frac{2}{a} \partial_c \partial^j \delta E_j^c \right)$$

$$D[\partial^a \xi] = \frac{1}{8\pi G} \int_\Sigma d^3x \partial^c \xi \left(a^2 \delta_k^d \partial_c \delta K_d^k - a^2 \partial_k \delta K_c^k - \mathcal{H}_{\text{conf}} \delta_c^k \partial_d \delta E_k^d \right).$$

The general gauge transformation is $\delta_{[\xi^0, \xi]} f = \{ f, H[\bar{N}\xi^0] + D[\partial^a \xi] \}$ for any phase-space function, corresponding to the Lie derivative along the vector field $(\xi^0, \partial^a \xi)$.

The factor of $\bar{N}$ in the Hamiltonian constraint used as a generator for gauge transformations is explained by the relationship between the space–time vector field written in different bases. In a coordinate system, the components ξ^μ refer to the basis t^a as the time-evolution vector field and s_i^a as three independent spatial vectors: $\xi^a = \xi^0 t^a + \xi^i s_i^a$. The gauge parameters used in the canonical constraints, on the other hand, refer to the vector field $\varepsilon^a = \varepsilon^0 n^a + \varepsilon^i s_i^a$ in a basis adapted to the spatial foliation, with unit timelike normal n^a. The relationship is determined by lapse and shift: $t^a = N n^a + N^a$, such that for a vanishing background shift $\bar{N}^a = 0$, $\varepsilon^a = \varepsilon^0 N^{-1} t^a + \varepsilon^i s_i^a = \xi^0 t^a + \xi^i s_i^a$, or $\varepsilon^0 = N\xi^0$ as the gauge parameter. See also [1, 23].

To derive gauge transformations of our perturbation fields, we compare the general triad perturbation (10.66) with its gauge transformation

$$\delta_{[\xi^0,\xi]}\delta E_i^a = 2\mathcal{H}_{\text{conf}}a^2\xi^0\delta_i^a + a^2\left(\frac{1}{3}\delta_i^a\nabla^2\xi - \partial^a\partial_i\xi\right).$$

Thus, $\delta_{[\xi^0,\xi]}\psi = -\mathcal{H}_{\text{conf}}\xi^0$ and $\delta_{[\xi^0,\xi]}E = \xi$. Similarly, from

$$\delta K_a^i = -\delta_a^i\left(\dot{\psi} + \mathcal{H}_{\text{conf}}(\psi + \phi)\right) + \partial_a\partial^i\left(\mathcal{H}_{\text{conf}}E - (B - \dot{E})\right)$$

(from the equation of motion for δE_i^a) and

$$\delta_{[\xi^0,\xi]}\delta K_a^i = \partial^i\partial_a(\xi^0 + \mathcal{H}_{\text{conf}}\xi) - \frac{\mathcal{H}_{\text{conf}}^2}{2}\xi^0\delta_a^i$$

we have $\delta_{[\xi^0,\xi]}\psi = -\mathcal{H}_{\text{conf}}\xi^0$, $\delta_{[\xi^0,\xi]}E = \xi$, $\delta_{[\xi^0,\xi]}(B - \dot{E}) = -\xi^0$, $\delta_{[\xi^0,\xi]}\phi = \dot{\xi}^0$ $+ \mathcal{H}_{\text{conf}}\xi^0$. In $\Psi := \psi - \mathcal{H}_{\text{conf}}(B - \dot{E})$, all gauge transformations cancel out, showing its gauge invariance.

Equations of motion can then be expressed fully in terms of gauge-invariant quantities with all gauge-dependent terms cancelling:

$$\nabla^2\Psi - 3\mathcal{H}_{\text{conf}}(\mathcal{H}_{\text{conf}}\Phi + \dot{\Psi}) = -4\pi Ga^2\delta T_0^{0(\text{GI})} \tag{10.67}$$

$$\partial_a\left(\mathcal{H}_{\text{conf}}\Phi + \dot{\Psi}\right) = -4\pi Ga^2\delta T_a^{0(\text{GI})} \tag{10.68}$$

$$\left(\ddot{\Psi} + \mathcal{H}_{\text{conf}}(2\dot{\Psi} + \dot{\Phi}) + (2\dot{\mathcal{H}}_{\text{conf}} + \mathcal{H}_{\text{conf}}^2)\Phi + \frac{1}{2}\nabla^2(\Phi - \Psi)\right)\delta_a^b$$
$$-\frac{1}{2}\partial^b\partial_a(\Phi - \Psi) = 4\pi Ga^2\delta T_a^{b(\text{GI})} \tag{10.69}$$

with stress-energy components in terms of gauge-invariant matter fields.

10.3.1.2 Evolution versus Gauge

Gauge transformations as well as equations of motion are generated by the constraints. One may then wonder why gauge-invariant combinations such as Ψ and Φ are subject to evolution at all, if evolution plays the same role as a gauge transformation. First, from the derivations it is clear that gauge and evolution are, in the perturbative setting, generated by different combinations of the constraints: $\delta_{[\xi^0,\xi]} = \{\cdot, H[\bar{N}\xi^0] + D[\partial^a\xi]\}$ and $\bullet = \{\cdot, H[\bar{N}(1 + \phi)]\}$. For linearized equations, $\delta_{[\xi^0,\xi]}$ is sensitive only to $H^{(1)}$ since the gauge generators are already of second order, while evolution, in a gauge determined by the background lapse, is sensitive to $H^{(2)}$. These are different terms in the expansion of the constraint, and so, at least formally, gauge and evolution are clearly kept separate.

In fact, by splitting the variables into background quantities and perturbations we have been able to isolate one global gauge degree of freedom as time, fixing the background gauge by choosing a specific background lapse. Gauge invariance is then only imposed for perturbations, not for the background values. Gauge-invariant perturbations are still subject to the background evolution.

10.3.2 Quantum Corrections

In Sect. 10.1.2.2, we have not derived a complete quantum-corrected Hamiltonian, but rather gauge-fixed derivations have shown that corrections of certain types must arise. It is then a highly non-trivial test of the whole consistency to see whether consistent equations can result. If this is the case, one might also obtain a covariant space–time picture for the corrections computed in a canonical derivation.

Explicit corrections are available in particular for inverse-triad effects. These terms were derived for special simple states, and with a gauge fixing assumed. In this form, corrections are incomplete and likely to be inconsistent from the point of view of anomaly-freedom. But both issues, the incompleteness of corrections and consistency, can be combined to get a much wider picture: we start with corrections of the form seen so far and test what extra terms they require for a consistent formulation. If a consistent formulation can be found in this way, it suggests terms that one must be able to see emerge from a more general derivation of quantum corrections. The full theory is thus being tested, going beyond the simple states used initially. On the other hand, if no consistent formulation incorporating the corrections seen exists, they are ruled out. Also in this case feedback for the full theory can be obtained since one would somehow have to explain why a crucial quantum-geometry effect, if it cannot be realized consistently, should disappear without a trace in effective equations. If there is no good reason why an inconsistent correction should cancel out, the theory suggesting this correction would itself be ruled out.

One example has been considered in quite some detail: consistent formulations of inverse-triad corrections for linear inhomogeneities on a spatially flat FLRW background [24]. We have computed the primary correction function α in (10.32) for lattice states (making use of longitudinal gauge) from eigenvalues of inverse-triad operators on regular lattice states, as they have been seen to come from a background-state expansion for effective equations. A characteristic functional behavior arose, but depending only on the fluxes, corresponding to the fact that inverse-triad operators in the (Abelianized) lattice setting have the same eigenstates as flux operators. In general, we should expect relevant semiclassical states not to be based on regular graphs, and certainly not to be triad eigenstates. Extra dependences on the (non-Abelian) connection or extrinsic curvature may then arise, which is more difficult to compute from expectation values at the state level. But the possible form of such terms, required for consistency, can be seen from considerations of anomaly-freedom.

For inverse-triad corrections we make the ansatz

$$H_{\text{grav}}^{(\alpha)} := \frac{1}{16\pi G} \int d^3x \left(\bar{N} \left(\bar{\alpha} H^{(0)} + \alpha^{(2)} H^{(0)} + \bar{\alpha} H^{(\alpha)(2)} \right) + \delta N \bar{\alpha} H^{(\alpha)(1)} \right)$$

for a corrected Hamiltonian constraint expanded to second order in inhomogeneities, with $H^{(0)} = -6\mathscr{H}_{\text{conf}}^2 a$ and

$$H^{(\alpha)(1)} = -4(1+f)\mathscr{H}_{\text{conf}} a \delta_j^c \delta K_c^j - (1+g)\frac{\mathscr{H}_{\text{conf}}^2}{a}\delta_c^j \delta E_j^c + \frac{2}{a}\partial_c\partial^j \delta E_j^c$$

$$H^{(\alpha)(2)} = a\delta K_c^j \delta K_d^k \delta_k^c \delta_j^d - a(\delta K_c^j \delta_j^c)^2 - \frac{2\mathscr{H}_{\text{conf}}}{a}\delta E_j^c \delta K_c^j$$

$$-\frac{\mathscr{H}_{\text{conf}}^2}{2a^3}\delta E_j^c \delta E_k^d \delta_c^k \delta_d^j + \frac{\mathscr{H}_{\text{conf}}^2}{4a^3}(\delta E_j^c \delta_c^j)^2 - (1+h)\frac{\delta^{jk}}{2a^3}(\partial_c\delta E_j^c)(\partial_d\delta E_k^d).$$

The primary correction function is $\bar{\alpha}$ (the function α evaluated in the background variables), as derived before, and $\alpha^{(2)}$ the term it provides when expanded to second order in inhomogeneities. In the additional terms, we have inserted counterterm coefficients f, g and h which have not been derived from the Abelianized quantum Hamiltonian but will be required for a consistent algebra. (Not all terms in the constraints need be amended by counterterm coefficients for consistency.) The counterterm f multiplies a δK-dependent term; it can thus be interpreted as representing a connection-dependence of inverse-triad corrections; see also Sect. 10.2.2.

Going through a lengthy analysis of the constraint algebra [24], one finds that it is first-class to second order in inhomogeneities if $2f + g = 0$ and

$$-h - f + \frac{a}{\bar{\alpha}}\frac{\partial\bar{\alpha}}{\partial a} = 0$$

$$f - g - 2a\frac{\partial f}{\partial a} - \frac{a}{\bar{\alpha}}\frac{\partial\bar{\alpha}}{\partial a} = 0$$

$$-f + g - a\frac{\partial g}{\partial a} + \frac{a}{\bar{\alpha}}\frac{\partial\bar{\alpha}}{\partial a} = 0$$

$$\frac{1}{6}\frac{\partial\bar{\alpha}}{\partial a}\frac{\delta E_j^c}{a^3} + \frac{\partial\alpha^{(2)}}{\partial(\delta E_i^a)}(\delta_j^a\delta_i^c - \delta_j^c\delta_i^a) = 0.$$

Moreover, there is a condition $\bar{\nu}\bar{\sigma} = \bar{\alpha}^2$ (in the case of a scalar Hamiltonian (10.54)) for matter correction functions in terms of $\bar{\alpha}^2$ if matter is present. The matter Hamiltonian, if present, contains counterterms as well. We will later refer to the matter counterterms f_1 and f_3 which appear in the scalar-field Hamiltonian constraint with a lapse perturbation, $H_{\text{scalar}}^{(\alpha)}[\delta N] = \int d^3x \delta N(\bar{\nu}\mathscr{H}_{\text{kin}}^{(\alpha)(1)} + \mathscr{H}_{\text{pot}}^{(\alpha)(1)})$ with a primary inverse-triad correction function $\bar{\nu}$ analogous to $\bar{\alpha}$, and

$$\mathscr{H}_{\text{kin}}^{(\alpha)(1)} = (1+f_1)\frac{\bar{\pi}\delta\pi}{a^3} - (1+f_2)\frac{\bar{\pi}^2}{2a^3}\frac{\delta_a^i\delta E_i^a}{2a^2} \tag{10.70}$$

Fig. 10.4 Illustration of the hypersurface-deformation algebra of two Hamiltonian constraints

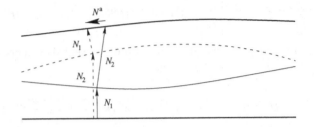

$$\mathcal{H}_{\text{pot}}^{(\alpha)(1)} = a^3 \left((1 + f_3) W_{,\varphi}\,(\bar{\varphi})\delta\varphi + W(\bar{\varphi})\frac{\delta_a^i \delta E_i^a}{2a^2} \right) \qquad (10.71)$$

For the matter counterterms we have additional relations with $\bar{\alpha}$ and $\bar{v}$ by consistency, such as

$$f_1 = f - \frac{a}{6\bar{v}}\frac{\partial \bar{v}}{\partial a}, \quad f_2 = 2f_1, \quad a\frac{\partial f_3}{\partial a} + 3f_3 - 3f = 0 \qquad (10.72)$$

All coefficients are fixed in terms of $\bar{\alpha}$ and $\bar{v}$, which can be derived in isotropic models or in longitudinal gauge as before.

There is thus an anomaly-free system of constraints including quantum corrections, or in other words: inverse-triad corrections constitute a consistent deformation of the classical theory. The conditions for the counterterms do allow non-trivial solutions; moreover, the amount of quantization ambiguities is reduced by relating matter correction functions to $\bar{\alpha}$. The underlying discreteness, which via the flux spectrum is responsible for the presence of inverse-triad corrections, does not destroy general covariance. (For holonomy corrections, which are also a consequence of discreteness, this consistency has not yet been verified for cosmological perturbations. Consistent examples for these corrections exist in models of spherical symmetry [25] and for $2 + 1$ gravity [26].)

The hypersurface-deformation algebra obtained from the corrected constraints is first class:

$$\{H^{(\alpha)}[N_1], H^{(\alpha)}[N_2]\} = D\left[\bar{\alpha}^2 \bar{N} a^{-1/2}\partial^a(\delta N_2 - \delta N_1)\right]. \qquad (10.73)$$

But it is not identical to the classical one: the correction function $\bar{\alpha}$ appears in its structure functions. This means that the classical space–time picture does not apply, because any foliated space–time must lead to the same classical algebra which follows simply from geometry as illustrated in Fig. 10.4. Even higher-curvature terms do not change the algebra [27].

10.3.3 Scalar Modes

Cosmologically, anomaly-freedom provides consistent cosmological perturbation equations: the Hamiltonian constraint, the diffeomorphism constraint and evolution equations. In addition, quantum corrections to constraints do change gauge invariant variables, and only those quantities, but no gauge artefacts, appear in quantum-corrected perturbation equations [22, 28]

$$\partial_c \left(\dot{\Psi} + \mathscr{H}_{\mathrm{conf}}(1+f)\Phi \right) = \pi G \frac{\bar{\alpha}}{\bar{v}} \dot{\bar{\varphi}} \partial_c \delta\varphi^{\mathrm{GI}} \tag{10.74}$$

$$\Delta(\bar{\alpha}^2 \Psi) - 3\mathscr{H}_{\mathrm{conf}}(1+f)\left(\dot{\Psi} + \mathscr{H}_{\mathrm{conf}}\Phi(1+f) \right) \tag{10.75}$$

$$= 4\pi G \frac{\bar{\alpha}}{\bar{v}}(1+f_3)\left(\dot{\bar{\varphi}}\delta\dot{\varphi}^{\mathrm{GI}} - \dot{\bar{\varphi}}^2(1+f_1)\Phi + \bar{v}a^2 V_{,\varphi}(\bar{\varphi})\delta\varphi^{\mathrm{GI}} \right) \tag{10.76}$$

$$\ddot{\Psi} + \mathscr{H}_{\mathrm{conf}}\left(2\dot{\Psi}\left(1 - \frac{a}{2\bar{\alpha}}\frac{\mathrm{d}\bar{\alpha}}{\mathrm{d}a} \right) + \dot{\Phi}(1+f) \right) \tag{10.77}$$

$$+ \left(2\dot{\mathscr{H}}_{\mathrm{conf}} + \mathscr{H}_{\mathrm{conf}}^2 \left(1 + \frac{a}{2}\frac{\mathrm{d}f}{\mathrm{d}a} - \frac{a}{2\bar{\alpha}}\frac{\mathrm{d}\bar{\alpha}}{\mathrm{d}a} \right) \right)\Phi(1+f) \tag{10.78}$$

$$= 4\pi G \frac{\bar{\alpha}}{\bar{v}}\left(\dot{\bar{\varphi}}\delta\dot{\varphi}^{\mathrm{GI}} - a^2 \bar{v} V_{,\varphi}(\bar{\varphi})\delta\varphi^{\mathrm{GI}} \right) \tag{10.79}$$

with the corrected gauge-invariant variables

$$\Psi = \psi - \mathscr{H}_{\mathrm{conf}}(1+f)\frac{B - \dot{E}}{\bar{\alpha}^2} \tag{10.80}$$

$$\Phi = \phi + \left(\frac{B - \dot{E}}{\bar{\alpha}^2} \right)^{\boldsymbol{\cdot}} + \mathscr{H}_{\mathrm{conf}}\frac{B - \dot{E}}{\bar{\alpha}^2}. \tag{10.81}$$

Moreover, the classical relationship between the two scalar modes is corrected to $\Phi = \Psi(1+h)$.

In the presence of a scalar matter degree of freedom, a gravitational gauge-invariant scalar mode such as Ψ or Φ can be combined with the matter mode to form a gauge-invariant perturbation with an expression independent of E and B in any gauge-fixing. In this way, one defines the curvature perturbation

$$\mathscr{R} = \psi + \frac{\mathscr{H}_{\mathrm{conf}}}{\dot{\bar{\varphi}}}(1 + f - f_1)\delta\varphi \tag{10.82}$$

depending on a matter counterterm f_1.

Classically, this perturbation obeys an equation of motion that in an expanding universe has a constant mode and a decaying mode if spatial derivatives can be ignored, that is for long wave lengths larger than the Hubble radius [29, 30]. The existence of a conserved quantity can be seen based on general properties of Hamiltonian analysis: For scalar modes, there is a single local degree of freedom on the reduced phase space obtained by implementing all first-class constraints. There must thus be a second-order evolution equation for the degree of freedom, which can be chosen to be $\mathscr{R}$. By an appropriate redefinition of $\mathscr{R}$ to $y\mathscr{R}$ with some function y depending on the background geometry, one can eliminate the term in this equation lacking time derivatives. The resulting equation will have a constant solution for $y\mathscr{R}$, and there are reduced phase-space techniques which can be used to compute $y\mathscr{R}$ systematically [31]. A shorter method has been applied in [16] also in the case of inverse-triad corrections, and the result is that $\mathscr{R}$ itself is conserved on large scales irrespective of the corrections.

Thus, the curvature perturbation remains conserved on large scales. One can efficiently formulate its evolution in terms of a rescaled perturbation, the Mukhanov variable $u = z\mathscr{R}$ with a background function z which classically equals $z = a\dot{\bar{\varphi}}/\mathscr{H}_{\text{conf}}$ but is also quantum corrected. The new perturbation, unlike the original Φ and Ψ, satisfies an evolution equation decoupled from all others, the Mukhanov equation

$$-\ddot{u} + \left(s^2\Delta + \frac{\ddot{z}}{z}\right)u = 0 \tag{10.83}$$

with $s^2 = \alpha^2(1 - f_3)$ [16] (with a matter counterterm f_3). Despite first appearance, with corrections only in spatial derivatives but not in time derivatives of the Laplacian, this equation is covariant: it is obtained from a system of first-class constraints deforming the classical notion of space–time symmetries.

10.3.4 Vector Modes

Vector modes are subject to a smaller number of independent gauge transformations than scalar modes and are thus easier to make consistent. In conformal time, they come from metric components of the form

$$-N^2 = -a^2, \quad N_a = a^2 F_a, \quad h_{ab} = a^2(\delta_{ab} + \partial_a f_b + \partial_b f_a). \tag{10.84}$$

whose mode functions F_a and f_a are divergence-free: $\partial_a F^a = 0$ and $\partial_a f^a = 0$. In a canonical formulation, the function F_a determines the shift perturbation, while f_a enters the spatial metric, or the densitized triad

$$\delta E_i^a = -\bar{p}(c_1\partial^a f_i + c_2\partial_i f^a) \tag{10.85}$$

with constants satisfying $c_1 + c_2 = 1$. Observables only depend on the symmetrized triad, which is independent of the choice of c_1 and c_2. The equations in triad variables plus corrections have been worked out in [32].

10.3.4.1 Inverse-Triad Corrections

The canonical variables are subject to the diffeomorphism constraint

$$D_{\text{grav}}[N^a] = \frac{1}{8\pi G} \int_\Sigma d^3x \delta N^c \left(-\bar{p}(\partial_k \delta K_c^k) - \bar{k}\delta_c^k(\partial_d \delta E_k^d)\right) \quad (10.86)$$

and the Hamiltonian constraint

$$H_{\text{grav}}^{(\alpha)}[N] = \frac{1}{16\pi G} \int_\Sigma d^3x \bar{N}\alpha(\bar{p}, \delta E_i^a)\left(-6\bar{k}^2\sqrt{\bar{p}} - \frac{\bar{k}^2}{2\bar{p}^{\frac{3}{2}}}(\delta E_j^c \delta E_k^d \delta_c^k \delta_d^j)\right.$$
$$\left. + \sqrt{\bar{p}}(\delta K_c^j \delta K_d^k \delta_k^c \delta_j^d) - \frac{2\bar{k}}{\sqrt{\bar{p}}}(\delta E_j^c \delta K_c^j)\right) \quad (10.87)$$

for inverse-triad corrections. It turns out that all counterterms required for scalar modes vanish identically for vector modes. Going through the canonical procedure of deriving gauge-invariant variables and equation of motion, one finds that the combination $\sigma^a = F^a - \dot{f}^a$ is gauge invariant (without quantum corrections in terms of metric components) and must satisfy the corrected equation of motion

$$\frac{1}{\bar{\alpha}}\left(-\frac{1}{2}\frac{d}{d\eta}(\partial^i \sigma_a + \partial_a \sigma^i) - \bar{k}(\bar{\alpha} - \bar{\alpha}'\bar{p})(\partial^i \sigma_a + \partial_a \sigma^i)\right) = 8\pi G\bar{p}\delta T^{(v)ia} \quad (10.88)$$

with a vectorial matter perturbation $\delta T_{ab}^{(v)}$.

10.3.4.2 Holonomy Corrections

For vector modes, with their simpler gauge structure, it is possible to find consistent deformations resulting from holonomy corrections. Since no such consistent deformation is known for scalar modes, it is still unclear whether holonomy corrections can be consistent altogether. But the example of vector modes is instructive in that it shows possible properties of perturbations around background solutions arising from holonomy corrections, such as bouncing ones.

An example for a consistent holonomy-corrected Hamiltonian constraint, when evaluated for vector-mode perturbations, is

$$H_{\text{grav}}^{(\alpha)}[N] = \frac{1}{16\pi G} \int_\Sigma d^3x \bar{N}\left(-6\sqrt{\bar{p}}\left(\frac{\sin \delta\gamma\bar{k}}{\delta\gamma}\right)^2\right.$$
$$-\frac{1}{2\bar{p}^{\frac{3}{2}}}\left(\frac{\sin \delta\gamma\bar{k}}{\delta\gamma}\right)^2(\delta E_j^c \delta E_k^d \delta_c^k \delta_d^j)$$
$$\left. + \sqrt{\bar{p}}(\delta K_c^j \delta K_d^k \delta_k^c \delta_j^d) - \frac{2}{\sqrt{\bar{p}}}\left(\frac{\sin 2\delta\gamma\bar{k}}{2\delta\gamma}\right)(\delta E_j^c \delta K_c^j)\right). \quad (10.89)$$

Now,

$$\sigma^a = F^a - \dot{f}^a + \bar{k}\left(1 - \frac{\sin 2\delta\gamma\bar{k}}{2\delta\gamma\bar{k}}\right)f^a \qquad (10.90)$$

is gauge invariant and satisfies the equation of motion

$$-\frac{1}{2}\frac{d}{d\eta}(\partial^i\sigma_a + \partial_a\sigma^i) - \frac{1}{2}\bar{k}\left(1 + \frac{\sin 2\delta\gamma\bar{k}}{2\delta\gamma\bar{k}}\right)(\partial^i\sigma_a + \partial_a\sigma^i) = 8\pi G\bar{p}\delta T^{(v)ia}.$$
$$(10.91)$$

Using the corrected background equation

$$\dot{\bar{p}} = \bar{p}\frac{\sin 2\delta\gamma\bar{k}}{\delta\gamma} \qquad (10.92)$$

the vacuum equation of motion for vector modes can quite simply be written as

$$\frac{d\log\sigma_k^i}{d\log a} = -\left(1 + \frac{2\delta\gamma\bar{k}}{\sin 2\delta\gamma\bar{k}}\right). \qquad (10.93)$$

This last form of the equation is not suitable for perturbations through the bounce where a turns around. (Indeed, at the bounce the right-hand side of (10.93) diverges.) But it does show that the rate of decay of vector modes in an expanding universe is modified by holonomy corrections.

10.3.5 Gravitational Waves

Tensor modes are easiest to discuss since they are not subject to gauge transformations or overdeterminedness. Still, from the propagation of gravitational waves compared to light one obtains interesting consistency tests [33]. For inverse-triad corrections, the Hamiltonian implies the linearized wave equation

$$\frac{1}{\bar{\alpha}}\ddot{h}_a^i + 2\frac{\dot{a}}{a}\left(1 - \frac{2ad\bar{\alpha}/da}{\alpha}\right)\dot{h}_a^i - \bar{\alpha}\nabla^2 h_a^i = 16\pi G\Pi_a^i$$

for the transverse-traceless part of metric perturbations and with a source-term Π_a^i.

One can rewrite this equation in simpler form and in Mukhanov style analogously to (10.83):

$$-\ddot{w} + \left(\bar{\alpha}^2\Delta + \frac{\ddot{a}}{a}\right)w = 0 \qquad (10.94)$$

where w is proportional to h [16, 34]. Also this equation is covariant despite appearance, and it is corrected differently than equation (10.83) for scalar modes because

$\bar{\alpha}$, not s, appears as the coefficient of spatial derivatives. Both terms differ by counterterm correction functions, which are now seen to be important for the relative evolution of scalar and tensor modes. As an important consequence of the different equations satisfied by scalar and tensor modes, the tensor-to-scalar ratio is corrected in a characteristic way; see [16] for the power spectra and indices.

From this, we have the dispersion relation $\omega^2 = \bar{\alpha}^2 k^2$ for gravitational waves. In the perturbative range of inverse-triad corrections, we have $\bar{\alpha} > 1$ (see Fig. 10.2) and thus face the threat of super-luminal motion combined with possible violations of causality. To test this issue, we must compare the gravitational dispersion relation with that of electromagnetic waves. The speed of gravitational waves should be compared to the physical speed of light, which receives corrections from the same effect of inverse triads in the Maxwell Hamiltonian. Here, we have

$$H_{\text{EM}} = \int_\Sigma d^3 x \left(\alpha_{\text{EM}}(h_{cd}) \frac{2\pi}{\sqrt{\det h}} E^a E^b h_{ab} + \beta_{\text{EM}}(h_{cd}) \frac{\sqrt{\det h}}{16\pi} F_{ab} F_{cd} h^{ac} h^{bd} \right)$$

(see also [35]) which on an FLRW background implies the wave equation

$$\partial_t \left(\bar{\alpha}_{\text{EM}}^{-1} \partial_t A_a \right) - \bar{\beta}_{\text{EM}} \nabla^2 A_a = 0$$

with dispersion relation $\omega^2 = \bar{\alpha}_{\text{EM}} \bar{\beta}_{\text{EM}} k^2$. Also light appears "super-luminal", when compared to the classical speed of light. But the ratio of $\bar{\alpha}^2$ to $\bar{\alpha}_{\text{EM}} \bar{\beta}_{\text{EM}}$ will determine whether there is truly super-luminal motion.

Initially, there is no clear relationship between the correction factors, which are subject to quantization ambiguities and could lead to ratios less than or larger than one. But implementing anomaly-freedom for the corrected system of constraints requires, analogously to the condition in Sect. 9.1.6.1, that

$$\bar{\alpha}^2 = \bar{\alpha}_{\text{EM}} \bar{\beta}_{\text{EM}}.$$

Gravitational waves and light travel at the same speed. Anomaly-freedom ensures that there are no violations of causality.

10.3.6 Comparison with Gauge-Fixed Treatments

For scalar modes, holonomy corrections have been implemented only in gauge-fixed treatments of different forms. Different gauge choices exist, such as the longitudinal [36–38] or the uniform one [39], and quantum corrections can be implemented in different ways without having a chance to see restrictions from anomaly cancellation. The available equations are thus rather ambiguous and it is not clear whether they can make significant predictions. When applied to inverse-triad corrections, gauge-fixed equations would not reveal the effects seen here by the counterterms.

10.4 Cosmological Applications

Even though the very early universe goes through several phases of high density, anything we can currently probe is still orders of magnitude removed from the Planck scale. In such a situation, one has to look very closely if one wants to find potentially observable phenomena within quantum gravity or quantum cosmology. Loop quantum gravity, by the deformed space–time structures it leads to, provides new ingredients in addition to the usual corrections in powers of the density. These new corrections may help to detect quantum-gravity effects, but for now they mainly pose an important consistency check which loop quantum gravity still has to pass. Some coarse models have already been ruled out, others show a remarkable balance between being close to detectability and barely not being ruled out yet. In this section we describe the current (and of course incomplete) status of observational applications in loop quantum cosmology, preceded by a more general discussion of inhomogeneities in quantum cosmology.

10.4.1 Quantum-to-Classical Transition

If one quantizes the perturbed Hamiltonian constraint, one obtains a quantum equation correcting the Wheeler–DeWitt or loop equation by back-reaction terms, as well as an infinite number of constraints for the perturbative modes. The whole system poses a complicated consistency problem after quantization because it must remain first class. This problem has not been solved yet even in the Wheeler–DeWitt context, which has a smaller number of corrections compared to loop quantum cosmology. But if one ignores the constraints for the modes and analyzes the perturbed Wheeler–DeWitt equation, one gains access to back-reaction properties [40].

The infinitely many perturbation modes not only provide degrees of freedom to describe the formation of structure, many of them do not play any role for observations and thus can be interpreted as a large environment of unaccessed degrees of freedom. In this situation, one can develop decoherence scenarios in order to explain how the observed degrees of freedom turn from quantum fluctuations into classical perturbations by coupling to a large environment of small disturbances [41–44]. However, decoherence in cosmology has a different status than decoherence of matter systems because of interpretational issues. Accordingly, several aspects of the quantum-to-classical transition have been questioned [45], and replaced by modified or alternative scenarios which may result in new options for observations [46–48].

In general, however, quantum effects in cosmology originating from the behavior of gravity are expected to be of the tiny size $\ell_{\mathrm{P}}\mathcal{H}$ if primarily the curvature scale determines their value. This expectation is no longer correct if the space–time structure changes, which is where loop quantum cosmology comes in.

10.4.2 Big-Bang Nucleosynthesis

Before using perturbation equations for the development of structure, already the modified background equations as they arise in the simplest isotropic models of loop quantum cosmology have some effects on cosmology. Especially the phase of big-bang nucleosynthesis is sensitive to the relative expansion rates of relativistic fermions (protons and neutrons being interconverted by exchanging electrons and neutrinos) to that of radiation. If the dilution rates are subject to quantum corrections, production rates of light elements are affected.

From such an analysis one can infer the bound $\mathcal{N}/a^3\mathcal{V} < 3/\ell_{\mathrm{P}}^3$ on the density of discrete patches [49]. This result is based on a correction to the equation of state parameter $w = \frac{1}{3}(1 - \mathrm{d}\log\bar{\alpha}/\mathrm{d}\log a)$ for relativistic fermions and radiation, deviating from the classical conformally invariant nature $w = 1/3$. One could expect different dilution rates for fermions and radiation, which could produce larger effects and stronger bounds. But this does not arise from inverse-triad corrections, which correct the equations of state of fermions [50] and radiation [51], but do so in the same way.

In fact, the upper limit $\mathcal{N}/a^3\mathcal{V} < 3/\ell_{\mathrm{P}}^3$ obtained for the patch density is not unexpected if there should be at most one discrete patch per Planck cube. An observational analysis of big-bang nucleosynthesis is hard to make more constrained in order to produce a tighter bound, but it is encouraging that the bound obtained is not removed from the expectation by orders of magnitude. Perturbation equations for structure formation, on the other hand, can provide more specific restrictions of quantum-geometry corrections and still show room for improvements.

10.4.3 Super-Inflation

Inverse-triad corrections as well as holonomy corrections give rise to a phase of super-inflation, with an increasing Hubble rate $\dot{\mathcal{H}} > 0$ [52]. The phase is normally not long in terms of the scale factor, which may change only mildly. However, during super-inflation the number of e-foldings is to be determined by factoring in the change of the Hubble rate: $N = \log((a\mathcal{H})_{\mathrm{final}}/(a\mathcal{H})_{\mathrm{initial}})$, and this number can more easily be large [53].

Another inflationary effect occurs when loop quantum cosmology is combined with standard inflation. The usual conditions for the potential then do not change, but it is easier to provide suitable initial values for the inflaton, far up its potential. During the brief super-inflation phase of loop quantum cosmology, the inflaton acquires an anti-friction term which can drive it up the potential [52, 54, 55]. Also bounce models show this effect, and can set the conditions of inflation from a preceding collapse phase [56].

10.4.4 Scalar Modes

Several crucial and characteristic effects have been recognized for scalar modes which may provide clear observational signatures.

10.4.4.1 Effective Anisotropic Stress

Even in the absence of anisotropic stress in matter, $\Phi = \Psi(1 + h)$ corrects the classical relationship $\Phi = \Psi$. This can be interpreted as quantum space–time implying an effective anisotropic stress even in the absence of sources. As another consequence, scalar modes can have propagation speeds different from that of light.

10.4.4.2 Curvature Perturbation

The curvature perturbation is an important quantity for an analysis of structure formation. Its behavior is changed by inverse-triad corrections in loop quantum cosmology in two different ways: Its gauge-invariant value has a different expression (10.82) in terms of metric and matter perturbations than classically; and its equation of motion, the Mukhanov equation (10.83), is corrected. With these modifications, as well as the loss of the classical space–time structure, it is initially not clear whether the curvature perturbation remains conserved on large scales. (A modified constraint algebra implies a modified Bianchi identity and corrections to the conservation law of matter, from which conservation of power can be derived classically.) If it is not conserved, a growing mode would signal an opportunity for quantum-gravity corrections larger than expected [57].

Modifications to space–time structure also imply that no effective line element can be used to describe the expanding geometry. A priori, there is thus no contradiction between non-conservation of power on large scales and the fact that large-scale scalar metric perturbations ψ, whose spatial dependence can be ignored, simply rescale the scale factor in the line element and should not be physically dynamical. In fact, Sect. 9.1.6.4 contains independent arguments that inverse-triad corrections change the space–time structure so significantly that the form of flat FLRW line elements can no longer be taken for granted. An analysis of cosmological structure formation without the benefit of effective line elements is difficult to perform, even with the tantalizing prospect of large quantum-gravity corrections. The situation that turns out to be realized strikes a convenient balance: Thanks to unexpected cancellations the curvature perturbation is conserved [16], and yet interesting new effects do happen [58] because of the form in which the Mukhanov equation is corrected. For an analysis it is helpful that effective perturbed FLRW line elements can be used.

10.4.4.3 Large-Scale Modes

If power is conserved for modes outside the Hubble scale, one may expect that the complicated perturbation analysis within quantum gravity, dealing with the thorny anomaly problem, could be skipped. Those modes simply follow the background dynamics, which can be derived in quantum-corrected form by using isotropic models devoid of any anomaly problems.

However, this expectation is incorrect. First, the relationship between gauge-invariant perturbations and the original metric and matter perturbations is important. Moreover, for a complete analysis of the power spectrum one must consider the phases of Hubble exit and re-entry as well, in which modes no longer are super-Hubble and become more sensitive to the perturbation dynamics. Indeed, examples show the existence of quantum corrections in power spectra even in cases in which power is conserved on large scales [16]. Corrections depend sensitively on some of the counterterms that arise in the anomaly analysis, and which could not possibly be seen in a pure background treatment. Corrections to power spectra thus cannot be mere background effects.

10.4.5 Tensor Modes

Tensor modes have a characteristic blue-tilt, which is enhanced if $x > -1/2$ for the lattice-refinement parameter x. For $x = -1/2$, only small correction of the size $8\pi G\rho\ell_{\mathrm{P}}^2$ arise [59–62]. It turns out that the power is increased for large wave lengths, where strong quantum-geometry effects show up. Normally, quantum gravity is expected to make itself noticeable in the UV for high frequencies [63, 64]. The discrepancy is explained by the fact that the quantum-geometry corrections are primarily a propagation, not a production effect. Modes of large wave length spend the longest time outside the Hubble radius, where deviations from the classical behavior are most characteristic.

The power spectrum of tensor modes is especially interesting in combination with the scalar spectrum. As already mentioned, the respective Mukhanov equations are corrected differently, in a way that depends sensitively on counterterms. Stronger corrections to the tensor-to-scalar ratio than in other scenarios are thus expected.

10.4.6 Indications for Evolution Through a Bounce

Inverse-triad corrections in their anomaly-free form have been found so far only for regimes in which the corrections are small, $\alpha \sim 1$. The strong quantum regime in which α increases from zero and thereby provide inflation [52] is much more difficult to control in the presence of inhomogeneities. Thus, consistent deformations of the classical perturbation equations can so far be analyzed only in combination with standard inflation [36, 37, 53, 56, 65–67], but not in the possibly more interesting case in which they may provide alternative scenarios of structure formation and perhaps solve some of the fine-tuning problems of inflation.

The same comments apply to evolution of structure through a bounce, which would suggest several ingredients for an alternative scenario of structure formation [68]. The situation of bounce models in loop quantum cosmology is thus more incomplete than the situation of inflationary models in the same framework. Here we provide a brief overview.

10.4.6.1 Matching

Bounce cosmology can be analyzed to a large degree based on classical equations amended by matching conditions for modes at the bounce [69]. Quantum gravity is then important only, but crucially so, to tell how structure is transferred from the collapse phase to expansion. There are always two independent solutions for the evolution of the scalar mode, subject to a second-order equation in time, and one must know how the dominant mode of the collapse phase, generating structure, is matched with the two modes in the expanding phase. Classically, there is a singularity instead of the bounce, and so one cannot simply perturb a classical equation by quantum corrections. The gauge-invariant modes and their dynamics must be derived and analyzed, again requiring one to consider and solve the anomaly problem. If one tries to avoid a full treatment, for instance by resorting to gauge-fixing choices, one loses access to the full linear dynamics. While some results may still be approximately correct in some sense, even a small error in the matching conditions can have a significant effect on the structure that arises at later times after the modes are evolved. On the other hand, the matching can also provide enhancement effects of quantum-gravity corrections for the same reason: even if matching coefficients change by tiny amounts of the usual size expected for quantum gravity, their implications can have sizeable effects if they are responsible for mixing in a growing mode.

10.4.6.2 Gauge Fixing

Bounce models based on holonomy corrections of loop quantum cosmology have been extended to linear perturbations by employing gauge fixing [39]. While tensor modes are not subject to gauge transformations or strong consistency requirements from anomaly freedom, their quantum corrections are restricted if one demands that they follow from a consistent system of equations that also includes scalar modes. If tensor-mode equations are obtained without paying attention to the consistency of accompanying scalar-mode equations, their analysis must be considered as based on gauge-fixing as well. In this spirit, tensor modes have been analyzed with strong inverse-triad corrections (super-inflation) or holonomy corrections (bounce) in [56, 70–73].

Another form to fix the gauge is the hybrid approach [74]; see Sect. 9.2.3. Also here the gauge-fixing removes the anomaly problem and allows one to study the evolution of structure through a bounce, at least numerically. One analysis performed in this way had provided another cautionary note: Linear perturbations evolved through the bounce seem to be enhanced significantly just at the holonomy bounce, suggesting that perturbation theory in this regime may be unstable [75].

References

1. Bojowald, M.: Canonical Gravity and Applications: Cosmology. Black Holes, and Quantum Gravity. Cambridge University Press, Cambridge (2010)
2. Pinho, E.J.C., Pinto-Neto, N.: Phys. Rev. D **76**, 023506 (2007). arXiv:hep-th/0610192
3. Falciano, F.T., Pinto-Neto, N.: Phys. Rev. D **79**, 023507 (2009). arXiv:0810.3542
4. Bojowald, M., Hernández, H., Kagan, M., Skirzewski, A.: Phys. Rev. D **75**, 064022 (2007). gr-qc/0611112
5. Rovelli, C., Smolin, L.: Phys. Rev. D **52**(10), 5743 (1995). gr-qc/0611112
6. Bojowald, M., Skirzewski, A.: Int. J. Geom. Meth. Mod. Phys. In: Borowiec, A., Francaviglia, M. (ed.) Proceedings of "Current Mathematical Topics in Gravitation and Cosmology" (42nd Karpacz Winter School of Theoretical Physics), 4, 25 (2007). hep-th/0606232
7. Gaul, M., Rovelli, C.: Class. Quantum Grav. **18**, 1593 (2001). gr-qc/0011106
8. Vandersloot, K.: Phys. Rev. D **71**, 103506 (2005). gr-qc/0502082
9. Perez, A.: Phys. Rev. D **73**, 044007 (2006). gr-qc/0509118
10. Bojowald, M.: Class. Quantum Grav. **19**, 5113 (2002). gr-qc/0206053
11. Giesel, K., Thiemann, T.: Class. Quantum Grav. **23**, 5667 (2006). gr-qc/0507036
12. Thiemann, T.: Class. Quantum Grav. **15**, 1281 (1998). gr-qc/9705019
13. Sahlmann, H., Thiemann, T.: Class. Quantum Grav. **23**, 867 (2006). gr-qc/0207030
14. Bojowald, M., Lidsey, J.E., Mulryne, D.J., Singh, P., Tavakol, R.: Phys. Rev. D **70**, 043530 (2004). gr-qc/0403106
15. Thiemann, T.: Class. Quantum Grav. **15**, 839 (1998). gr-qc/9606089
16. Bojowald, M., Calcagni, G.: JCAP **1103**, 032 (2011). arXiv:1011.2779
17. Bojowald, M.: Class. Quantum Grav. **23**, 987 (2006). gr-qc/0508118
18. Brunnemann, J., Thiemann, T.: Class. Quantum Grav. **23**, 1395 (2006). gr-qc/0505032
19. Giesel, K., Thiemann, T.: Class. Quantum Grav. **24**, 2499 (2007). gr-qc/0607100
20. Giesel, K., Thiemann, T.: Class. Quantum Grav. **24**, 2565 (2007). gr-qc/0607101
21. Bardeen, J.M.: Phys. Rev. D **22**, 1882 (1980)
22. Bojowald, M., Hossain, G., Kagan, M., Shankaranarayanan, S.: Phys. Rev. D **79**, 043505 (2009). gr-qc/0607101
23. Pons, J.M., Salisbury, D.C., Shepley, L.C.: Phys. Rev. D **55**, 658 (1997). gr-qc/9612037
24. Bojowald, M., Hossain, G., Kagan, M., Shankaranarayanan, S.: Phys. Rev. D **78**, 063547 (2008). arXiv:0806.3929
25. Reyes, J.D.: Spherically symmetric loop quantum gravity: connections to 2-dimensional models and applications to gravitational collapse. Ph.D. thesis, The Pennsylvania State University (2009)
26. Perez, A., Pranzetti, D.: arXiv:1001.3292
27. Deruelle, N., Sasaki, M., Sendouda, Y., Yamauchi, D.: Prog. Theor. Phys. **123**, 169 (2009). arXiv:0908.0679
28. Bojowald, M., Hossain, G., Kagan, M., Shankaranarayanan, S.: Phys. Rev. D **82**, 109903 (2010)
29. Salopek, D.S., Bond, J.R.: Phys. Rev. D **42**, 3936 (1990)
30. Wands, D., Malik, K.A., Lyth, D.H., Liddle, A.R.: Phys. Rev. D **62**, 043527 (2000). astro-ph/0003278
31. Langlois, D.: Class. Quantum Grav. **11**, 389 (1994)
32. Bojowald, M., Hossain, G.: Class. Quantum Grav. **24**, 4801 (2007). arXiv:0709.0872
33. Bojowald, M., Hossain, G.: Phys. Rev. D **77**, 023508 (2008). arXiv:0709.2365

34. Calcagni, G., Hossain, G.: Adv. Sci. Lett. **2**, 184 (2009). arXiv:0810.4330
35. Kozameh, C.N., Parisi, M.F.: Class. Quantum Grav. **21**, 2617 (2004). gr-qc/0310014
36. Hossain, G.M.: Class. Quantum Grav. **22**, 2511 (2005). gr-qc/0411012
37. Hofmann, S., Winkler, O.: astro-ph/0411124
38. Wu, J.P., Ling, Y.: JCAP **1005**, 026 (2010). arXiv:1001.1227
39. Artymowski, M., Lalak, Z., Szulc, L.: JCAP **0901**, 004 (2009). arXiv:0807.0160
40. Halliwell, J.J., Hawking, S.W.: Phys. Rev. D **31**(8), 1777 (1985)
41. Polarski, D., Starobinsky, A.A.: Class. Quantum Grav. **13**, 377 (1996). gr-qc/9504030
42. Lesgourgues, J., Polarski, D., Starobinsky, A.A.: Nucl Phys B **497**, 479 (1997). gr-qc/9611019
43. Kiefer, C., Polarski, D., Starobinsky, A.A.: Int J Mod Phys D **7**, 455 (1998). gr-qc/9802003
44. Kiefer, C., Lohmar, I., Polarski, D., Starobinsky, A.A.: Class. Quantum Grav. **24**, 1699 (2007). astro-ph/0610700
45. Perez, A., Sahlmann, H., Sudarsky, D.: Class. Quantum Grav. **23**, 2317 (2006). gr-qc/0508100
46. Burgess, C.P., Holman, R., Hoover, D.: astro-ph/0601646
47. De Unánue, A., Sudarsky, D.: Phys. Rev. D **78**, 043510 (2008). arXiv:0801.4702
48. León, G., Sudarsky, D.: Class. Quantum Grav. **27**, 225017 (2010). arXiv:1003.5950
49. Bojowald, M., Das, R., Scherrer, R.: Phys. Rev. D **77**, 084003 (2008). arXiv:0710.5734
50. Bojowald, M., Das, R.: Phys. Rev. D **78**, 064009 (2008). arXiv:0710.5722
51. Bojowald, M., Das, R.: Phys. Rev. D **75**, 123521 (2007). arXiv:0710.5721
52. Bojowald, M.: Phys. Rev. Lett **89**, 261301 (2002). gr-qc/0206054
53. Copeland, E.J., Mulryne, D.J., Nunes, N.J., Shaeri, M.: Phys. Rev. D **77**, 023510 (2008). arXiv:0708.1261
54. Tsujikawa, S., Singh, P., Maartens, R.: Class. Quantum Grav. **21**, 5767 (2004). astro-ph/0311015
55. Lidsey, J.E., Mulryne, D.J., Nunes, N.J., Tavakol, R.: Phys. Rev. D **70**, 063521 (2004). gr-qc/0406042
56. Mielczarek, J.: Phys. Rev. D **81**, 063503 (2010). arXiv:0908.4329
57. Bojowald, M., Hernández, H., Kagan, M., Singh, P., Skirzewski, A.: Phys. Rev. Lett **98**, 031301 (2007). astro-ph/0611685
58. Bojowald, M., Calcagni, G., Tsujikawa, S.: arXiv:1101.5391
59. Barrau, A., Grain, J.: arXiv:0805.0356
60. Barrau, A., Grain, J.: Phys. Rev. Lett **102**, 081301 (2009). arXiv:0902.0145
61. Grain, J., Cailleteau, T., Barrau, A., Gorecki, A.: Phys. Rev. D **81**, 024040 (2010). arXiv:0910.2892
62. Mielczarek, J., Szydłowski, M.: Phys Lett B **657**, 20 (2007). arXiv:0705.4449
63. Allen, B.: Relativistic Gravitation and Gravitational Radiation, Marck, J.A., Lasota, J.P., pp. 373–417. Cambridge University Press, Cambridge (1997)
64. Afonso, C.M., Henriques, A.B., Moniz, P.V.: arXiv:1005.3666
65. Mulryne, D.J., Nunes, N.J.: Phys. Rev. D **74**, 083507 (2006). astro-ph/0607037
66. Calcagni, G., Cortês, M.V.: Class. Quantum Grav. **24**, 829 (2007). gr-qc/0607059
67. Shimano, M., Harada, T.: Phys. Rev. D **80**, 063538 (2009). arXiv:0909.0334
68. Novello, M., Bergliaffa, S.E.P.: Phys. Rep. **463**, 127 (2008)
69. Wands, D.: Adv. Sci. Lett. **2**, 194 (2009). arXiv:0809.4556
70. Mielczarek, J., Szydłowski, M.: arXiv:0710.2742
71. Mielczarek, J.: JCAP **0811**, 011 (2008). arXiv:0807.0712
72. Mielczarek, J., Cailleteau, T., Grain, J., Barrau, A.: Phys. Rev. D **81**, 104049 (2010). arXiv:1003.4660
73. Copeland, E.J., Mulryne, D.J., Nunes, N.J., Shaeri, M.: Phys. Rev. D **79**, 023508 (2009). arXiv:0810.0104
74. Martín-Benito, M., Garay, L.J., Mena Marugán, G.A.: Phys. Rev. D **78**, 083516 (2008). arXiv:0804.1098
75. Brizuela, D., Mena Marugán, G.A., Pawlowski, T.: Class. Quantum Grav. **27**, 052001 (2010). arXiv:0902.0697

Part IV
Mathematical Issues

Some issues of mathematical interest arise in studies of the diverse models and equations encountered in loop quantum cosmology. These regard difference equations and their properties of specific interest for quantum space-time structures, as well as general techniques of deriving effective descriptions of quantum systems. The latter have contact with Poisson geometry and non-commutative algebra. From a physical point of view, such techniques are important for semiclassical approximations and the derivation of physical Hilbert-space properties.

Chapter 11
Difference Equations

Difference equations of a specific type play a large role in loop quantum cosmology. All equations initially encountered are linear, as usual for a dynamical equation for wave functions. But their coefficients are in general non-constant and depend on the discrete variables, which implies several subtleties in the analysis and solution of such equations. Moreover, in models less symmetric than isotropic ones or with matter fields, partial difference equations, difference-differential equations or highly coupled "functional" differential equations in inhomogeneous contexts arise. Several examples for such classes of difference equations have been derived in the models seen in preceding chapters. In this chapter, some mathematical properties related to questions of physical interest are discussed. Finally, we will sketch how non-linear difference equations may arise as a result of inhomogeneity.

11.1 Singularities and Dynamical Initial Conditions

In a mathematical sense, the difference equations of loop quantum cosmology are singular: their coefficients of highest (or lowest) order may vanish for some values of the recurrence parameters. Transition matrices then become degenerate and non-invertible, which may upset the whole recurrence scheme and prevent one from computing a complete wave function starting with initial values. If the recurrence does indeed break down, the system would be considered singular also in a physical sense. But mathematically singular behavior does not always imply physically singular behavior; the difference between these concepts is exactly loop quantum cosmology's way to resolve classical singularities.

A transition matrix can most easily be formulated if a difference equation of some given order is, without loss of generality, reformulated as a first-order difference equation in vector form. A second-order difference equation $a(n)\psi_{n-1} + b(n)\psi_n + c(n)\psi_{n+1} = 0$, for instance, can be reformulated as a first-order equation

$$A_n \begin{pmatrix} \psi_{n-1} \\ \psi_n \end{pmatrix} := \begin{pmatrix} a(n) & 0 \\ 0 & 1 \end{pmatrix} \begin{pmatrix} \psi_{n-1} \\ \psi_n \end{pmatrix} = \begin{pmatrix} -b(n) & -c(n) \\ 1 & 0 \end{pmatrix} \begin{pmatrix} \psi_n \\ \psi_{n+1} \end{pmatrix} =: B_n \begin{pmatrix} \psi_n \\ \psi_{n+1} \end{pmatrix} \tag{11.1}$$

M. Bojowald, *Quantum Cosmology*, Lecture Notes in Physics 835,
DOI: 10.1007/978-1-4419-8276-6_11, © Springer Science+Business Media, LLC 2011

written here as backward evolution which is of interest when approaching a classical singularity (at $n = 0$, say) from large positive values of n. The transition matrix

$$T_m^n := A_{m+1}^{-1} B_{m+1} A_{m+2}^{-1} B_{m+2} \cdots A_{n-1}^{-1} B_{n-1} A_n^{-1} B_n \qquad (11.2)$$

then determines how the value $\psi_m = (T_m^n)_{11} \psi_n + (T_m^n)_{12} \psi_{n+1}$ is obtained from initial values that may be chosen at n and $n + 1$, provided all the matrices A_n involved are in fact invertible.

For the type of difference equation (4.15) considered in isotropic loop quantum cosmology, A_1 is not invertible and T_0^n does not exist for any $n \neq 0$. With A_1 having a non-trivial kernel, the recurrence does not completely determine the values of (ψ_0, ψ_1) which T_0^n was supposed to provide in terms of the initial data. This can be a problem if (ψ_0, ψ_1) is required for further evolution, which would have to give a unique value for ψ_{-1} by applying $\mathscr{B}_0$ to (ψ_0, ψ_1), and so on.

Example 11.1 The difference equation of isotropic loop quantum cosmology has the general form analyzed here, more specifically $a(n) = X(n-1)$ and $c(n) = X(n+1)$ with $X(0) = 0$. Moreover, $b(0) = 0$. We obtain the matrices

$$A_n = \begin{pmatrix} X(n-1) & 0 \\ 0 & 1 \end{pmatrix}, \quad B_n = \begin{pmatrix} -b(n) & -X(n+1) \\ 1 & 0 \end{pmatrix}.$$

The only degenerate case of A_n is for $n = 1$, in which case we have a kernel spanned by $(1, 0)$, corresponding to the freedom of choosing ψ_0 (the first component of (ψ_{n-1}, ψ_n) at $n = 1$ in (11.1)). In attempting to follow the recurrence further, we next apply B_0 to a vector which, with the undetermined ψ_0 (the first component of (ψ_n, ψ_{n+1}) at $n = 0$ in (11.1)), is known only up to multiples of $(1,0)$. While B_0 does not annihilate this vector, the recurrence is not sensitive to the action of B_0 on it. (The undetermined ψ_0 is simply carried through from the first component of the right-hand side of (11.1) to the second component of the left-hand side.) Also the next matrix, B_{-1}, which has a kernel spanned by $(0, 1)$, is insensitive to the undetermined ψ_0 and the recurrence progresses unhindered.

As seen in the example, the finite-dimensional formulation of the recurrence, with ψ_n appearing at different places of the vectors involved in (11.1), makes the discussion of the recurrence difficult to organize. An infinite-dimensional formulation has an advantage at this stage: We consider whole sequences $\Psi_i \in \mathbb{C}^{\mathbb{Z}}$, fix initial values $\Psi_0 := (\ldots, \psi_n, \psi_{n+1}, \ldots)$ with specifically assigned initial data $(\psi_{n_0}, \psi_{n_0+1})$ for some fixed n_0 (and with entries ψ_k other than ψ_{n_0} and ψ_{n_0+1} as dummy variables), and evolve by applying suitable matrices to the sequences. In particular, we now define the infinite-dimensional matrices

$$
\mathscr{A}_n = \begin{pmatrix}
\ddots & & & & & & & \\
& \begin{matrix} 1 & 0 \\ 0 & 1 \end{matrix} & & \cdots & & \begin{matrix} 0 & 0 \\ & 0 \end{matrix} & \\
& & \ddots & & & & \\
& & & 1 & & & \\
\vdots & & & a(n) & & \vdots & \\
& & & & 1 & & \\
& & & & & \ddots & \\
& \begin{matrix} 0 & \\ 0 & 0 \end{matrix} & & \cdots & & \begin{matrix} 1 & 0 \\ 0 & 1 \end{matrix} & \\
& & & & & & \ddots
\end{pmatrix} \tag{11.3}
$$

$$
\mathscr{B}_n = \begin{pmatrix}
\ddots & & & & & & & \\
& \begin{matrix} 1 & 0 \\ 0 & 1 \end{matrix} & & \cdots & & \begin{matrix} 0 & 0 \\ & 0 \end{matrix} & \\
& & \ddots & & & & \\
& & & 1 & & & \\
\vdots & & & 0 \; -b(n) \; -c(n) & & \vdots & \\
& & & 1 & & & \\
& & & & \ddots & & \\
& \begin{matrix} 0 & \\ 0 & 0 \end{matrix} & & \cdots & & \begin{matrix} 1 & 0 \\ 0 & 1 \end{matrix} & \\
& & & & & & \ddots
\end{pmatrix} \tag{11.4}
$$

where the entry $a(n)$ in the diagonal $\mathscr{A}_n$ and the entries $-b(n)$ and $-c(n)$ in $\mathscr{B}_n$ appear in row $n-1$ (with $-b(n)$ in column n, $-c(n)$ in column $n+1$). Then, we iterate by solving the equations

$$
\mathscr{A}_{n_0-i} \Psi_{i+1} = \mathscr{B}_{n_0-i} \Psi_i
$$

for $i = 0, 1, \ldots$. Most of these infinitely many components of each equation are of the trivial form $\psi_k = \psi_k$, except for one equation that determines ψ_{n_0-i-1} from the two preceding values (if $a(n_0 - i) \neq 0$). After N steps, the sequence Ψ_N contains $N + 2$ specific values for the wave function in terms of initial data.

In this form, we can generally formulate the condition of quantum hyperbolicity: If one of the $\mathscr{A}_n$ is degenerate, such as $\mathscr{A}_1$ in the case of isotropic loop quantum cosmology, we require that all $\mathscr{B}_n$ map $\ker \mathscr{A}_1$ into itself. If this is the case, values for ψ_n that remain undetermined owing to the non-invertability of $\mathscr{A}_1$ do not spoil the further recurrence; those undetermined values will remain undetermined, but they will not be required to compute the rest of the wave function uniquely from initial values. Generalizations to higher-order difference equations are easily done in this form, as are generalizations to the case of multiple kernels: If several $\mathscr{A}_m$ are degenerate, we require that all $\mathscr{B}_n$ map $\bigcup_m \ker\mathscr{A}_m$ into itself.

In a physical sense, the evolution is thus non-singular: initial values posed on one side of $n = 0$ can uniquely be extended throughout the domain where a general wave function ψ_n is defined. No extra input is needed to determine the wave function once the classical singularity is passed. In loop quantum cosmology, this comes about because the state $|0\rangle$ of vanishing volume decouples from the dynamics: the value the wave function takes there remains undetermined, but it is not required for knowing the wave function anywhere else. The singularity is eliminated, but not by excluding states of vanishing volume by hand. Instead, it is the dynamics itself which decides to step over such states, rejecting the singularity.

> What is important here is the property that evolution across the classical singularity is unique and happens without extra input. Even classically it is sometimes possible to extend space–time solutions through a singularity in a distributional sense if the space–time metric is not twice differentiable everywhere but leads to a diverging curvature tensor which can be made sense of as a distribution on space–time. While this would allow one to define a formal solution to Einstein's equation at both sides of the singularity, retaining the singular divergence but removing the singular boundary, such an extension is not unique. There is no deterministic removal of classical singularities in this way. The difference equations of loop quantum cosmology, on the other hand, uniquely extend the wave function across regions of superspace which classically would be singular. This condition is similar in spirit to using generalized hyperbolicity as a condition substituting geodesic incompleteness in classical singularity theorems [1, 2].

> Note also that quantum hyperbolicity applies to all wave functions solving the difference equation. Thus, singularity removal is independent of properties of the physical inner product, providing a very general mechanism. Specific pictures of non-singular space–times, such as bounces, may require further details and then tend to be less general.

The kernel of $\mathscr{A}_1$ has a further consequence: imposing the difference equation, and now defining the transition matrices $\mathscr{T}^{n_0-1}: \Psi_0 \to \Psi_{n_0-1}$ by

$$\mathscr{T}^{n_0-1} = \mathscr{A}_2^{-1}\mathscr{B}_2 \cdots \mathscr{A}_{n_0}^{-1}\mathscr{B}_{n_0},$$

leads to a state which must satisfy $v^T \mathscr{B}_1 \mathscr{T}^{n_0-1}\Psi_0 = 0$ for any $v \in \ker\mathscr{A}_1$ since $\mathscr{A}_1\Psi_{n_0} = \mathscr{B}_1\Psi_{n_0-1} = \mathscr{B}_1\mathscr{T}^{n_0-1}\Psi_0$. Thus for a kernel of dimension k we obtain k independent linear conditions on the initial values in Ψ_0. Without specifying initial conditions, initial data are restricted by the dynamics. These are the dynamical initial conditions [3, 4] discussed in specific cases before. In general, the number of dynamical initial conditions on a wave function is determined by $\dim \bigcup_m \ker\mathscr{A}_m$.

Similar considerations hold in anisotropic and even midisuperspace models, but the analysis is more involved since ψ_n would not just be complex-valued but take values in ℓ^2-spaces or even in products of such spaces with a large number of factors.

11.2 Properties

For interpretations and physical applications, specific mathematical properties of difference equations as they arise in loop quantum cosmology are of interest. The main ones are stability, which is a statement about oscillating versus exponential behavior of solutions, and asymptotic boundedness, which provides conditions for an exponentially decaying rather than increasing solution. These properties are related to physical Hilbert space issues, which will be discussed in a dedicated way in the next chapter.

11.2.1 Stability

Difference equations in loop quantum cosmology are formulated for wave functions on minisuperspace. There are regions, typically of large values of densitized-triad components, where one expects all solutions to the constraint equation (or at least a large set for certain initial values) to behave semiclassically. Thus, they should be oscillatory and not increase (or decrease) exponentially. Even if one is using a Hamiltonian constraint operator which is formally self-adjoint and has the correct continuum limit (for instance if holonomy corrections vanish for $\delta \to 0$) stable behavior in this sense is not guaranteed if one actually uses a finite non-zero step-size. Analyzing stability and making sure that it is realized in all required regimes provides stringent tests for the viability of a quantization scheme [5]. In loop quantum cosmology, this is closely related to lattice refinement.

One can probe stability numerically by solving the tree-level equations obtained by replacing the classically quadratic connection terms in the Hamiltonian constraint with periodic functions of the connection as indicated by holonomy corrections. Such a modified Hamiltonian in general does not correspond to an effective Hamiltonian but only constitutes the tree-level approximation. (This means that there are no corrections from quantum dynamics, but only from quantum geometry.) Normally, it would be difficult to justify using these equations, for instance to analyze the fate of classical singularities where quantum dynamics and quantum back-reaction should be significant, too. But one can use these equations to test whether quantum geometry remains tame enough in regimes which should be very nearly semiclassical. Here, quantum back-reaction cannot be strong, and deviations of tree-level equations from classical equations would indicate that the quantization is incorrect, or unstable in the terminology of difference equations.

These methods, which have often been applied, can be used as tests of stability in various cases, such as for Bianchi models [6–8] or the black-hole interior [9–13]. Their drawback is that they rely on numerical evolutions which require an analysis of a large number of different solutions to probe the full parameter space. An elegant procedure to analyze stability in general terms, in a way insensitive to picking specific initial values for solutions, is provided by generating-function techniques. This method also allows one to compute specific initial conditions for a wave function of desired behavior. Another technique which applies more generally and is more closely related to numerical tools is von-Neumann stability.

11.2.1.1 Generating Functions

Properties of solutions to a difference equation for coefficients ψ_m can conveniently be collected in a generating function, defined as the function

$$G(x) = \sum_{m=0}^{\infty} \psi_m x^m \tag{11.5}$$

of one variable x such that $m!\psi_m$ is its m-th Taylor coefficient. Recurrence relations for ψ_m with non-constant coefficients are mapped to differential equations for $G(x)$. Each monomial m-dependent coefficient of ψ_m of some order k in the difference equation will contribute a k-th order derivative to the differential equation for $G(x)$. At first order, we have the useful equation

$$\frac{\partial(x^k G(x))}{\partial x} = \sum_{m=0}^{\infty} (m+k)\psi_m x^{m+k-1}. \tag{11.6}$$

These properties clearly show that generating functions are most suitable for difference equations with polynomial coefficients of low order. In loop quantum cosmology, difference equations often have non-polynomial coefficients (in particular, square roots), but they can be approximated by polynomial ones in certain regimes such as for large m. This is in fact the region where a stability analysis is most important.

One manifestation of unstable behavior can often be found in rapidly oscillating solutions of the form of a modulated $(-1)^m$. This means that differences of adjacent values of a solution ψ_m would be of the same order as ψ_m and converge to zero only if the ψ_m themselves converge to zero (which would be acceptable since the oscillation would be suppressed for large m). Such differences can be analyzed in terms of the generating function [14] by looking at $(1-x)G(x)$ since

$$(1-x)G(x) = \psi_0 + \sum_{m=0}^{\infty} (\psi_{m+1} - \psi_m)x^{m+1}.$$

If this function is free of singularities at $x = -1$, the sequence ψ_m converges to a finite value without oscillation. Knowing the generating function and studying its poles thus allows one to find initial values for solutions ψ_m which evolve to a converging value without short-scale oscillations.

Example 11.2 In the Taylor expansion of $G(x) := (1+x)^{-1}$, which satisfies $d((1+x)G(x))/dx = 0$ and may accordingly be viewed as a generating function for solutions of a difference equation $\psi_{m+1} + \psi_m = 0$, the coefficients of x^m in

$$\frac{1}{1+x} = 1 - x + x^2 - x^3 + \cdots \tag{11.7}$$

have alternating sign, in accordance with the fact that there is a simple pole at $x = -1$. Poles at $x = 1$ can imply different behaviors. Two examples of this are seen in the expansions for $(1 - x)^{-1}$,

$$\frac{1}{1-x} = 1 + x + x^2 + x^3 + \cdots,$$

and $\log(1 - x)$,

$$-\log(1 - x) = x + \frac{1}{2}x^2 + \frac{1}{3}x^3 + \cdots$$

In the first, the coefficients of the sequence are of constant value (solving $\psi_{m+1} - \psi_m = 0$); in the second, they converge to zero. A function $G(x) := -\log(1 - x)$ solves $\mathrm{d}((1 - x)\mathrm{d}G(x)/\mathrm{d}x)\mathrm{d}x = 0$, qualifying it as the generating function of $(m + 1)\psi_{m+1} - m\psi_m = 0$.

As we will discuss later, what is more relevant for the overall growth of coefficients is the pole at $x = 1$ of $(1 - x)G(x)$.

11.2.1.2 Von-Neumann Stability

Von-Neumann stability analysis in general applies to partial difference equations, but also here one can reduce discussions essentially to an ordinary difference equation by singling out one parameter as evolution variable and decomposing the rest by orthogonal functions. Evolution is then determined by a time-dependent matrix, whose eigenvalues reveal stability properties.

Higher-order equations can be reduced to a first-order equation for vector-valued functions: $\sum_{k=-M}^{M} a_{n+k}\psi_{n+k} = 0$ is equivalent to a vector equation of the form $\mathbf{v}_n = B(n)\mathbf{v}_{n-1}$ for $\mathbf{v}_n = (\psi_{n+M}, \psi_{n+M-1}, \ldots, \psi_{n-M+1})^T$. The evolution of an eigenvector $\mathbf{w}_{n-1}$ with eigenvalue $\lambda(n)$ of the matrix $B(n)$ is given by $\mathbf{w}_n = \lambda(n)\mathbf{w}_{n-1}$. Thus, when the size of the corresponding eigenvalue is $|\lambda(n)| > 1$, the values in the sequence associated to $\mathbf{w}_{n-1}$ will grow as well. In this way, one can provide a general analysis of the difference equations of loop quantum cosmology and test in which regions of minisuperspace solutions with generic non-exponential behavior are possible. If this overlaps with the whole range where semiclassical behavior is expected, the model is consistent. This method was one of the first to indicate that lattice refinement is necessary for consistent models of loop quantum cosmology [15].

11.2.1.3 Bianchi I LRS Behavior

We first look at an LRS version of the Bianchi I model, a homogeneous model which allows one rotational symmetry axis as also seen for black-hole interior models in

Sect. 8.3. (See [15], and [16–18] for further examples.) This reduction leaves two independent parameters in an invariant triad. After loop quantization (without lattice refinement to be specific) and rescaling the wave function by the volume to simplify coefficients, $t_{m,n} = V(m,n)\psi_{m,n}$, we can bring (8.91)— without the γ^2-term in one coefficients which arises from the intrinsic curvature of the Kantowski–Sachs model but is absent in Bianchi I LRS—to the form of the partial difference equation

$$c(n)(t_{m-2,n}-2t_{m,n}+t_{m+2,n})+2d(m)(t_{m+1,n+1}-t_{m+1,n-1}-t_{m-1,n+1}+t_{m-1,n-1}) = 0 \tag{11.8}$$

in two independent variables. Here,

$$c(n) = \begin{cases} 0 & \text{if } n = 0 \\ \sqrt{1+1/2n} - \sqrt{1-1/2n} & \text{otherwise} \end{cases} \tag{11.9}$$

and

$$d(m) = \begin{cases} 0 & \text{if } m = 0 \\ 1/m & \text{otherwise.} \end{cases} \tag{11.10}$$

Since $t_{m,n}$ includes the volume of each basis state, it must vanish at the minisuperspace boundaries: $t_{0,n} = 0 = t_{m,0}$ for all n and m. With $\tilde{t}_{m,n} = t_{m+1,n} - t_{m-1,n}$ ($m \geq 1$), we simplify the recursion relation further:

$$c(n)(\tilde{t}_{m+1,n} - \tilde{t}_{m-1,n}) + 2d(m)(\tilde{t}_{m,n+1} - \tilde{t}_{m,n-1}) = 0. \tag{11.11}$$

There is no loss of information while going from t to $\tilde{t}$ because we can compute $t_{m+1,n} = \tilde{t}_{m,n} + t_{m-1,n}$ once we know $\tilde{t}$ and use the boundary conditions for t. Boundary values of $\tilde{t}$ are free.

We first reduce the partial difference equation (11.11) for $\tilde{t}_{m,n}$ to a set of ordinary difference equations by separation, looking for solutions of the form $\tilde{t}_{m,n} = a_m b_n$. Then a_m and b_n must satisfy

$$a_{m+1} - a_{m-1} = \frac{2\lambda}{m}a_m, \quad b_{n+1} - b_{n-1} = -\lambda c(n)b_n \tag{11.12}$$

with a separation parameter λ. Any pair of solutions for these two sequences will provide a solution to the original recursion relation.

The equation for a_m has polynomial coefficients (after multiplying with m) and can easily be studied by generating functions (as in the preceding section). Oscillatory behavior of solutions is indeed likely: For negative λ, for instance, solutions generically alternate in the signs of a_m and also grow unstably in size. Such a behavior, realized even in regimes of large m which are supposed to result in semiclassical behavior, cannot be allowed for acceptable solutions. In fact, for large m the right-hand side of the first equation in (11.12) is usually small compared to the left-hand side such that $a_{m+2} \approx a_m$. Going in even multiples of the basic step thus does not show strong oscillations. The relation between neighboring values a_{m+1} and a_m is then crucial to see whether or not oscillations arise. This behavior is determined by the initial values a_0 and a_1 which give $a_2 = 2\lambda a_1 + a_0$. For negative λ, a_2 can easily have the opposite sign of a_1 which translates to a_m having the opposite sign of a_{m+1} for large m and thus alternating behavior. (For positive λ it is easy to suppress oscillations. However, since λ enters the equation for b_n with the sign reversed, the b-sequence will then develop alternating behavior.)

To determine the generating function for a_m, we multiply the difference equation by mx^{m-1} and sum over all $m \geq 0$:

$$\sum_{m=0}^{\infty} ((m+1)a_{m+2} - 2\lambda a_{m+1} - (m+1)a_m)\, x^m = 0. \tag{11.13}$$

Multiplications with m can then be related to derivatives by x, such that

$$\frac{d}{dx}\left(\frac{G(x) - a_0}{x} - xG(x)\right) - 2\lambda \frac{G(x) - a_0}{x}$$
$$= \frac{d}{dx}\left(\frac{1 - x^2}{x}G(x)\right) - 2\lambda \frac{G(x)}{x} + a_0 \frac{1 + 2\lambda x}{x^2} = 0 \tag{11.14}$$

must be satisfied. This equation has singularities at $x = -1$, $x = 0$ and $x = 1$. The pole at $x = -1$ will be of interest for the oscillatory behavior, as seen before in general terms. Strictly speaking, the pole at $x = 0$ makes it difficult to Taylor expand around it, although this procedure was used to introduce the generating function. But this does not affect properties of formal expansion coefficients.

In fact, for the form of resulting equations it is convenient to eliminate the pole at $x = 0$ by redefining

$$H(x) = \frac{G(x) - a_0}{x} \tag{11.15}$$

which must satisfy

$$\frac{d}{dx}\left((1 - x^2)H(x)\right) - 2\lambda H(x) = a_0. \tag{11.16}$$

We have eliminated one singularity in the differential equation and the new function is still related to solutions of our difference equation: Expanding

$$H(x) = \sum_{m=0}^{\infty} \alpha_m x^m, \tag{11.17}$$

and comparing with $G(x)$ shows that

$$a_m = \alpha_{m-1}, \quad \text{for } m \geq 1. \tag{11.18}$$

This simple shift clearly does not affect the oscillating behavior. Moreover, once we know the two initial values of the α_m-sequence, α_0 (equalling a_1) and α_1 (equalling a_2), we can find those of the a_m sequence using

$$\alpha_1 - a_0 = 2\lambda\alpha_0. \tag{11.19}$$

The question of main interest now is how to avoid alternating oscillatory behavior. Because of the $(1-x^2)$ factor appearing in the differential equation (11.16) for $H(x)$, the function will in general have poles at $x = \pm 1$, and is regular for $|x| < 1$. If there is no singularity at $x = 1$, we have $(1 - x)H(x)|_{x=1} = \alpha_0 + \sum_{m=0}^{\infty}(\alpha_{m+1} - \alpha_m) = \lim_{m\to\infty} \alpha_m$ such that the sequence α_m has a finite limit. Moreover, if there is no pole at $x = -1$, we have $\lim_{m\to\infty}(\alpha_{m+1} - \alpha_m) = 0$ and alternating oscillations are suppressed. Obviously, $(1 - x)H(x)$ will have the same behavior at $x = -1$ as $H(x)$. The absence of a pole requires a relation between the two initial values α_0 and α_1.

We first solve the homogeneous equation (11.16) for $a_0 = 0$ by $H(x) = c(1 + x)^{\lambda-1}(1 - x)^{-\lambda-1}$. By varying the constant c we obtain the general solution to the inhomogeneous equation as

$$H(x) = c(x)(1 + x)^{\lambda-1}(1 - x)^{-\lambda-1} \tag{11.20}$$

with

$$c(x) = a_0 \int^x \left(\frac{1-t}{1+t}\right)^\lambda dt = c_0 - \frac{2^\lambda a_0}{\lambda - 1}(1+x)^{1-\lambda} {}_2F_1(1-\lambda, -\lambda; 2-\lambda; (1+x)/2) \tag{11.21}$$

for $\lambda \neq 1$ in terms of the hypergeometric function ${}_2F_1$. (If $\lambda = 1$, the equation can be integrated in a manner similar to the $\lambda = -1$ case in the example below. As mentioned, the case of negative λ is more interesting as regards oscillating behavior in the a_m-solutions.) This gives

$$H(x) = c_0(1+x)^{\lambda-1}(1-x)^{-\lambda-1} - \frac{2^\lambda a_0}{\lambda - 1}(1-x)^{-\lambda-1} {}_2F_1(1-\lambda, -\lambda; 2-\lambda; (1+x)/2) \tag{11.22}$$

where only the first term is relevant for the singularity structure at $x = -1$ since the hypergeometric function ${}_2F_1(a, b; c; z)$ is regular at $z = 0$ (taking the value ${}_2F_1(a, b; c; 0) = 1$ for all a, b, c). Thus, the singularity at $x = -1$ can always be removed by choosing $c_0 = 0$. Since

$$a_1 = \alpha_0 = H(0) = c_0 - 2^\lambda a_0/(\lambda - 1){}_2F_1(1 - \lambda, -\lambda; 2 - \lambda; 1/2)$$
$$= c_0 + a_0 - \lambda a_0(\psi(1/2 - \lambda/2) - \psi(1 - \lambda/2))$$

with the digamma function $\psi(z) = d\Gamma(z)/dz$, we have the condition

$$a_1 = a_0(1 - \lambda(\psi(1/2 - \lambda/2) - \psi(1 - \lambda/2))). \tag{11.23}$$

This expression is finite for all λ that are not positive integers since the digamma function is analytic except for simple poles at $-z \in \mathbb{N}$. For $\lambda = -1$, for instance, we can use $\psi(1) - \psi(3/2) = 2\log 2 - 2$ and obtain the special case studied in more detail below.

At $x = 1$, $_2F_1(a, b; c; (1 + x)/2)$ has a branch point which is logarithmic for $c - a - b \in \mathbb{Z}$ or $c - a - b \notin \mathbb{Q}$. Thus, $(1 - x)G(x)$ always has a singularity at $x = 1$, for positive λ enhanced by the factor $(1 - x)^{-\lambda-1}$. For $\lambda > 0$ the sequence α_m is unbounded.

Example 11.3 To be specific, we now choose $\lambda = -1$. With the initial condition $H(0) = \alpha_0$ and the relation (11.19) between a_0 and the initial values α_0 and α_1, we obtain the solution

$$H(x) = \frac{\alpha_0 - (2\alpha_0 + \alpha_1)x - (4\alpha_0 + 2\alpha_1)\log(1 - x)}{(1 + x)^2}. \tag{11.24}$$

For generic α_0 and α_1, this function has singularities at $x = \pm 1$. To ensure that the singularity at $x = -1$ does not give rise to oscillatory behavior at large m, we require that

$$\lim_{x \to -1} \left((1 - x)(1 + x)^2 H(x) \right) = 0. \tag{11.25}$$

Solving this relation implies

$$\alpha_1 = \frac{4\log 2 - 3}{2\log 2 - 1}\alpha_0$$

and so $(1-x)H(x)$ is regular also at $x = 1$. (At $x = 1$, a singularity remains in $H(x)$ because of the $\log(1 - x)$ term, but $(1 - x)H(x)$ is regular. With the logarithmic behavior, coefficients of the Taylor series go as $1/m$.) One can check that the pole at $x = 1$ for $\lambda > 0$ is of higher order. Thus, those solutions will not be bounded, but they do not show oscillations.

This example illustrates how precise conditions for initial values of sequences can be found that give rise to solutions whose oscillations are damped. The same model is also an interesting example for applying von-Neumann stability analysis. In fact, the model used so far turns out to be unstable unless lattice refinement is taken into account [15]. Difference equations then become more complicated with step-sizes no longer being constant. In such a case, generating functions are more difficult to find, but von-Neumann stability analysis can still be used. We will see this in the following example, which turns out to be more restrictive than Bianchi I models.

11.2.1.4 Kantowski–Sachs

The strongest restrictions on refinement analyzed so far arise for the Kantowski–Sachs model describing a non-rotating vacuum black hole interior as in Sect. 8.3. Since this is an anisotropic model, much a-priori freedom exists for possible refinement schemes. However, the possibilities are strongly restricted by stability conditions. In fact, so far no scheme has been formulated that would provide good behavior in all semiclassical regimes, especially near the horizon. But even in regimes far from the horizon, where implications of the coordinate singularity present in the Kantowski–Sachs form could be ignored, many possible refinements are ruled out.

Stability problems in the loop-quantized Kantowski–Sachs model were first seen using generating functions [18]. It turns out that they can be resolved only with non-trivial lattice refinements which require non-equidistant difference equations. Generating functions are then less suitable, but von-Neumann analysis can still be used.

We consider two choices of $\mathcal{N}_x$ and $\mathcal{N}_\vartheta$ to determine the lattice refinement as discussed in Chap. 8. The simpler case is $\mathcal{N}_x \propto \sqrt{\tau}$ and $\mathcal{N}_\vartheta \propto \sqrt{\mu}$ which provides a difference equation transformable to an equidistant one. (In the context of stability it is sufficient to consider regimes of fixed sign of τ and μ; we thus drop absolute values, assuming positive signs.) For large values of μ, τ, the coefficients (8.95), (8.96) of the Hamiltonian constraint become

$$C_\pm(\mu, \tau) \sim 4\delta\sqrt{\frac{\tau}{\mu}}, \qquad C_0(\mu, \tau) \sim \frac{\delta}{\tau}.$$

Asymptotically, the coefficients of the $\psi_{\mu\pm2\delta/\sqrt{\mu},\tau}$ and $\psi_{\mu,\tau}$ terms in the difference equation become $C_0(\mu, \tau)\mu$, which we insert into the Hamiltonian constraint equation (8.94). We also change variables to $\tilde{\mu}(\mu) = \mu^{3/2}$ and $\tilde{\tau}(\tau) = \tau^{3/2}$, which makes the lattice-refined steps almost equidistant: we can expand

$$\mu(\tilde{\mu} + 3\delta) = (\tilde{\mu} + 3\delta)^{2/3} = \tilde{\mu}^{2/3} + 2\delta\tilde{\mu}^{-1/3} + \cdots = \mu + 2\delta/\sqrt{\mu} + \cdots$$

In a continuum approximation, the values of ψ_μ evaluated at $\mu + 2\delta/\sqrt{\mu}$ are thus well approximated by the values of $\tilde{\psi}_{\tilde{\mu}} := \psi_{\mu(\tilde{\mu})}$ evaluated at $\tilde{\mu} + 3\delta$, up to higher derivatives of the wave function which can be attributed to quantization ambiguities and are irrelevant for semiclassical stability. We obtain the equidistant difference equation [19]

$$4\tilde{\tau}(\psi_{\tilde{\mu}+3\delta,\tilde{\tau}+2\delta} - \psi_{\tilde{\mu}-3\delta,\tilde{\tau}+3\delta} + \psi_{\tilde{\mu}-3\delta,\tilde{\tau}-3\delta} - \psi_{\tilde{\mu}+3\delta,\tilde{\tau}-3\delta})$$
$$+ \tilde{\mu}(\psi_{\tilde{\mu}+6\delta,\tilde{\tau}} - 2\psi_{\tilde{\mu},\tilde{\tau}} + \psi_{\tilde{\mu}-6\delta,\tilde{\tau}}) = 0.$$

(In general, a non-equidistant difference equation in one variable μ with steps evaluated at $\mu + k\delta\mu^x$ is transformed to nearly-equidistant form by $\tilde{\mu} = \mu^{1-x}$, with step-size $k\delta(1 - x)$.)

The asymptotic behavior can be seen from evolution on the integer lattice given by varying m and n in $\tilde{\mu} = 3m\delta$ and $\tilde{\tau} = 3n\delta$. Using n as our evolution parameter, we make the ansatz $\psi_{3m\delta,3n\delta} = u_m \exp(in\omega)$ in order to test stability for evolution in n. We obtain a recurrence relation

$$2in(u_{n+1} - u_{n-1}) - m\sin(\theta)u_n = 0$$

for u_n, equivalent to the vector equation

$$\begin{pmatrix} u_{n+1} \\ u_n \end{pmatrix} = \begin{pmatrix} -\frac{1}{2}im\sin(\theta)/n & 1 \\ 1 & 0 \end{pmatrix}\begin{pmatrix} u_n \\ u_{n-1} \end{pmatrix}. \tag{11.26}$$

Stability is determined by the size of eigenvalues of the evolution matrix, which are

$$\lambda_\pm = \frac{-im\sin\theta \pm \sqrt{16n^2 - m^2\sin^2\theta}}{4n}.$$

For $16n^2 - m^2\sin^2\theta \geq 0$, $|\lambda| = 1$ and the solution is stable; unstable modes arise when $16n^2 - m^2\sin^2\theta < 0$. The most unstable mode corresponds to the choice $\sin\theta = 1$, giving instabilities in terms of the original variables when $\mu > 4\tau$. In this regime, all solutions behave exponentially rather than oscillatory. It includes parts of the solutions for the Schwarzschild interior near the singularity, but also parts for values of μ and τ where one expects classical behavior to be valid. Especially near the horizon, but not just at the horizon with $\mu = 0$, the stability condition is violated. Irrespective of the physical inner product which would be used to evaluate such wave functions for their observable information, instability implies that quantum solutions in those regions cannot be wave packets following the classical trajectory. The correct classical limit is not realized for a quantization based on the refinement scheme used here.

For the choices $\mathcal{N}_\vartheta \propto \sqrt{\tau}$ and $\mathcal{N}_x \propto \mu/\sqrt{\tau}$ we will find that no instability arises for large μ and τ. Now, there is no choice of variables that allows us to asymptotically approach an equidistant recursion relation because of the mixing of the μ and τ variables in the step-size functions. Instead, we will make the assumption that in the limit of large μ, τ the solution does not change much between lattice points separated by steps of the size $\delta\mathcal{N}_x^{-1}$ and $\delta\mathcal{N}_\vartheta^{-1}$.

In the limit of large μ and τ the coefficient functions of the recursion relation to leading order are now

$$C_\pm(\mu,\tau) \sim 4\delta, \qquad C_0(\mu,\tau) \sim \frac{\delta}{\mu}.$$

The difference equation to be analyzed then is

$$4(\psi_{\mu+2\delta/\sqrt{\tau},\tau+2\delta\sqrt{\tau}/\mu} - \psi_{\mu-2\delta/\sqrt{\tau},\tau+2\delta\sqrt{\tau}/\mu}$$
$$- \psi_{\mu+2\delta/\sqrt{\tau},\tau-2\delta\sqrt{\tau}/\mu} + \psi_{\mu-2\delta/\sqrt{\tau},\tau-2\delta\sqrt{\tau}/\mu})$$
$$+ (\psi_{\mu+4\delta/\sqrt{\tau},\tau} - 2\psi_{\mu,\tau} + \psi_{\mu-4\delta/\sqrt{\tau},\tau}) = 0.$$

We assume that we have a solution to this relation that does not vary much between increments of μ by $\pm 2\delta/\sqrt{\mu}$, and similarly for τ. Both $\mathcal{N}_x$ and $\mathcal{N}_\vartheta$ are constant to first order in shifts $\mu \pm 2\delta \mathcal{N}_x^{-1}$ and similarly for τ, in the asymptotic limit. Thus, we assume that $\alpha = 2\delta \mathcal{N}_x^{-1}$ and $\beta = 2\delta \mathcal{N}_\vartheta^{-1}$ are constants, and use the scalings $\mu = \alpha m$ and $\tau = \beta n$. We get an equation similar to the case of $\mathcal{N}_x \propto \sqrt{\tau}$ and $\mathcal{N}_\vartheta \propto \sqrt{\mu}$, but with constant coefficients. Using the decomposition $\psi_{\alpha m, \beta n} = u_n \exp(im\theta)$, we arrive at the matrix equation

$$
\begin{pmatrix} u_{n+1} \\ u_n \end{pmatrix} = \begin{pmatrix} -\frac{i}{2}\sin\theta & 1 \\ 1 & 0 \end{pmatrix} \begin{pmatrix} u_n \\ u_{n-1} \end{pmatrix}. \tag{11.27}
$$

The relevant matrix has eigenvalues λ with $|\lambda| = 1$ for all m and n, and the solution is stable.

Still, the lattice refinement found to be stable for large μ and τ cannot be a fully satisfactory quantization of Kantowski–Sachs models. For classical solutions, small curvature is realized not just for large densitized-triad components but also for small μ, a regime which in the interpretation of a Schwarzschild black-hole interior would be near the horizon. Quantum corrections should certainly be small in this regime, but for the refinement scheme used we have a small vertex number $\mathcal{N}_x \propto \mu/\sqrt{|\tau|}$ for small μ. With a small number of vertices, discreteness corrections are large. Near the horizon the refinement scheme has to deviate from what is given here, not because of instability but to ensure small quantum-geometry corrections. In particular, we cannot have a total vertex number $\mathcal{N}_\vartheta^2 \mathcal{N}_x$ proportional to volume (or a positive power of volume) since the volume $\mu\sqrt{|\tau|}$ drops off to zero near the horizon. Using the relations

$$
\mathcal{N}_x(\lambda)v_x(\lambda) = \frac{|\tilde{p}_b(\lambda)|}{\sqrt{|\tilde{p}_c(\lambda)|}}\mathcal{L}, \quad \mathcal{N}_\vartheta(\lambda)v_\vartheta(\lambda) = \sqrt{|\tilde{p}_c(\lambda)|}
$$

relating discrete and continuous geometries, the patch sizes v_x have to decrease sufficiently fast when the horizon is approached for $\mathcal{N}_x$ to remain large while $\tilde{p}_b$ is decreasing. Note that we need not consider the horizon itself for this argument, where the coordinates used for the homogeneous Schwarzschild slicing would break down. It is sufficient to consider a near-horizon regime, which should be accessible within the model and semiclassically.

11.2.2 Boundedness

In regimes which are either expected to be strongly quantum, such as those around classical singularities, or classically forbidden, such as the large-volume region in recollapsing models, solutions behave exponentially. This may happen in an unrestricted way—exponentially increasing or decreasing—in strong quantum regimes, but in classically forbidden regimes one must pick the correctly decaying solution. With two independent exponential solutions for a difference equation, picking the decaying one out of all solutions most of which increase exponentially is a tough

numerical task. It is thus important to find analytical tools to select initial values for
wave functions that evolve exactly into the decaying branch. Generating functions
can be used also here, but continued fractions provide an alternative powerful method
[20]. A numerical discussion in the case of the closed isotropic model can be found
in [21].

For an example, let us look at the difference equation

$$a_{m+1} - a_{m-1} = 2\lambda m^{-1} a_m .$$

As used before, it is not the difference equation of isotropic models, but arises when
the partial difference equation of an anisotropic model is separated, $\lambda \in \mathbb{R}$ being the
separation parameter. The generating function for solutions is

$$G(x) = \sum_{m=0}^{\infty} a_{m+1} x^m = c_0 (1+x)^{\lambda-1} (1-x)^{-\lambda-1}$$

$$- \frac{2^\lambda a_0}{\lambda - 1} (1-x)^{-\lambda-1} {}_2 F_1 (1-\lambda, -\lambda; 2-\lambda; (1+x)/2)$$

which was used before for stability but is also useful to determine asymptotic properties. For solutions to have oscillations with shrinking amplitude, $G(x)$ must be
regular at $x = -1$ to guarantee that $\sum_m (-1)^m a_m$ is convergent. This is realized
only for special initial values of the solution to the difference equation, satisfying

$$a_1/a_0 = 1 - \lambda \psi(1/2 - \lambda/2) + \lambda \psi(1 - \lambda/2)$$

with the digamma function $\psi(z) = \mathrm{d} \log \Gamma(z)/\mathrm{d}z$. Moreover, the parameter c_0 in
the generating function must vanish in this case. The initial values are determined
through $a_1 = G(0)$ while a_0 already appears in the generating function.

Bounded solutions to the isotropic difference equations can also be controlled.
For asymptotic boundedness it is sufficient to consider large values of n such that
coefficients of the general equation simplify to

$$\psi_{n+4} - 2\psi_n + \psi_{n-4} = -\Lambda n \psi_n \tag{11.28}$$

written here for the flat vacuum model with a cosmological constant Λ, or

$$s_{n+4} + 2s_n + s_{n-4} = \Lambda n \psi_n \tag{11.29}$$

for $s_n := (-1)^{n/4} \psi_n$.

In these equations we implicitly use the lattice-refinement parameter $x = 0$ for illustrative
purposes. Physically, the refinement is not sufficiently strong at large volume n where oscillations of the wave function stop in spite of classically unending expansion. The model
(11.28) is thus not realistic for all n, but it does provide an interesting mathematical example. For the stronger refinement of $x = -1/2$, for instance, we obtain a difference equation
$\psi_{n+4} - (2 - \Lambda)\psi_n + \psi_{n-4} = 0$ with constant coefficients, easily solved by $\psi_n = \exp(ikn)$
with $\cos 4k = 1 - \Lambda/2$. For small positive Lambda, two independent solutions oscillating
for all n, no matter how large, result.

Generic solutions exponentially increase for large n and numerically it is difficult to pick the exponentially decaying and thus bounded one. Also here, we have to find conditions on initial values of the solution ensuring boundedness, which are easier to determine from the recurrence

$$h(n+4) = \Lambda n - 2 - \frac{1}{h(n)} \tag{11.30}$$

for $h(n) := s_n/s_{n-4}$. If we successively insert values of $h(n-4k)$, $k = 0, 1, \ldots$ using the difference equation, we obtain an expression for $h(n+4)$ in terms of initial values at small n. For a bounded solution, we want $h(n)$ to converge, which by the considerations given is equivalent to the convergence of a continued fraction in terms of initial values [20]. For s_n, this translates to the condition

$$\frac{s_0}{s_1} = \Lambda - \frac{1}{2\Lambda - \cdots}.$$

Continued-fraction methods are more widely applicable and can also be used for other models. In the particular case considered here, or rather a Euclidean version obtained by switching the sign of Λ, the ratio is of interest in comparison with exact solutions of the equation in terms of Bessel functions [22]: the Euclidean analog of (11.28) is solved by

$$\psi_n = C_1 J_{n/4+1/2\Lambda}(1/2\Lambda) + C_2 Y_{n/4+1/2\Lambda}(1/2\Lambda)$$

which is bounded at large n only for $C_2 = 0$. The integration constant is related to initial values in the recurrence, for which we have the continued-fraction condition of boundedness. Combining the conditions results in

$$\frac{J_{m-1+1/2\Lambda}(1/2\Lambda)}{J_{m+1/2\Lambda}(1/2\Lambda)} = 2 + 4\Lambda m - \frac{1}{2 + 4\Lambda(m+1) - \frac{1}{2+4\Lambda(m+2)-\cdots}}.$$

11.3 Non-Linear Loop Quantum Cosmology

Consider a spatial slice with a perturbatively inhomogeneous geometry around an FLRW model, approximated as a collection of isotropic patches whose individual geometries are all similar and nearly uncorrelated. In an idealization of identical isotropic geometries, independently characterized by $(\bar{c}_i, \bar{p}_i)$, we write the inhomogeneous state as

$$\psi_{\text{inhom}}(\bar{c}_1, \bar{p}_2, \bar{c}_2, \bar{p}_2, \ldots) \approx \prod_i \psi_{\text{iso}}(\bar{c}_i, \bar{p}_i). \tag{11.31}$$

In this case, the state should not differ from an exactly isotropic one, but as an inhomogeneous state it is subject to different dynamics. We view the isotropic geometries as a collection of points in minisuperspace, in this picture having a many-particle state. If we expand the spatially integrated inhomogeneous Hamiltonian constraint around the isotropic one,

$$C(A, E) = \sum_i C_{\text{iso}}(\bar{c}, \bar{p}) + C^{(2)}(\delta A, \delta E) \qquad (11.32)$$

to second order with $\delta E \sim \bar{p}_{i+1} - \bar{p}_i$ and $\delta A \sim \bar{c}_{i+1} - \bar{c}_i$ (or some other differencing scheme whose precise form is not important here), we obtain the isotropic (one-particle) Hamiltonian to leading order and a (two-particle) interaction term $C^{(2)} = \sum_{i,j} C_{i,j}$ between different isotropic geometries in minisuperspace. (There is no interaction in space. For the strength of interaction terms, the difference in geometries is important, not the difference in spatial positions of the patches.)

With the inhomogeneous state and a quantization of the expanded Hamiltonian we compute the quantum Hamiltonian

$$C_Q = \langle \psi_{\text{inhom}} | \hat{C} | \psi_{\text{inhom}} \rangle = \sum_i C_Q^{\text{iso}} + \sum_{ij} \int \int (\psi_{\text{iso}}^*)^2 \hat{C}_{ij} \psi_{\text{iso}}^2. \qquad (11.33)$$

In the second term, two factors of ψ_{iso} remain from ψ_{inhom} due to the two-particle interaction. Hamiltonian evolution of the inhomogeneous system with the coupling term of higher order in the wave function, after picking an internal time, is then equivalent to non-linear Schrödinger evolution. For instance, if the coupling term is a delta-function, the discrete non-linear Schrödinger equation

$$i\hbar \frac{\partial \psi_n}{\partial t} = \psi_{n+1} - 2|\psi_n|^2 \psi_n + \psi_{n-1} \qquad (11.34)$$

is obtained. (In this case, the derivation is analogous to the Gross–Pitaevski equation following for a Bose–Einstein condensate.) A delta-function interaction is not realistic in the cosmological case, which rather provides polynomial interaction terms. Then, the non-linearity will be different and spans several increments in the difference equation, but is still strictly related to the form of the interactions.

This example illustrates how inhomogeneous features could be captured by non-linear effects in the homogeneous description, rather than providing a fundamentally non-linear wave equation as for instance in [23, 24].

References

1. Clarke, C.J.S.: Class. Quantum Grav. **15**, 975 (1998). gr-qc/9702033
2. Vickers, J.A., Wilson, J.P.: Class. Quantum Grav. **17**, 1333 (2000). gr-qc/9907105
3. Bojowald, M.: Phys. Rev. Lett. **87**, 121301 (2001). gr-qc/0104072

4. Bojowald, M.: Gen. Rel. Grav. **35**, 1877 (2003). gr-qc/0305069
5. Bojowald, M., Date, G.: Class. Quantum Grav. **21**, 121 (2004). gr-qc/0307083
6. Chiou, D.W.: Phys. Rev. D **75**, 024029 (2007). gr-qc/0609029
7. Chiou, D.W.: arXiv:gr-qc/0703010
8. Chiou, D.W.: Phys. Rev. D **76**, 124037 (2007). arXiv:0710.0416
9. Modesto, L.: Adv. High. Energy Phys. **2008**, 459290 (2008). gr-qc/0611043
10. Chiou, D.W., Vandersloot, K.: Phys. Rev. D **76**, 084015 (2007). arXiv:0707.2548
11. Chiou, D.W.: Phys. Rev. D **78**, 044019 (2008). arXiv:0803.3659
12. Böhmer, C.G., Vandersloot, K.: Phys. Rev. D **76**, 104030 (2007). arXiv:0709.2129
13. Campiglia, M., Gambini, R., Pullin, J.: AIP Conf. Proc. **977**, 52 (2008). arXiv:0712.0817
14. Cartin, D., Khanna, G., Bojowald, M.: Class. Quantum Grav. **21**, 4495 (2004). gr-qc/0405126
15. Rosen, J., Jung, J.H., Khanna, G.: Class. Quantum Grav. **23**, 7075 (2006). gr-qc/0607044
16. Cartin, D., Khanna, G.: Phys. Rev. Lett. **94**, 111302 (2005). gr-qc/0501016
17. Date, G.: Phys. Rev. D **72**, 067301 (2005). gr-qc/0505030
18. Cartin, D., Khanna, G.: Phys. Rev. D **73**, 104009 (2006). gr-qc/0602025
19. Bojowald, M., Cartin, D., Khanna, G.: Phys. Rev. D **76**, 064018 (2007). arXiv:0704.1137
20. Bojowald, M., Rej, A.: Class. Quantum Grav. **22**, 3399 (2005). gr-qc/0504100
21. Green, D., Unruh, W.: Phys. Rev. D **70**, 103502 (2004). gr-qc/0408074
22. Noui, K., Perez, A., Vandersloot, K.: Phys. Rev. D **71**, 044025 (2005). gr-qc/0411039
23. Singh, T.P.: Bulg. J. Phys. **33**, 217 (2006). gr-qc/0510042
24. Singh, T.P.: Int. J. Mod. Phys. D **17**, 611 (2008). arXiv:0705.2357

Chapter 12
Physical Hilbert Spaces

The Wheeler–DeWitt equation or the difference equation of loop quantum cosmology present a constraint that states have to satisfy, analogous to the Friedmann equation which is a constraint in canonical relativity. Zero eigenvalues of the constraint operator are thus to be found. For the difference equation (4.15) encountered for isotropic models there is in fact a kinematically normalizable eigenstate of zero eigenvalue, given by $\psi_\mu = \delta_{\mu,0}$. This state, however, is supported only on the classical singularity and of no interest to describe an expanding universe.

Sometimes also interesting states can correspond to normalizable zero-eigenstates, for instance if they belong to a model in which all solutions recollapse, requiring exponentially decaying wave functions at large volume. But in general one has to deal with states that belong to a zero eigenvalue as part of the continuous spectrum, and which cannot be normalizable in the kinematical inner product as it is defined on the Bohr Hilbert space. Then, solutions to the constraint equation cannot form a subspace of the kinematical Hilbert space but constitute a new physical Hilbert space to be constructed by endowing the solution space with a physical inner product. Several techniques to do so exist, but explicit constructions are complicated in general models. For a reparameterization–invariant system as realized by models of general relativity, the physical Hilbert space issue is often related to the problem of time.

12.1 Group Averaging

One method to derive a physical Hilbert space for a given constrained system is group averaging [1]. It applies in particular if all the constraints to be solved generate a unitary group action on the kinematical Hilbert space. One can then integrate over the group to ensure that states considered are invariant under the action.

For a single self-adjoint constraint $\hat{C}$, the unitary group action is Abelian: $\exp(it\hat{C})$, $t \in \mathbb{R}$. Starting with an arbitrary kinematical state $|\psi\rangle$, an averaged state is obtained by integrating $\exp(it\hat{C})|\psi\rangle$ over all values of t. The integration does

M. Bojowald, *Quantum Cosmology*, Lecture Notes in Physics 835,
DOI: 10.1007/978-1-4419-8276-6_12, © Springer Science+Business Media, LLC 2011

not necessarily exist for all states, and the result may not be another normalizable state. However, for states $|\psi\rangle$ in a suitable dense set $\mathscr{D} \subset \mathscr{H}$ the integration can be made sense of as a distribution: a linear functional $\langle\eta_\psi| : \mathscr{D} \to \mathbb{C}$ mapping every state $|\phi\rangle \in \mathscr{D}$ to a complex number. This number is defined by

$$\langle\eta_\psi|\phi\rangle = \int\limits_{-\infty}^{\infty} dt\,\langle\psi|\exp(it\hat{C})|\phi\rangle. \tag{12.1}$$

Heuristically, one can interpret $\int_{-\infty}^{\infty} dt\exp(it\hat{C}) = \delta(\hat{C})$ as a delta-function whose insertion ensures that the action of $\hat{C}$ on states vanishes. The distributional, group-averaged state may then be written as $\eta_\psi = \delta(\hat{C})\psi$ which is not normalizable in the kinematical inner product because one would have to multiply two delta-functions before integrating. The group-averaging inner product makes sense of such an expression by elegantly removing one of the delta-functions, making the integral well-defined and invariant under the group action.

In a similar way, symmetric states, obtained not by implementing first-class constraints but by imposing symmetry, can be interpreted as ordinary states multiplied with delta-functions to make a state vanish on non-symmetric configurations. Distributional techniques similar to group averaging have been used to make sense of the reduced state spaces, giving rise to the models of loop quantum cosmology as in Sects. 8.2.5 and 10.1.2.2.

The space of distributions on $\mathscr{D}$, called $\mathscr{D}'$, does not carry a natural inner product. But on its subspace obtained by group averaging one can easily introduce one. Given two such distributions $\langle\eta_\psi|$ and $\langle\eta_\phi|$, we define the bilinear form

$$\langle\eta_\phi|\eta_\psi\rangle_{\text{phys}} := \int\limits_{-\infty}^{\infty} dt\,\langle\phi|\exp(it\hat{C})|\psi\rangle. \tag{12.2}$$

On the right-hand side we use kinematical states $|\phi\rangle$ and $|\psi\rangle$ averaged to $|\eta_\phi\rangle$ and $|\eta_\phi\rangle$, respectively. Such states are not unique, but thanks to unitarity of the group action the integral in the definition of $\langle\eta_\phi|\eta_\psi\rangle_{\text{phys}}$ does not depend on which representative is chosen: Any other states averaging to the same distributions must be of the form $|\phi'\rangle = \exp(iu\hat{C})|\phi\rangle$ and $|\psi'\rangle = \exp(iv\hat{C})|\psi\rangle$ with real u and v, such that

$$\int\limits_{-\infty}^{\infty} dt\,\langle\phi'|\exp(it\hat{C})|\psi'\rangle = \int\limits_{-\infty}^{\infty} dt\,\langle\phi|\exp(i(t+v-u)\hat{C})|\psi\rangle = \int\limits_{-\infty}^{\infty} dt'\,\langle\phi|\exp(it'\hat{C})|\psi\rangle.$$

Factoring out a possible kernel of the bilinear form and Cauchy-completing the space provides the physical Hilbert space. Uniqueness properties and examples can be found in [2–4].

Alternatively, one can understand the procedure by first introducing a parameter λ along the unitary flow generated by the constraint operator on arbitrary states:

$$\hat{C}|\phi_\lambda\rangle = i\frac{d}{d\lambda}|\phi_\lambda\rangle. \tag{12.3}$$

One thus deals with the constraint in a way similar to a Hamiltonian providing a Schrödinger-type equation. A family of states solving the Schrödinger equation (12.3) is a solution to the constraint only if it is actually λ-independent such that $\hat{C}|\phi_\lambda\rangle = 0$ follows from $d|\phi_\lambda\rangle/d\lambda = 0$. In the solution space, this can be achieved by integrating over λ and defining $|\phi\rangle = \int_{-\infty}^{\infty} d\lambda |\phi_\lambda\rangle$. The state $|\phi\rangle$ is then annihilated by the constraint, and noting that we can always write $|\phi_\lambda\rangle = \exp(-i\lambda\hat{C})|\phi_0\rangle$ thanks to the Schrödinger equation $|\phi_\lambda\rangle$ satisfies, we recognize $|\phi\rangle$ as the group average of $|\phi_0\rangle$. In general one has to be careful with commuting the action of $\hat{C}$ and the λ-integration. This is taken care of in the detailed procedure of group averaging by selecting an appropriate dense set $\mathscr{D}$.

Example 12.1 Self-adjointness is not always required for group averaging integrals to exist, but violations can have undesired consequences. First taking the simple self-adjoint case $\hat{C} = i\partial_x$, the flow equation $-\partial_x\phi_\lambda(x) = \partial_\lambda\phi_\lambda(x)$ has the general solution $\phi_\lambda(x) = \phi(\lambda + x)$ which easily integrates over λ to a constant independent of x. Now looking at the constraint $\hat{C} = \partial_x$ which is not self-adjoint, we have $\phi_\lambda(x) = \phi(\lambda + ix)$ as the general solution of the flow equation. Interpreting this general solution as an arbitrary holomorphic function on the complex plane with coordinate $\lambda + ix$, the λ-integration is performed along a line parallel to the real axis but shifted by an amount of x. Clearly, ϕ must fall off at infinity for the integration to exist. Since any bounded entire function is a constant according to Liouville's theorem, the only pole-free function allowed is the trivial zero function. Otherwise, ϕ must have poles on the complex plane. But then, changing x and shifting one of the integration contours past a pole with non-vanishing residue will change the value of the integration. Except for the zero solution, averaged solutions cannot be constant and independent of x, as one would expect it for the constraint ∂_x. They are only piecewise constant and include discontinuous steps at values of x corresponding to the imaginary parts of poles of ϕ.

Example 12.2 ([5]). Let the constraint be $\hat{C} = -a\hat{x}\hat{p} + b$ with two constants a and b which may be complex-valued. To implement group averaging we first solve the flow equation

$$(axi\partial_x + b)\phi_\lambda(x) = i\partial_\lambda\phi_\lambda(x)$$

in the x-representation. The general solution is

$$\phi_\lambda(x) = e^{\frac{1}{2}ib(a^{-1}\log(x)-\lambda)}f(\log(x)/a + \lambda)$$

with an arbitrary differentiable function f. This solution realizes the flow of the constraint on the state space. To proceed further and integrate over λ, it is useful to Fourier transform f provided that a is real: $f(u) = (2\pi)^{-1}\int_{-\infty}^{\infty} d\omega e^{-i\omega u}\tilde{f}(\omega)$. Then,

$$\int\limits_{-\infty}^{\infty} d\lambda \phi_\lambda(x) = e^{\frac{1}{2}ib\log(x)/a} \int\limits_{-\infty}^{\infty} \frac{d\omega}{2\pi} e^{-i\omega\log(x)/a} \tilde{f}(\omega) \int\limits_{-\infty}^{\infty} d\lambda e^{-i(\omega+b/2)\lambda}$$

if we are allowed to commute the integrations. If b is real, the λ-integration results in $\delta(\omega + b/2)$ and we have invariant solutions $\phi(x) \propto e^{ib\log(x)/a}$. If b is not real, the integration would have to be done more carefully, and appropriate fall-off conditions for $\tilde{f}$ would be required for convergence.

12.2 Observables

Physical observables are self-adjoint operators on the physical Hilbert space. If one knows a kinematical operator $\hat{O}$ which commutes with the constraint $\hat{C}$, then called a Dirac observable, its action can directly be taken over to group-averaged states by the dual action:

$$\langle \hat{O} \eta_\psi | \phi \rangle := \langle \eta_\psi | \hat{O}^\dagger | \phi \rangle = \int\limits_{-\infty}^{\infty} dt \langle \psi | \exp(it\hat{C}) \hat{O}^\dagger | \phi \rangle \tag{12.4}$$

provides an action on physical states. It is independent of the representative chosen: Another one is related to $|\phi\rangle$ by $|\phi'\rangle = \exp(iu\hat{C})|\phi\rangle$ with a real number u that just adds to t in the integration without changing the result. If $\hat{O}$ does not commute with $\hat{C}$ one may still define its action on physical states in this way, but the notion is ambiguous: it depends on which representative $|\psi\rangle$ is chosen for the physical state $|\eta_\psi\rangle$ before averaging the right-hand side.

A natural action of operators not commuting with the constraint on physical states is obtained for evolving observables, a special class of relational observables [6–10], provided one manages to write the constraint as $\hat{C} = -i\hbar\partial_\varphi + \hat{H}$ with a suitable phase–space variable φ (called internal time) and a φ-independent $\hat{H}$. Then,

$$\hat{\mathcal{O}}(\varphi) = e^{-i\varphi\hat{H}/\hbar} \hat{O} e^{i\varphi\hat{H}/\hbar} \tag{12.5}$$

provides parameterized families of operators commuting with the constraint $\hat{H} - i\hbar\partial_\varphi$:

$$[\hat{\mathcal{O}}(\varphi), \hat{C}] = i\hbar\partial_\varphi\hat{\mathcal{O}}(\varphi) + \exp(-i\varphi\hat{H}/\hbar)[\hat{O}, \hat{H}]\exp(i\varphi\hat{H}/\hbar) = 0.$$

Here, we crucially use the φ-independence of $\hat{H}$, which implies $i\hbar\partial_\varphi\hat{\mathcal{O}}(\varphi) = \exp(-i\varphi\hat{H}/\hbar)\hat{H}\hat{O}\exp(i\varphi\hat{H}/\hbar) - \exp(-i\varphi\hat{H}/\hbar)\hat{O}\hat{H}\exp(i\varphi\hat{H}/\hbar)$. A physical state is thus mapped into a physical state by $\hat{\mathcal{O}}(\varphi)$.

Virtues and disadvantages of evolving observables have been discussed in [11]. Evolving observables satisfy the equation

$$\frac{\partial \langle \hat{\mathcal{O}}(\varphi) \rangle}{\partial \varphi} = -\frac{\langle [\hat{\mathcal{O}}, \hat{H}] \rangle}{i\hbar}. \tag{12.6}$$

Notice the minus sign, to be commented on at the end of Sect. 12.3.1.

12.3 Internal Time

Solving quantum constraints and computing observable quantities is often facilitated
if global internal-time variables exist; most of the explicit techniques even rely on
that feature: the existence of simple phase–space functions that can be used as global
parameters in dynamical solutions. We have already seen this in the preceding section,
where the existence of a phase–space variable φ allowing us to write the constraint
as a linear momentum of time φ plus a time-independent Hamiltonian, providing
general expressions for observables. This procedure is called deparameterization.

12.3.1 Non-Relativistic Parameterized Systems

Constraints are rather simple to solve if a phase–space variable φ exists such that they
take the form $\hat{C} = \hat{p}_\varphi + \hat{H}$ where H is independent of φ and its momentum p_φ. One
can then replace the constraint equation $\hat{C}|\psi\rangle = 0$ by a conventional Schrödinger
flow in the internal time φ. If $\hat{H}$ in this decomposition is self-adjoint, a physical inner
product can be defined using the kinematical one at any fixed value of φ.

Physical states in such a case are solutions to the Schrödinger equation

$$i\hbar\partial_\varphi \psi(\varphi) = \hat{H}\psi(\varphi)$$

and thus clearly depend on the internal time φ. In contrast to the discussion of general
constraints where we introduced an identical equation in terms of λ instead of φ,
however, no further averaging is necessary: φ represents a variable on the phase
space of the system, not an auxiliary quantity to reformulate the equations. The
constraint equation $\hat{C}|\psi\rangle$ is solved by the Schrödinger flow, and we can easily define
the physical inner product

$$\langle \psi_1(\varphi)|\psi_2(\varphi)\rangle_{\text{phys}} = \langle \psi_1(\varphi_0)|\psi_2(\varphi_0)\rangle_{\varphi_0}$$

in terms of the kinematical inner product of unconstrained states, assumed to take
the form $\langle \cdot|\cdot\rangle = \int d\varphi \langle \cdot|\cdot\rangle_\varphi$. We are thus dropping one φ-integration and replace it
with evaluation at some fixed φ_0. Since the inner product is preserved by any unitary
φ-evolution, the physical inner product is independent of the choice of φ_0.

A complete set of observables can formally be constructed in the evolving-
observable sense. First, $\hat{p}_\varphi$ is clearly an observable since we assumed $\hat{H}$ to be φ-
independent: $[\hat{p}_\varphi, \hat{C}] = 0$. Any kinematical operator $\hat{O}$ other than $\hat{\varphi}$ can be made
into an evolving observable

$$\hat{O}(\varphi) = \exp(-i(\varphi - \varphi_0)\hat{H}/\hbar)\hat{O}\exp(i(\varphi - \varphi_0)\hat{H}/\hbar) \qquad (12.7)$$

mapping physical solutions into other physical solutions. Properties such as expectation values in physical states can simply be computed by the restricted inner product $\langle\cdot|\cdot\rangle_{\varphi_0}$.

Example 12.3 (*Linear Hamiltonian*) For $\hat{H} = \hat{p}$, the Schrödinger equation $\partial_\varphi \psi = -\partial_q \psi$ has the general solution $\psi(\varphi - q)$. We immediately see that the motion of the expectation value $\langle\hat{q}\rangle$ must be linear in internal time φ, and that the initial shape of the state is preserved during the motion. While $\hat{p}$ is already an observable, the position observable must be an evolving one:

$$\hat{\mathcal{Q}} = \exp(-i(\varphi - \varphi_0)\hat{p}/\hbar)\hat{q}\exp(i(\varphi - \varphi_0)\hat{p}/\hbar) = \hat{q} - (\varphi - \varphi_0) \qquad (12.8)$$

(using $[\hat{q}, f(\hat{p})] = i\hbar f'(\hat{p})$). Thus, $\partial\langle\hat{\mathcal{Q}}\rangle/\partial\varphi = -1$, in accordance with (12.6). At this stage, we can explain the possibly unexpected minus sign in (12.6), opposite of what one normally has for Schrödinger evolution, for instance in (5.9). We invert (12.8) to write $\hat{q}(\varphi) = \hat{\mathcal{Q}} + \varphi - \varphi_0$, where $\hat{\mathcal{Q}}$, as a Dirac observable, is a constant of motion. For the time-dependent expectation value $\langle\hat{q}\rangle(\varphi)$ we then have $d\langle\hat{q}\rangle/d\varphi = 1$, with the expected sign. In general, the inversion involved in the transition between observables and internal-time dependent operators is the origin of the sign difference between (12.6) and (5.9).

12.3.2 Relativistic Parameterized Systems

Relativistic systems have constraints of the form $\hat{C} = \hat{p}_\varphi^2 - \hat{H}^2$, and are deparameterizable if $\hat{H}$ does not depend on φ. They can thus be brought into the non-relativistic parameterized form only by taking a square root, which requires sign choices. To view the resulting equations as those corresponding to a system with definite frequency sign of φ-evolution, one factorizes

$$\hat{p}_\varphi^2 - \hat{H}^2 = (\hat{p}_\varphi + |\hat{H}|)(\hat{p}_\varphi - |\hat{H}|).$$

The two sectors of solutions annihilated by either of the two factors correspond to positive and negative frequency, respectively. Within these sectors, solutions can also be classified by the sign of $\hat{H}$, which are called right-moving and left-moving by analogy with the case $\hat{H} = \hat{p}$ for a free, massless relativistic Klein–Gordon particle.

For a simple form of $\hat{H}$, such as $\hat{p}^2 + m^2$ for a free relativistic particle of mass m, one can formulate a complete inner product, find all physical solutions and derive properties such as the φ-dependence of expectation values and moments. A useful formulation employs a Dirac-style procedure to take the square root in terms of matrices [12]. Others are based on Fourier transformation.

Example 12.4 (*Klein–Gordon inner product*) For a free, massless relativistic particle we consider the constraint $E^2 - p^2 = 0$, quantized to

$$\frac{\partial^2 \psi}{\partial t^2} - \frac{\partial^2 \psi}{\partial x^2} = 0$$

Solutions can easily be found as $\psi(t, x) = \psi_+(x + t) + \psi_-(x - t)$ with arbitrary ψ_+ and ψ_-. The inner product is easiest to discuss after a Fourier transformation, for which we consider wave functions satisfying $(\omega^2 - k^2)\tilde{\psi}(\omega, k) = 0$. Thus, $\tilde{\psi}$ is supported only on $\omega = |k|$ (positive frequency) or $\omega = -|k|$ (negative frequency).

The Klein–Gordon bilinear form

$$(\psi_1, \psi_2) = i \int dx \left(\psi_1^* \frac{\partial \psi_2}{\partial t} - \frac{\partial \psi_1^*}{\partial t} \psi_2 \right)$$

is preserved in time for solutions ψ_1 and ψ_2 to the Klein–Gordon equation. However, it does not directly provide an inner product because it is not positive definite:

$$(\psi_1, \psi_2) = \int dx \left(\int dk_1 \tilde{\psi}_1^* e^{i(\omega_1 t - k_1 x)} \int dk_2 \omega_2 \tilde{\psi}_2 e^{-i(\omega_2 t - k_2 x)} \right.$$
$$\left. + \int dk_1 \omega_1 \tilde{\psi}_1^* e^{i(\omega_1 t - k_1 x)} \int dk_2 \tilde{\psi}_2 e^{i(\omega_2 t - k_2 x)} \right)$$
$$= \int dk (\omega_1(k) + \omega_2(k)) \tilde{\psi}_1^* \tilde{\psi}_2 e^{-i(\omega_1(k) - \omega_2(k))t}$$

shows that there are three different cases:

- $\psi_1 = \psi_2$, positive frequency: $\tilde{\psi}_1$ supported on $\omega = |k|$. Then, $(\psi_1, \psi_1) = 2 \int dk |k| \tilde{\psi}_1^* \tilde{\psi}_1 > 0$ and we have positive Klein–Gordon norm.
- $\psi_1 = \psi_2$, negative frequency: $\tilde{\psi}_1$ supported on $\omega = -|k|$. Then, $(\psi_1, \psi_1) = -2 \int dk |k| \tilde{\psi}_1^* \tilde{\psi}_1 < 0$ and we have negative Klein–Gordon norm.
- ψ_1 and ψ_2 of opposite frequency signs, e.g. $\omega_1 = |k_1|$, $\omega_2 = -|k_2|$. Then, $(\psi_1, \psi_2) = 0$ and the solutions are orthogonal.

A Hilbert space with a definite inner product can be defined as $(\mathscr{H}, \langle \cdot, \cdot \rangle) = (\mathscr{H}_+, (\cdot, \cdot)) \oplus (\mathscr{H}_-, -(\cdot, \cdot))$ [13].

Example 12.5 (Relativistic harmonic oscillator [14]). We consider a system with two degrees of freedom q and φ with momenta p and p_φ, subject to the constraint

$$\hat{C} = \hat{p}_\varphi^2 - \hat{p}^2 - \hat{q}^2.$$

Physical states must solve the equation

$$i\hbar \frac{\partial}{\partial \varphi} \psi(q, \varphi) = \pm \sqrt{\hat{p}^2 + \hat{q}^2} \psi(q, \varphi)$$

and can be written as

$$\psi_\pm(q, \varphi) = \sum_{n=0}^{\infty} c_n \phi_n(q) \exp(\mp i\lambda_n \varphi / \hbar)$$

in terms of harmonic-oscillator eigenstates $\phi_n(q)$, $\lambda_n = \sqrt{(2n+1)\hbar}$. The constants c_n in this general solution are determined by "initial" values chosen at a fixed value of φ. For instance, using the well-known harmonic-oscillator coherent states at $\varphi = 0$ (in the present context playing the role of a kinematical coherent state), we have

$$c_n = \exp\left(-\frac{|z|^2}{2}\right)\frac{z^n}{\sqrt{n!}}, \quad z \in \mathbb{C}$$

such that

$$\psi(q,0) = \left(\frac{2}{\pi}\right)^{1/4}\exp\left(-\frac{1}{2}|z|^2 - z^2 + 2q^2 - 4izq\right)$$

The corresponding evolving state has coefficients

$$c_n e^{-i\lambda_n\varphi/\hbar} = \frac{1}{\sqrt{n!}}e^{-|z|^2/2}z^n\exp(-i\sqrt{2n+1}\varphi/\sqrt{\hbar})$$

and is non-coherent for $\varphi \neq 0$, but may be considered as a physical semiclassical state for some time of the evolution [14].

In general, the construction of physical semiclassical states via wave functions can be surprisingly subtle. For instance, in the physical Hilbert space underlying loop quantum cosmology sourced be a free massless scalar [15], one may choose Gaussian states at a fixed value of φ and evolve with φ as internal time. If the Gaussian is sharp enough, one expects to have a good semiclassical state with small volume and curvature fluctuations. However, an actual computation of the fluctuations reveals that in this innocent-looking state they are infinite [16]. One can define semiclassical states via the moments, as already used in Chap. 5, but reconstructing the form of a wave function from given fluctuations or higher moments is highly non-trivial.

12.4 Examples in Quantum Cosmology

If φ is taken as the free massless scalar field in a model of quantum cosmology, one can use the procedures seen here to derive physical Hilbert spaces and their properties. As already used for the solvable models of Chap. 5, one can rewrite the Friedmann equation as a relativistically deparameterizable constraint

$$p_\varphi^2 - \frac{16\pi G}{3}(1-x)V^2 P^2 = 0 \tag{12.9}$$

in the canonical variables (5.3). ; see (5.4). In a similar form, physical Hilbert spaces for the Wheeler–DeWitt quantization of this constraint have been introduced in [17]. The φ-Hamiltonian is then the one used in the discussion of solvable models. Analogously, holonomy corrections and inverse-triad corrections of loop quantum cosmology can be included in the Hamiltonian before applying the same techniques to compute the physical Hilbert space [15, 18]. In addition to this extension of the methods of [17] to loop quantum cosmology, [15, 18] have provided detailed numerical tools to compute physical states and extract observables. An alternative derivation of the physical Hilbert space for such systems, based on a Dirac-style first-order formulation of the constraint operator quadratic in $\hat{p}_\varphi$ can be found in [12].

For meaningful physical evolution in internal time it is important that the φ-Hamiltonian is self-adjoint. As shown by [19] and [20], the φ-Hamiltonian of isotropic loop quantum cosmology is essentially self-adjoint for the flat and closed models with vanishing (or negative) cosmological constant. With a positive cosmological constant, however, the Hamiltonian is not essentially self-adjoint [21]. This property is related to the fact, mentioned in Sect. 5.4.1.3, that the classical volume diverges at a finite φ. A semiclassical state thus reaches the boundary at infinity in a finite amount of internal time, and must be reflected back in order to preserve probability. Reflection conditions for wave functions are not unique and do not follow from classical physics; thus, there is no unique self-adjoint extension of the quantum Hamiltonian. Physical interpretations of the reflection and possible extensions of the dynamics are questionable, however, because infinite volume requires an infinite amount of proper time to be reached.

Another approach to construct a physical inner product is based on spin–foam models [23], a suggested covariant, path-integral like version of loop quantum gravity which has branched out into quantum cosmology. The first construction of spin–foam motivated physical inner products in quantum cosmology was done for a model without matter but a positive cosmological constant [22]. More recently, a more general formulation of spin–foam cosmology has been attempted [24, 25] and is still being developed. Well-known results of loop quantum cosmology regarding physical Hilbert spaces can then be used to shed light on more complicated constructions for general spin foams [26–31].

Constructions of physical Hilbert spaces in quantum cosmology are straightforward applications of standard methods if there is no non-trivial matter potential, such that the resulting Hamiltonians are indeed φ-independent. Otherwise the whole procedure complicates enormously to the degree of being intractable. If deparameterization is formally applied to models without good internal time, instead using a variable with turning point (a zero of its momentum along a physical solution) for a description local in time, the system freezes at the turning point [32, 33]. At this point, or even before the freeze-out [34], it is difficult to decide which aspects of the physical state can still be trusted. Transformations to a new time variable which remains valid around the turning point of the old time would have to change the physical Hilbert space, whose construction is based on the choice of time for deparameterization. Even if the physical Hilbert space could be derived completely, such transformations would be difficult to perform. Moreover, time must be chosen before quantization in such a setting, and different choices of time give rise to different quantum representations with possibly differing physics. Here, effective tools to address physical Hilbert space issues become essential to understand physical properties of generic models in loop quantum cosmology, as introduced in the next and final chapter.

References

1. Marolf, D. (1995). gr-qc/9508015
2. Giulini, D., Marolf, D.: Class. Quantum Grav. **16**, 2479 (1999). gr-qc/9812024
3. Giulini, D., Marolf, D.: Class. Quantum Grav. **16**, 2489 (1999). gr-qc/9902045
4. Gomberoff, A., Marolf, D.: Int. J. Mod. Phys. D **8**, 519 (1999). gr-qc/9902069
5. Bojowald, M., Singh, P., Skirzewski, A.: Phys. Rev. D **70**, 124022 (2004). gr-qc/0408094
6. Bergmann, P.G.: Rev. Mod. Phys. **33**, 510 (1961)
7. Rovelli, C.: Class. Quantum Grav. **8**, 297 (1991)
8. Rovelli, C.: Class. Quantum Grav. **8**, 317 (1991)
9. Dittrich, B.: Gen. Rel. Grav. **39**, 1891 (2007). gr-qc/0411013
10. Dittrich, B.: Class. Quant. Grav. **23**, 6155 (2006). gr-qc/0507106
11. Lawrie, I.D.: Phys. Rev. D **83**, 043503 (2011). arXiv:1011.4444
12. Kreienbühl, A.: Phys. Rev. D **79**, 123509 (2009). arXiv:0901.4730
13. Hartle, J.B., Marolf, D.: Phys. Rev. D **56**, 6247 (1997). gr-qc/9703021
14. Bojowald, M., Tsobanjan, A.: Class. Quantum Grav. **27**, 145004 (2010). arXiv:0911.4950
15. Ashtekar, A., Pawlowski, T., Singh, P.: Phys. Rev. D **73**, 124038 (2006). gr-qc/0604013
16. Varadarajan, M.: Class. Quantum Grav. **26**, 085006 (2009). arXiv:0812.0272
17. Blyth, W.F., Isham, C.J.: Phys. Rev. D **11**, 768 (1975)
18. Ashtekar, A., Pawlowski, T., Singh, P.: Phys. Rev. D **74**, 084003 (2006). gr-qc/0607039
19. Kaminski, W., Lewandowski, J.: Class. Quant. Grav. **25**, 035001 (2008). arXiv:0709.3120
20. Szulc, L., Kaminski, W., Lewandowski, J.: Class. Quantum Grav. **24**, 2621 (2007). gr-qc/0612101
21. Kaminski, W., Pawlowski, T.: Phys. Rev. D **81**, 024014 (2010). arXiv:0912.0612
22. Perez, A.: Class. Quantum Grav. **20**, R43 (2003). gr-qc/0301113
23. Noui, K., Perez, A., Vandersloot, K.: Phys. Rev. D **71**, 044025 (2005). gr-qc/0411039
24. Rovelli, C., Vidotto, F.: Class. Quantum Grav. **25**, 225024 (2008). arXiv:0805.4585
25. Bianchi, E., Rovelli, C., Vidotto, F.: Phys. Rev. D **82**, 084035 (2010). arXiv:1003.3483
26. Ashtekar, A., Campiglia, M., Henderson, A.: Phys. Lett. B **681**, 347 (2009). arXiv:0909.4221
27. Ashtekar, A., Campiglia, M., Henderson, A.: Class. Quantum Grav. **27**, 135020 (2010). arXiv:1001.5147
28. Rovelli, C., Vidotto, F.: arXiv:0911.3097
29. Campiglia, M., Henderson, A., Nelson, W.: Phys. Rev. D **82**, 064036 (2010). arXiv:1007.3723
30. Henderson, A., Rovelli, C., Vidotto, F., Wilson-Ewing, E.: Class. Quantum Grav. **28**, 025003 (2011). arXiv:1010.0502
31. Calcagni, G., Gielen, S., Oriti, D.: Class. Quantum Grav. **28**, 125014 (2011). arXiv:1011.4290
32. Wald, R.M.: Phys. Rev. D **48**, 2377 (1993). gr-qc/9305024
33. Higuchi, A., Wald, R.M.: Phys. Rev. D **51**, 544 (1995). gr-qc/9407038
34. Rovelli, C.: Phys. Rev. D **43**, 442 (1991)

Chapter 13
General Aspects of Effective Descriptions

Effective actions or equations are always a powerful tool to analyze general effects of a quantum theory. They can be difficult to derive, but once obtained at least in an approximate form they allow detailed studies and provide more intuition than what can be obtained from dealing with wave functions. Key for their derivation is the availability of a simple solvable model in which quantum properties can be obtained in an exact and compact form. In quantum mechanics, this sovable model is the harmonic oscillator, which in the form of free field theories also appears as the firm solvable basis for quantum field theory on a background space-time. Perturbation theory around the solvable model allows one to include anharmonicities, interactions or extra fields. The harmonic cosmology of a free massless scalar in a flat, isotropic universe plays the solvable role for quantum cosmology.

13.1 Canonical Effective Equations

For a canonical quantization such as loop quantum cosmology we have to use a Hamiltonian way of deriving effective equations. We start with a $*$-algebra $\mathscr{A}$ of operators, possibly together with a representation on a Hilbert space $\mathscr{H}$. A representation would help us in dealing with pure states, but our discussion here will be at the general level of density states which may be mixed. We will focus on operators whose algebra, rather than representation, turns out to be more important; most statements in this chapter are representation-independent. (The algebra of observables is crucial. We will be able to distinguish between a Wheeler–DeWitt quantization and a loop quantization because the latter contains only holonomies of connection components which obey different algebraic relations with fluxes. Once the algebra of operators is fixed, crucial consequences are captured independently of whether a Schrödinger or Bohr representation is introduced.)

M. Bojowald, *Quantum Cosmology*, Lecture Notes in Physics 835,
DOI: 10.1007/978-1-4419-8276-6_13, © Springer Science+Business Media, LLC 2011

13.1.1 States and Moments

A state is a positive linear functional $\omega \colon \mathscr{A} \to \mathbb{C}$ on the $*$-algebra, that is a linear functional satisfying $\omega(\hat{A}^* \hat{A}) \geq 0$ for all $\hat{A} \in \mathscr{A}$ as well as $\omega(\hat{A}^*) = \omega(\hat{A})^*$. (The latter property can be derived if the algebra is unital, as we normally assume.) One example is a pure state ψ in a Hilbert-space representation of $\mathscr{A}$, for which $\omega_{\psi}(\hat{A}) = \langle \psi, \hat{A}\psi \rangle$, or a density matrix ρ for which $\omega_{\rho}(\hat{A}) = \mathrm{tr}(\hat{A}\rho)$. The notion of states allows us to view observables in the algebra $\mathscr{A}$ in a "classical" way as functions on a phase space. We first introduce the space on which these functions are defined as the space $\mathscr{P}$ of all states ω on the algebra. Any element A of the algebra then defines a function $\langle \hat{A} \rangle \colon \mathscr{P} \to \mathbb{C}, \omega \mapsto \omega(\hat{A})$. As the notation indicates, this function is nothing but the expectation-value functional associated with $\hat{A}$.

13.1.2 Quantum Phase Space

Classical observables are real-valued functions on the phase space. Accordingly, we require that quantum observables $\hat{O}$ provide real expectation-value functionals $\langle \hat{O} \rangle$, a large class being self-adjoint ones: $\hat{O}^* = \hat{O}$. Moreover, classical observables are functions on a phase space, which is equipped with a Poisson bracket. For expectation-value functionals, we introduce

$$\{\langle \hat{A} \rangle, \langle \hat{B} \rangle\} = \frac{\langle [\hat{A}, \hat{B}] \rangle}{i\hbar}. \tag{13.1}$$

For functions obtained as products or other expressions in terms of expectation-value functions, we extend the Poisson bracket using linearity and the Leibniz rule. It follows from the properties of the commutator that this is indeed a Poisson bracket. (Note that we define the Poisson bracket by (13.1). It is not required to equal the classical Poisson bracket for all expressions corresponding to $\hat{A}$ and $\hat{B}$. In general, it will be identical to the classical Poisson bracket only if $\hat{A}$ and $\hat{B}$ are basic operators.)

For explicit calculations, it is often most useful to choose coordinates. There is a natural choice if we have a distinguished set of basic operators in our algebra, which form a closed subalgebra under taking commutators and which can be used to generate all other elements in the algebra by their products. For quantum mechanics, these basic operators would usually be taken as the position and momentum operators $\hat{q}$ and $\hat{p}$ with $[\hat{q}, \hat{p}] = i\hbar$, but one may use other versions such as $e^{i\mu c/2}$ and $\hat{p}$ with $[e^{i\mu c/2}, \hat{p}] = -\frac{4}{3}\pi \gamma \ell_{\mathrm{p}}^2 \mu e^{i\mu c/2}$ in loop quantum cosmology, or entire smeared field algebras in quantum field theory and the holonomy-flux algebra in loop quantum gravity.

Let us assume that we have a set of basic operators $\hat{J}_i$, $i = 1, 2, \ldots, N$, which is linear and closed under taking commutators: $[\hat{J}_i, \hat{J}_j] = i\hbar C_{ij}^k \hat{J}_k$. (For a canonical algebra, we allow the identity $\hat{J} = \hat{\mathbb{I}}$ as a possible basic operator.) Expectation-value functionals of the basic operators then satisfy the same Poisson algebra:

$\{\langle \hat{J}_i \rangle, \langle \hat{J}_j \rangle\} = C_{ij}^k \langle \hat{J}_k \rangle$. But they do not provide a complete set of coordinates on the space of all functions on $\mathscr{P}$ because in general we have $\langle \hat{J}_i \hat{J}_j \rangle \neq \langle \hat{J}_i \rangle \langle \hat{J}_j \rangle$, and no other general relation between $\langle \hat{J}_i \hat{J}_j \rangle$ or other expectation values of products and the basic expectation values of $\hat{J}_i$ exists. In fact, even if the basic algebra is finite, as in quantum mechanics or homogeneous models of quantum cosmology, the space of states is usually infinite-dimensional. (Exceptions are cyclic finitely generated algebras as they occur for spin systems; see the examples below.) Expectation-value functionals of products of the basic operators, called moments, provide the remaining coordinates. Unless there are non-trivial relations between operators in the algebra, all moments of density states are independent and in fact of infinite number.

It is convenient to define moments not directly as expectation-value functionals of products of basic operators but in the form already introduced in (5.23):

$$\Delta(O_1 O_2 \cdots O_n) := \left\langle (\hat{O}_1 - \langle \hat{O}_1 \rangle)(\hat{O}_2 - \langle \hat{O}_2 \rangle) \cdots (\hat{O}_n - \langle \hat{O}_n \rangle) \right\rangle_{\text{symm}}$$

$$= \frac{1}{n!} \sum_{\pi \in S_n} \left\langle (\hat{O}_{\pi(1)} - \langle \hat{O}_{\pi(1)} \rangle)(\hat{O}_{\pi(2)} - \langle \hat{O}_{\pi(2)} \rangle) \cdots (\hat{O}_{\pi(n)} - \langle \hat{O}_{\pi(n)} \rangle) \right\rangle$$

(13.2)

where we totally symmetrize to remove redundancy due to re-ordering. (Moreover, symmetrization makes the moments real provided the $\hat{O}_i$ used are self-adjoint.) These moments are non-trivial only for $n \geq 2$ since $\langle (\hat{O} - \langle \hat{O} \rangle) \rangle = 0$ for normalized states. At second order, we use the standard notation $(\Delta O)^2 = \Delta(O^2)$.

With this choice of moments, it is guaranteed that expectation-value functionals of the basic operators Poisson commute with their moments: $\{\langle \hat{O} \rangle, \Delta(O_1 O_2 \ldots)\} = 0$ provided the basic operators $\hat{O}$, $\hat{O}_1$ and $\hat{O}_2$ satisfy a canonical algebra. This property often provides computational advantages, but even for non-canonical basic operators the choice of moments in this form is useful because they are all expected to be small in a semiclassical state, or vanish in a classical correspondence. In this limit only the basic expectation values remain and correspond to classical variables. Accordingly, the moments (5.23) are sometimes called quantum variables. For the general Poisson bracket of moments for canonical basic operators, see [1].

13.1.3 Relations

With expectation-value functionals and the moments we have a complete set of coordinates on the state space. But the moments cannot take arbitrary real values: Fluctuations of the form $(\Delta O)^2 = \langle (\hat{O} - \langle \hat{O} \rangle)^2 \rangle$ must clearly be non-negative, but in general they are restricted even more strongly by uncertainty relations. For each pair $(\hat{O}_1, \hat{O}_2)$ of operators we have the Schwarz inequality

$$\langle \hat{\Delta} O_1^* \hat{\Delta} O_1 \rangle \langle \hat{\Delta} O_2^* \hat{\Delta} O_2 \rangle \geq |\langle \hat{\Delta} O_1^* \hat{\Delta} O_2 \rangle|^2$$

with $\hat{\Delta} O_i := \hat{O}_i - \langle \hat{O}_i \rangle$. If the operators are self-adjoint, this means

$$\langle (\hat{\Delta} O_1)^2 \rangle \langle (\hat{\Delta} O_2)^2 \rangle \geq |\langle \hat{\Delta} O_1 \hat{\Delta} O_2 \rangle|^2. \tag{13.3}$$

Writing

$$\hat{\Delta} O_1 \hat{\Delta} O_2 = \frac{1}{2}(\hat{\Delta} O_1 \hat{\Delta} O_2 + \hat{\Delta} O_2 \hat{\Delta} O_1) + i\frac{1}{2i}[\hat{\Delta} O_1, \hat{\Delta} O_2]$$

with

$$\frac{1}{2}\langle \hat{\Delta} O_1 \hat{\Delta} O_2 + \hat{\Delta} O_2 \hat{\Delta} O_1 \rangle = \frac{1}{2}\langle \hat{O}_1 \hat{O}_2 + \hat{O}_2 \hat{O}_1 \rangle - \langle \hat{O}_1 \rangle \langle \hat{O}_2 \rangle$$

and $[\hat{\Delta} O_1, \hat{\Delta} O_2] = [\hat{O}_1, \hat{O}_2]$ we have

$$|\langle \hat{\Delta} O_1 \hat{\Delta} O_2 \rangle|^2 = \frac{1}{4}\left(\langle \hat{O}_1 \hat{O}_2 + \hat{O}_2 \hat{O}_1 \rangle^2 - 2\langle \hat{O}_1 \rangle \langle \hat{O}_2 \rangle \right)^2 + \frac{1}{4}\langle -i[\hat{O}_1, \hat{O}_2] \rangle^2.$$

Again, we have used self-adjointness of the operators to compute the absolute square of the complex number $\langle \hat{\Delta} O_1 \hat{\Delta} O_2 \rangle$. For the moments, we thus have the inequalities

$$(\Delta O_1)^2 (\Delta O_2)^2 - \Delta(O_1 O_2)^2 \geq \frac{1}{4}\langle -i[\hat{O}_1, \hat{O}_2] \rangle^2 \tag{13.4}$$

as uncertainty relations whenever $\hat{O}_1$ and $\hat{O}_2$ are self-adjoint. Similar conditions exist for moments of higher order, but they mix the order and are more tedious to derive.

Usually, $[\hat{O}_1, \hat{O}_2] \sim \hbar$. Near the saturation of the uncertainty relation, second-order moments are thus of the order $\hbar$, as realized for instance for Gaussian states of canonical quantum systems (see the example below). In that case, $\Delta(O_1 \cdots O_n) \sim O(\hbar^{n/2})$ provides a suitable condition for semiclassicality much wider than one using Gaussians or any specific class of wave functions.

Example 13.1 For the basic operators $\hat{q}$ and $\hat{p}$ represented as the usual operators on the Hilbert space of square-integrable functions of $q \in \mathbb{R}$, pure Gaussian states have the general form $\psi(q) = \exp(-z_1 q^2 + z_2 q + z_3)$ with three complex numbers z_i such that $\mathrm{Re}\, z_1 > 0$ for normalizability. The parameter z_3 is irrelevant for the moments since its real part is fixed by normalization while its imaginary part contributes only a phase factor. Writing $z_1 = \alpha_1 + i\beta_1$ and $z_2 = \alpha_2 + i\beta_2$ with real α_i and β_i, we have

$$\langle \hat{q} \rangle = \frac{\alpha_2}{2\alpha_1}, \quad \langle \hat{p} \rangle = \hbar \frac{\alpha_1 \beta_2 - \alpha_2 \beta_1}{\alpha_1}, \tag{13.5}$$

$$(\Delta q)^2 = \frac{1}{4\alpha_1}, \quad (\Delta p)^2 = \hbar^2 \alpha_1 + \hbar^2 \frac{\beta_1^2}{\alpha_1}, \quad \Delta(qp) = -\hbar \frac{\beta_1}{2\alpha_1} \tag{13.6}$$

characterizing the peak position, fluctuations and correlations of the state. For all allowed values of the parameters, Gaussian states saturate the uncertainty relation

$$(\Delta q)^2 (\Delta p)^2 - \Delta(qp)^2 = \frac{1}{4}\hbar^2. \tag{13.7}$$

For $\beta_1 = \mathrm{Im} z_1 \neq 0$, the state is correlated. In this case, the uncertainty product $(\Delta q)^2 (\Delta p)^2$ is larger than the minimally possible value $\hbar^2/4$. Correlations $\Delta(qp)$ are bounded from above for given values of fluctuations Δq and Δp. These are all the moments that can be varied for a Gaussian state; such states are thus much more special than general semiclassical ones.

With (13.2), we are considering moments for general density states. If moments are desired only for pure states of some type, further conditions arise which are complicated to derive completely. Fortunately, such a restriction is rarely necessary. Especially for quantum cosmology which constitutes models obtained by averaging inhomogeneous configurations, density states should be expected to be relevant for sufficient generality.

The preceding example illustrates that none of the second-order moments for a canonical algebra of basic operators is redundant even for pure states: At saturation we have one relation between the three variables, and there are indeed two free parameters in a Gaussion once the expectation values of $\hat{q}$ and $\hat{p}$ are fixed. The third independent moment then enters as a free parameter to describe deviations from saturation. Such unsaturated states, however, are more difficult to write down in closed form with specific moments; they are certainly no longer Gaussian. The following examples illustrate the relationship between moments and density states.

Example 13.2 Take the algebra M_2 of complex 2×2-matrices, with an adjoint defined in the usual way by transposition combined with complex conjugation. Self-adjoint generators can be taken to be the Pauli matrices σ_i together with the identity id. There are relations $\mathrm{id}\sigma_i = \sigma_i = \sigma_i \mathrm{id}$ and $\sigma_i \sigma_j = \delta_{ij} + \varepsilon_{ijk}\sigma_k$ between them such that any product can be reduced to an expression linear in the generators. That means that expectation values of matrices in the set $\{\mathrm{id}, \sigma_i\}$ as basic operators are sufficient to parameterize all expectation values, and no independent moments exist. Disregarding the trivial unit which must always have the same expectation value one, the quantum phase space is thus three-dimensional—in accordance with the fact that 2×2 density matrices have three free parameters. (They are self-adjoint and of trace one.) An irreducible Hilbert-space representation $\mathcal{H}_2$, on the other hand, provides a two-dimensional state space: the usual spin-$1/2$ representation is complex two-dimensional, which with normalization and factoring out a total phase corresponds to two real dimensions. In this simple example, we thus see that the state space we obtain by expectation values and moments is larger than the space of pure states.

Example 13.3 Now take the tensor product $M_2 \otimes M_2$ of the algebra M_2 with itself. It has 16 generators $\{\mathrm{id} \otimes \mathrm{id}, \mathrm{id} \otimes \sigma_i, \sigma_j \otimes \mathrm{id}, \sigma_k \otimes \sigma_l\}$. Again, all products can be reduced to expressions linear in the generators, and disregarding the unit we obtain a 15-dimensional state space in accordance with the dimension of the space of self-adjoint, trace one, 4×4 density matrices. An irreducible Hilbert-space representation $\mathcal{H}_2 \otimes \mathcal{H}_2$, on the other hand, is complex four-dimensional, implying six real dimensions for the pure state space.

13.1.4 Casimir Conditions

Further conditions may arise if several irreducible representations exist for a given linear system $[\hat{J}_i, \hat{J}_j] = i\hbar C_{ij}^k \hat{J}_k$. For instance, one may specify values of Casimirs for group coherent states, such as $\hat{C} := \hat{J}_1^2 + \hat{J}_2^2 + \hat{J}_3^2 = \hbar^2 j(j+1)$ for SU(2) with $C_{ij}^k = \varepsilon_{ij}^k$. Taking an expectation value of the operator equation then relates expectation values of the basic operators to second-order moments (for a quadratic Casimir), and provides relations for higher moments from expectation values of the form

$$\langle \hat{J}_1^k \hat{J}_2^l \hat{J}_3^m \hat{C} \rangle = \hbar^2 j(j+1) \langle \hat{J}_1^k \hat{J}_2^l \hat{J}_3^m \rangle. \tag{13.8}$$

Evaluations of these relations then follow the lines as in the example of reality conditions for harmonic loop quantum cosmology used in Sect. 6.1, which amount to a Casimir condition for the algebra sl(2, $\mathbb{R}$).

One may view the Casimir operator as a constraint operator and deal with it along the lines of effective constraints to be described in Sect. 13.2 for first-class constraints. In contrast to the procedure described there, however, a Casimir constraint is second class, even if it is a single constraint. In such a case, it amounts to a constraint on a non-symplectic Poisson manifold (with degenerate Poisson tensor) for which constraints are classified in a generalized way compared to Dirac's well-known procedure; see [2]. A Casimir constraint, by definition, commutes with all basic operators: $[\hat{J}_i, \hat{C}] = 0$. In the quantum phase space, it does not generate any gauge transformations, a property that distinguishes second-class constraints.

13.1.5 Equations of Motion

On the state space parameterized by expectation values and moments, a flow is defined once a Hamiltonian operator in the algebra $\mathscr{A}$ has been chosen. For expectation values of arbitrary time-independent operators $\hat{O}$, we have the well-known equation

$$\frac{d\langle \hat{O} \rangle}{dt} = \frac{\langle [\hat{O}, \hat{H}] \rangle}{i\hbar} \tag{13.9}$$

which can be derived equally well in a Schrödinger or Heisenberg picture. Such an equation also applies to evolving observables in deparameterized systems, as seen in the preceding chapter; (12.6). Using linearity and the Leibniz rule of derivatives we extend the dynamics to all moments, for instance

$$\frac{d(\Delta O)^2}{dt} = \frac{d(\langle \hat{O}^2 \rangle - \langle \hat{O} \rangle^2)}{dt} = \frac{\langle [\hat{O}^2, \hat{H}] \rangle}{i\hbar} - 2\langle \hat{O} \rangle \frac{\langle [\hat{O}, \hat{H}] \rangle}{i\hbar}. \tag{13.10}$$

Example 13.4 (Harmonic oscillator) With $\hat{H} = \frac{1}{2m}\hat{p}^2 + \frac{1}{2}m\omega^2\hat{q}^2$ in the Schrödinger representation, we have

$$\frac{d}{dt}\langle\hat{q}\rangle = \frac{1}{m}\langle\hat{p}\rangle, \quad \frac{d}{dt}\langle\hat{p}\rangle = -m\omega^2\langle\hat{q}\rangle$$

which can be solved easily and in closed form independently of any moment. Solutions agree exactly with the classical ones, as is well known from the Ehrenfest theorem. Similarly, the second-order moments satisfy

$$\frac{d}{dt}(\Delta q)^2 = \frac{2}{m}\Delta(qp)$$

$$\frac{d}{dt}\Delta(qp) = -m\omega^2(\Delta q)^2 + \frac{1}{m}(\Delta p)^2$$

$$\frac{d}{dt}(\Delta p)^2 = -2m\omega^2\Delta(qp)$$

and can also be solved for in closed form without knowing any higher moment. Second-order moments, moreover, are subject to the (generalized) uncertainty relation

$$(\Delta q)^2(\Delta p)^2 - \Delta(qp)^2 \geq \frac{\hbar^2}{4}.$$

The second-order equations allow one to discuss coherent and squeezed states in simple terms. We have non-spreading solutions of constant second-order moments provided that $\Delta(qp) = 0$, $\Delta p = m\omega\Delta q$. All these solutions correspond to coherent states following exactly the classical trajectory. They satisfy the uncertainty relation if $\Delta q \geq \sqrt{\hbar/2m\omega}$. If the uncertainty relation is saturated, we obtain fluctuations that are fixed uniquely to be those of the ground state. If we saturate the uncertainty relation at all times, we have general dynamical coherent states. From the previous equations, this implies fluctuations changing in time unless we have exactly the harmonic oscillator ground state; such states are called squeezed states. See Fig. 13.1 for an illustration. In their most general form, squeezed states may also have non-vanishing correlations if $\Delta(qp) \neq 0$. For given fluctuations, however, the amount of squeezing is bounded by the uncertainty relation $|\Delta(qp)| \leq \sqrt{(\Delta q)^2(\Delta p)^2 - \hbar^2/4}$ even if it is not required to be saturated.

Dynamics is more involved for anharmonic systems, in which the equations of motion for all infinitely many moments no longer decouple to finitely coupled sets of linear equations. In this case, quantum back-reaction results and the precise behavior of an evolving state, including its spreading and deformations from a Gaussian, is important for knowing the time dependence of expectation values or other moments. It is generated by the quantum Hamiltonian

$$H_Q = \langle H(\hat{q}, \hat{p})\rangle = \langle H(q + (\hat{q} - q), p + (\hat{p} - p))\rangle \tag{13.11}$$

$$= H(\langle\hat{q}\rangle, \langle\hat{p}\rangle) + \sum_{a,b} \frac{1}{a!b!} \frac{\partial^{a+b} H(\langle\hat{q}\rangle, \langle\hat{p}\rangle)}{\partial\langle\hat{q}\rangle^a \partial\langle\hat{p}\rangle^b} \Delta(q^a p^b). \tag{13.12}$$

Fig. 13.1 Spreading of a
squeezed state compared to
an unsqueezed state. Plotted
are the time-dependent
q-expectation value and
q-fluctuations around it for
two different states, one
without correlations (*solid*)
and one with non-vanishing
correlations (*dashed*)

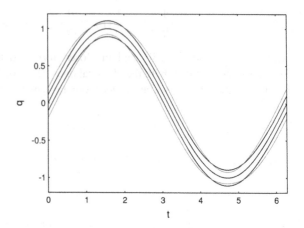

The expansion in terms of the moments, just as a semiclassical $\hbar$-expansion used for the WKB approximation, or like the Feynman expansion, is in general not convergent but asymptotic. The equations of motion $\dot{f} = \{f, H_Q\}$ it generates follow from the Poisson bracket in the quantum phase space. Its relationship to commutators then shows that H_Q generates the correct flow for expectation values and moments based on (13.9).

In canonical formulations, the Hamiltonian determines the dynamics together with the Poisson structure. One could therefore suspect that not only the Hamiltonian but also the Poisson brackets receive quantum corrections. This suspicion is correct: The Poisson structure receives corrections by the presence of moments as new quantum degrees of freedom, enlarging the classical phase space. However, by definition (13.1) there are no quantum corrections to the basic variables as long as they form a closed algebra (belonging to a linear quantum system). The Poisson brackets of expectation values of basic operators agree with the classical Poisson brackets, without quantum corrections in this relation.

If one does not work with the full Poisson manifold of expectation values and all moments, but rather embeds the classical phase space non-horizontally in the quantum phase space seen as a fiber bundle over the classical phase space, quantum corrections to Poisson brackets of expectation values do arise. For instance, in the following example we will see that under certain conditions (more precisely, when an adiabatic approximation can be applied for the moments) one can solve for the moments $\Delta(q^a p^b)(\langle \hat{q} \rangle, \langle \hat{p} \rangle)$ in terms of the expectation values. These solutions then describe an embedding $(q, p) \mapsto (q, p, \Delta(q^a p^b)(q, p))$ of the classical phase space in the quantum phase space. If one pulls back the symplectic form $d\langle \hat{q} \rangle \wedge d\langle \hat{p} \rangle + \Omega_{a,b;c,d}(\Delta) d\Delta(q^a p^b) \wedge d\Delta(q^c p^d)$ on the quantum phase space (with moment-dependent coefficients encoding the Poisson structure (13.1) for moments), one may obtain a quantum-corrected symplectic form

$$(1 + 2\Omega_{a,b;c,d}(\Delta(q, p)))(\partial\Delta(q^a p^b)/\partial q)(\partial\Delta(q^c p^d)/\partial p))dq \wedge dp.$$

(Such a correction does not arise in the following example because the moments solved for in terms of expectation values depend only on $\langle \hat{q} \rangle$, not on $\langle \hat{p} \rangle$.) Sometimes, corrections of this form also arise when one begins the derivation of effective equations by an assumption on, rather than derivation of, the dependence of moments on expectation values, as in [3].

Thus, if quantum corrections to the symplectic or Poisson structure arise, they are a consequence of a secondary step in solving effective equations or making assumptions about their solutions. If moments are kept independent of expectation values, which is the most general form of effective equations, no corrections to the Poisson brackets of expectation values arise, while Poisson brackets between moments as they follow from commutators have no classical analog whatsoever.

The infinitely-coupled dynamics of ordinary differential equations for moments is equivalent to the partial differential equation given by the Schrödinger flow. Usually, the Schrödinger equation is much more useful for solving the dynamics. But in certain regimes, equations for moments lend themselves more easily to approximation schemes.

Example 13.5 (Anharmonic oscillator) With a cubic anharmonicity in the Hamiltonian

$$\hat{H} = \frac{1}{2m}\hat{p}^2 + \frac{1}{2}m\omega^2\hat{q}^2 + \frac{1}{3}\lambda\hat{q}^3$$

we have equations of motion

$$\frac{d}{dt}\langle\hat{q}\rangle = \frac{1}{m}\langle\hat{p}\rangle, \quad \frac{d}{dt}\langle\hat{p}\rangle = -m\omega^2\langle\hat{q}\rangle - \lambda\langle\hat{q}\rangle^2 - \lambda(\Delta q)^2$$

with a non-linear coupling of expectation values and the q-fluctuation. To solve these equations, we must know the behavior of Δq, which is itself time dependent with a rate of change

$$\frac{d}{dt}(\Delta q)^2 = \frac{2}{m}C_{qp} \tag{13.13}$$

given by the covariance, obeying

$$\frac{d}{dt}C_{qp} = \frac{1}{m}C_{qp} + m\omega^2(\Delta q)^2 + 6\lambda\langle\hat{q}\rangle(\Delta q)^2 + 3\lambda\Delta(q^3). \tag{13.14}$$

Here, a higher moment $\Delta(q^3)$ of third order couples. Proceeding this way in the hope of finding a closed set of equations, one will have to include all infinitely many moments.

For a general anharmonic system, we consider a classical Hamiltonian $H = \frac{1}{2m}p^2 + \frac{1}{2}m\omega^2q^2 + U(q)$. In this case, additional approximation schemes open up which will allow us to compare results with other derivations based, for instance, on effective actions. Using the parameters of the anharmonic system, we first introduce dimensionless variables $\tilde{\Delta}(q^b p^a) := \hbar^{-(a+b)/2}(m\omega)^{b/2-a/2}\Delta(q^b p^a)$. The quantum Hamiltonian, including coupling terms between expectation values and moments, can then be expanded in powers of $\hbar^{1/2}$:

$$H_Q(\langle\hat{q}\rangle, \langle\hat{p}\rangle, \Delta(\cdots)) = \frac{1}{2m}\langle\hat{p}\rangle^2 + \frac{1}{2}m\omega^2\langle\hat{q}\rangle^2 + U(\langle\hat{q}\rangle) + \frac{\hbar\omega}{2}\left((\tilde{\Delta}q)^2 + (\tilde{\Delta}p)^2\right)$$

$$+ \sum_{n>2}\frac{1}{n!}\left(\frac{\hbar}{m\omega}\right)^{n/2} U^{(n)}(\langle\hat{q}\rangle)\tilde{\Delta}(q^n). \tag{13.15}$$

Here the dimensionless variables are useful to organize the formal expansion with an explicit expansion parameter $\hbar$. In general, such expansions can still be performed if one assumes the moments to obey the semiclassical hierarchy $\Delta(q^a p^b) \sim O(\hbar^{(a+b)/2})$. With the dimensionless moments, we have $\tilde{\Delta}(q^a p^b) \sim O(1)$ and all orders of $\hbar$ are indeed shown explicitly in (13.15). From this expression, we see the zero-point energy given just by the fluctuations, $(\tilde{\Delta}q)^2 = (\tilde{\Delta}p)^2 = 1/2$ for the ground state, and quantum back-reaction from coupling terms between expectation values and moments.

Equations of motion then follow in the usual Hamiltonian way as $\dot{f} = \{f, H_Q\}$ for any phase-space function f, using the Poisson brackets as defined in (13.1) together with linearity and the Leibniz rule. One can easily see that the resulting equations agree with what we had earlier obtained by the background-state method in Sect. 5.4.1, expanding the Hamiltonian operator directly but formally. For the anharmonic oscillator, we have

$$\langle\dot{q}\rangle = \frac{\langle\hat{p}\rangle}{m} \tag{13.16}$$

$$\langle\dot{p}\rangle = -m\omega^2\langle\hat{q}\rangle - U'(\langle\hat{q}\rangle) - \sum_n\frac{1}{n!}\left(\frac{\hbar}{m\omega}\right)^{n/2} U^{(n+1)}(\langle\hat{q}\rangle)\tilde{\Delta}(q^n) \tag{13.17}$$

$$\dot{\tilde{\Delta}}(q^{n-a}p^a) = -a\omega\tilde{\Delta}(q^{n-a+1}p^{a-1}) + (n-a)\omega\tilde{\Delta}(q^{n-a-1}p^{a+1})$$

$$- a\frac{U''(\langle\hat{q}\rangle)}{m\omega}\tilde{\Delta}(q^{n-a+1}p^{a-1})$$

$$+ \frac{\sqrt{\hbar}aU'''(\langle\hat{q}\rangle)}{2(m\omega)^{\frac{3}{2}}}\tilde{\Delta}(q^{n-a}p^{a-1})\tilde{\Delta}(q^2) + \frac{\hbar aU''''(\langle\hat{q}\rangle)}{3!(m\omega)^2}\tilde{\Delta}(q^{n-a}p^{a-1})\tilde{\Delta}(q^3)$$

$$- \frac{a}{2}\left(\frac{\sqrt{\hbar}U'''(\langle\hat{q}\rangle)}{(m\omega)^{\frac{3}{2}}}\tilde{\Delta}(q^{n-a+2}p^{a-1}) + \frac{\hbar U''''(\langle\hat{q}\rangle)}{3(m\omega)^2}\tilde{\Delta}(q^{n-a+3}p^{a-1}) + \cdots\right)$$

$$+ \frac{a(a-1)(a-2)}{24}\left(\frac{\sqrt{\hbar}U'''(\langle\hat{q}\rangle)}{(m\omega)^{\frac{3}{2}}}\tilde{\Delta}(q^{n-a}p^{a-3})\right.$$

$$\left. + \frac{\hbar U''''(\langle\hat{q}\rangle)}{(m\omega)^2}\tilde{\Delta}(q^{n-a+1}p^{a-3}) + \cdots\right). \tag{13.18}$$

As expected, these are infinitely many coupled equations for infinitely many variables. (Incidentally, note that the expansion is in terms of $\sqrt{\hbar}$, not of $\hbar$ as one might have expected. Half-integer powers of $\hbar$ only drop out if all odd-order moments vanish, which is the case for harmonic oscillator coherent states but not necessarily in general.)

Solving these equations even approximately requires truncations that allow one to disregard all but finitely many moments. The set of moments indeed has the hierarchy of finite-dimensional spaces $\mathscr{P}_n$ defined as the subsets obtained by allowing moments $\Delta(q^a p^b)$ only up to a finite order $a + b \leq n$. Commutators do not directly provide a closed Poisson bracket on these spaces; for instance, at third order we have $\{\Delta(q^3), \Delta(p^3)\} = 9(\Delta(q^2 p^2) - (\Delta q)^2(\Delta p)^2)$ which includes a moment of order four. (Only second-order moments have a closed Poisson algebra among themselves.) But truncating not only the Hamiltonian but also the Poisson brackets at a finite order n provides the dynamics on a closed Poisson manifold. That this truncation also of the Poisson brackets is in fact required for a consistent approximation of equations of motion can be seen from the observation that it results in a Hamiltonian flow corresponding exactly to the equations of motion truncated at the same order n.

There is thus a consistent truncation procedure which in terms of dimensionless moments disregards all terms beyond a certain order $\hbar^{n/2}$. Here, the expansion is a semiclassical one, but formally one can do similar truncations in other regimes. To that end, we would replace each moment $\Delta(q^a p^b)$ with $\mu^{(a+b)/2}\Delta(q^a p^b)$ and $\hbar$ with $\mu\hbar$, expand all equations up to terms of order $\mu^{n/2}$ (irrespective of the value of $\hbar$). After setting $\mu = 1$, we obtain the truncated equations.

In general, the truncation results in a higher-dimensional dynamical system in which the expectation values are accompanied by some of the moments as true quantum degrees of freedom. By the coupling terms among the equations, quantum back-reaction is realized. Sometimes one can further reduce the system to one of the classical dimension, involving only expectation values, such that quantum back-reaction is realized in a simpler way, for instance by an effective potential. Such an additional step requires solutions to some of the equations for moments in terms of expectation values, which can then be inserted into the expectation-value equations. For the anharmonic oscillator, this is possible by an adiabatic approximation for the moments which indeed allows one to solve the moment equations without knowing the behavior of expectation values, and then insert the solutions into the coupling terms of expectation-value equations.

Performing an adiabatic approximation again makes use of a formal parameter λ to arrange expansions, and in the end setting $\lambda = 1$. Now, a small value of the parameter should mean that the time dependence of moments is only weak. We thus rescale $\mathrm{d}/\mathrm{d}t$ to $\lambda\mathrm{d}/\mathrm{d}t$ and expand $\Delta(q^a p^b) = \sum_e \Delta_e(q^a p^b)\lambda^e$. After solving the equations order by order in λ, we will finally set $\lambda = 1$. There is no claim that the λ-expansion of $\Delta(\cdots)$ is a converging sum, and the procedure might seem questionable. Instead, the expansion merely serves to organize different kinds of time dependences of the moments. To n-th adiabatic order, comparing λ-coefficients in the adiabatic expansion of the general Hamiltonian equation of motion leads to

$$\{\Delta_n(q^a p^b), H_Q\} = \dot{\Delta}_{n-1}(q^a p^b). \tag{13.19}$$

(Writing this equation as $\{\Delta_n(q^a p^b) - \Delta_{n-1}(q^a p^b), H_Q\} = 0$ shows that at n-th adiabatic order the highest-order term is assumed to provide a contribution of negligibly small time dependence to the expansion of moments.) Iterating (13.19) over n, with the left-hand side computed as a Poisson bracket on phase space rather than interpreted as a time derivative, this procedure provides algebraic equations to all orders. We will see this explicitly when we apply these equations to the anharmonic oscillator.

To first order in $\hbar$ and zeroth in λ, we have from (13.18) without the $\hbar$-terms and without the time derivative (which is at least of first order in λ) [4]

$$0 = (n - a)\tilde{\Delta}_0(q^{n-a-1} p^{a+1}) - a\left(1 + \frac{U''(\langle\hat{q}\rangle)}{m\omega^2}\right)\tilde{\Delta}_0(q^{n-a+1} p^{a-1}). \quad (13.20)$$

As a step-2 recurrence in a for $\tilde{\Delta}_0(q^{n-a} p^a)$ starting with $\tilde{\Delta}_0(q^n)$, (13.20) must stop at $a = n$ since the sequence of n-th order moments is finite. Non-zero solutions of this form can arise only if $n = a$ is the last step in the recursion, which implies that a must be even for even n and odd for odd n. For even n and a we have the general solution

$$\tilde{\Delta}_0(q^{n-a} p^a) = \binom{n/2}{a/2}\binom{n}{a}^{-1}\left(1 + \frac{U''(\langle\hat{q}\rangle)}{m\omega^2}\right)^{a/2}\tilde{\Delta}_0(q^n), \quad (13.21)$$

and we will not need the odd-order solutions. To this order, $\tilde{\Delta}_0(q^n)$ remains free. To first order in λ, we have

$$(n - a)\tilde{\Delta}_1(q^{n-a-1} p^{a+1}) - a\left(1 + \frac{U''(\langle\hat{q}\rangle)}{m\omega^2}\right)\tilde{\Delta}_1(q^{n-a+1} p^{a-1}) = \frac{1}{\omega}\dot{\tilde{\Delta}}_0(q^{n-a} p^a)$$

which implies

$$\sum_{a \text{ even}}\binom{n/2}{a/2}\left(1 + \frac{U''(\langle\hat{q}\rangle)}{m\omega^2}\right)^{(n-a)/2}\dot{\tilde{\Delta}}_0(q^{n-a} p^a) = 0$$

for consistency. This condition requires

$$\tilde{\Delta}_0(q^n) = C_n(1 + U''(\langle\hat{q}\rangle)/m\omega^2)^{-n/4} \quad (13.22)$$

with constants C_n. At this stage, we see that the zeroth-order adiabatic solutions are not t-independent but implicitly depend on t via $\langle\hat{q}\rangle$ if the potential is not quadratic. The equation (13.20) we solved therefore cannot provide an exact solution to (13.18): Higher orders in the adiabatic approximation correct for this error in the time dependence; this reorganization of the t-dependence is the actual meaning of the formal λ-expansion.

To obtain moments of the ground state for the harmonic limit $U = 0$, we must have $C_n = 2^{-n} n!/(n/2)!$ (as well as $\tilde{\Delta}(q^{n-a} p^a) = 0$ for a and n odd, which is why we do not need the solution of (13.20) for such values). The zeroth adiabatic order for second-order moments is then fixed completely, and we obtain the first correction

$$\dot{p} = -m\omega^2 q - U'(q) - \frac{\hbar}{2m\omega} U'''(q)(\tilde{\Delta}_0 q)^2 + \cdots$$

$$= -m\omega^2 q - U'(q) - \frac{\hbar}{4m\omega} \frac{U'''(q)}{\sqrt{1 + U''(q)/m\omega^2}} + \cdots$$

to the classical equations of motion. The last term shows the leading correction in the effective force as compared to the classical force $-U'(q)$ from the effective potential in H_Q with the zero-point energy.

To second adiabatic order [4] we obtain additional corrections, which imply the second-order equation of motion

$$\left(m + \frac{\hbar U'''(\langle \hat{q} \rangle)^2}{32m^2\omega^5(1 + U''(\langle \hat{q} \rangle)/m\omega^2)^{5/2}} \right) \langle \ddot{\hat{q}} \rangle$$

$$+ \frac{\hbar \langle \dot{\hat{q}} \rangle^2 \left(4m\omega^2 U'''(\langle \hat{q} \rangle) U''''(\langle \hat{q} \rangle)(1 + U''(\langle \hat{q} \rangle)/m\omega^2) - 5U'''(\langle \hat{q} \rangle)^3 \right)}{128m^3\omega^7(1 + U''(\langle \hat{q} \rangle)/m\omega^2)^{7/2}}$$

$$+ m\omega^2 \langle \hat{q} \rangle + U'(\langle \hat{q} \rangle) + \frac{\hbar U'''(\langle \hat{q} \rangle)}{4m\omega(1 + U''(\langle \hat{q} \rangle)/m\omega^2)^{1/2}} = 0.$$

for the position expectation value. There is a new velocity-dependent contribution to the effective force at this order, and the classical mass has received a position-dependent correction. One can verify that this equation results from an action

$$\Gamma_{\text{eff}}[q(t)] = \int dt \left(\frac{1}{2} \left(m + \frac{\hbar U'''(q)^2}{2^5 m^2 \left(\omega^2 + m^{-1} U''(q) \right)^{5/2}} \right) \dot{q}^2 \right. \tag{13.23}$$

$$\left. - \frac{1}{2} m\omega^2 q^2 - U(q) - \frac{\hbar\omega}{2} \sqrt{1 + \frac{U''(q)}{m\omega^2}} \right) \tag{13.24}$$

in agreement, to this order, with the low-energy effective action derived for instance via path integration [5]. (Although the equations of motion coincide, the meaning of q in the low-energy effective action is different from that of $\langle \hat{q} \rangle$ in canonical effective equations: q is related to off-diagonal matrix elements of $\hat{q}$ [5] and not even guaranteed to be real. Conceptually, the low-energy effective action is thus problematic in a semiclassical context.)

13.1.6 General Properties

The examples of anharmonic systems illustrate that the solvable nature of the harmonic oscillator, not surprisingly, is very special. Solvability does not just mean that one is lucky enough to find closed solutions for a wave function. We even have closed solutions for all the moments (which would require additional integrations if derived from a wave function) and the dynamical behavior of moments of any given order is independent of the others. This strong harmonic sense of solvability is realized only rarely, in explicitly treatable but most often unrealistic ideal models. In addition to the harmonic oscillator in quantum mechanics, it can be found in the same form for free quantum field theories—or in the harmonic cosmologies of Chap. 6.

In other systems, quantum corrections arise from the back-reaction of fluctuations and higher moments. In the language of quantum field theory, these are loop corrections. The scheme presented here also produces such terms, but it is more encompassing: we obtain not only quantum-correction terms but compute state properties such as fluctuations along the way. Moments are not fixed completely, but their dynamical behavior is solved for, starting from an initial state through the initial values required to be posed for the differential equations we have. In this way, if we start with a harmonic oscillator ground state, we derive properties of the interacting ground state in the anharmonic system. For instance, from (13.21) and (13.22) we derive $(\tilde{\Delta}_0 q)^2 = \frac{1}{2}(1 + U''(\langle \hat{q} \rangle)/m\omega^2)^{-1/2}$ and $(\tilde{\Delta}_0 p)^2 = \frac{1}{2}(1 + U''(\langle \hat{q} \rangle)/m\omega^2)^{1/2}$ in the anharmonic ground state, which still saturate the uncertainty relation to the orders considered. This property is crucial for the general consistency of the scheme: At times other than the initial time, state properties are not put in from the outset, as it is sometimes suggested in other derivations based for instance on a geometrical formulation of quantum mechanics [3]. Having to prescribe the state at different times would require too much bias about the dynamical behavior especially if one evolves toward high quantum regimes.

We have also seen, for instance in the application to harmonic cosmology, that the equations shed much light on the availability and behavior of dynamical coherent states, especially in combination with the uncertainty relation. Thus, not only the interacting ground state is accessible, but a much larger class of states can be analyzed by choosing different initial values.

In comparison with the low-energy effective action we notice several new features. First, effective equations are state dependent; only when a specific state to expand around is selected, even if this is done implicitly as in most derivations of effective actions, do unique quantum-correction terms result. Not all regimes of interest clearly distinguish a unique state to expand around, such as the vacuum state, and so more freedom to parameterize state-dependent properties is present. The canonical procedure allows this freedom; for instance we may decide not to fix C_2 in the adiabatic solutions, on which only a lower bound is then given by the uncertainty relation. Secondly, general effective equations may not allow the application of an adiabatic expansion, or some other approximation in addition to the semiclassical one. In the example of anharmonic oscillators, it was the adiabaticity assumption that allowed us to solve for the moments in terms of expectation values. When there is no substitute for this assumption, moments remain a-priori undetermined and must be kept as additional independent variables. One is then dealing with a higher-dimensional effective system with true quantum degrees of freedom. This enlargement of the dimensionality is reminiscent of higher-derivative effective actions which require additional initial values to be posed for higher time derivatives of the fields. However, there is no straightforward mapping between the degrees of freedom, and in fact not all the freedom in perturbative higher-derivative effective actions is consistent [6]. The degrees of freedom of higher-dimensional canonical effective systems, on the other hand, are all consistent and have clear physical interpretations as the moments of a state.

The procedure is manageable in explicit terms if a free system is available. For canonical basic variables, this requires the Hamiltonian operator $\hat{H}$ to be quadratic. More generally, however, there are additional options formulated by a linear system with basic operators $\hat{J}_i$ forming a closed commutator algebra with the Hamiltonian: $[\hat{J}_i, \hat{H}] = \sum_j a_{ij} \hat{J}_j + b_i \hat{H}$. It is easy to see that such a linear algebra ensures decoupling properties of the Ehrenfest equations (13.9), which then close for the expectation values. Once a free system has been identified within a larger class of models, perturbation theory can be used to test robustness and to derive more general properties. An example for a linear system without a quadratic Hamiltonian is harmonic cosmology.

Positivity of the Hamiltonian is another important aspect. It is guaranteed for the harmonic oscillator, but it becomes an issue if systems need be reformulated to bring them in the usual Hamiltonian form, for instance by deparameterizing a constrained system. Especially if the system is relativistic, a square root must be taken and combined with a judicious sign choice. We may for instance have a constraint of the form $p_\varphi^2 - H^2 = 0$ as in relativistic deparameterizable systems (Sect. 12.3.2). Taking a square root $p_\varphi = \pm|H|$ means that the absolute value may destroy strict linearity, as already discussed in the context of cosmology. As we will see in more detail in the next section, quantum constraints can be dealt with just like quantum Hamiltonians, by taking the expectation value as a function on the state space and expanding in the moments. There are several differences compared to a non-relativistic system; for instance, additional moments of p_φ may contribute to quantum constraints unlike in the linearized deparameterized version. For now, we are interested only in the fact that we will have to take a square root when solving for p_φ and bringing the system to deparameterized form.

A sign ambiguity thus arises in solving for $p_\varphi = \pm|H|$ to obtain the two sectors of positive-frequency and negative-frequency modes. No difference occurs between classical and effective constraints from this perspective. But the nature of a linear quantum system may be spoiled by the absolute value around H. Even if H is quadratic in canonical variables, such as $H = cp$ in the solvable model of Wheeler–DeWitt cosmology, $|H|$ is not strictly quadratic. However, on states supported only on the positive or negative part of the spectrum of H, respectively, the absolute value can be dropped and the system will be linear. If H is time-independent, the required spectral property of states will be preserved and it is necessary to restrict only initial values to the spectral property. This can easily be done without too strong restrictions on the moments and the accessible states.

In a Hilbert-space representation, $|\widehat{cp}|$ is usually a non-local operator if wave functions such as $\psi(c)$ or $\psi(p)$ are used. However, this kind of non-locality is representation dependent (and thus unphysical). An operator of this form would certainly be local in a representation based on eigenstates of the operator $\widehat{cp}$. Since effective equations are representation-independent, they are insensitive to this non-locality problem.

13.2 Effective Constraints

As already indicated, a constraint operator gives rise to the quantum constraint $C_Q = \langle \hat{C} \rangle = C_{\text{class}}(q, p) + \cdots$ on the state space since it must vanish when the expectation value is taken in physical states. Quantum corrections to the classical constraint surface thus arise. However, a single constraint on the quantum phase space only removes two degrees of freedom (by constraining and factoring out a gauge, provided the constraint is part of a first-class system). Corresponding quantum variables would remain unconstrained, leaving infinitely many spurious variables which would never arise in a quantization of the classical reduced phase space.

Removing all degrees of freedom as appropriate requires additional constraints. In fact, there are infinitely many ones

$$C_{f(q,p)} := \langle f(\hat{q}, \hat{p}) \hat{C} \rangle$$

on the state space, which all have to vanish in physical states and which in general are independent of C_Q. Practically, one can usually work with polynomials $f(q, p)$, as we will also see in examples below. As with the truncations in the discussion of effective dynamics, solutions of the constraints become feasible to any given order in moments, where only finitely many C_f have to be considered.

At this general level, several properties important for the consistency of the constrained system can be derived [7]. The system of constraints is first-class if the ordering $\langle f(\hat{q}, \hat{p}) \hat{C} \rangle$ is indeed used as indicated in the definition. This means that we are dealing with operator products not symmetric in general. Naive reorderings either spoil the first-class nature, or lead to functions that are not constraints if the constraint operator no longer stands at the very right or left of the operator product. Alternatively, one may define quantum constraints via the generating functional $C_\alpha = \langle e^{i\hbar^{-1}\alpha_i \cdot \hat{x}^i} C(\hat{x}^i) \rangle$ (where $(\hat{x}^i)_{i=1,2} = (\hat{q}, \hat{p})$) for Weyl-ordered and thus symmetric constraints, which also provides a first-class system. This procedure does include certain reorderings of our constraints above, obtained by all possible derivatives of C_α by the components α_i, and then setting $\alpha = 0$. But practically it is not always easy to derive the Weyl-ordered constraints and we will continue with the non-symmetric constraints.

In fact, we do not require reality of kinematical quantum variables $\Delta(q^a p^b)$, realized before constraints are solved. Usually the kinematical inner product has to be changed to the physical one after the constraints are solved, and requiring self-adjointness or reality before this step would not guarantee the correct reality properties. Then there is no reason either to require constraint functions to be real, or to come from symmetrically ordered operator products including the constraint. We will impose reality conditions only after solving the constraints to ensure physical normalization. In this way, physical observables are accessible without deriving the full physical Hilbert space or a specific integral form of the physical inner product. Also here, the procedure is very well amenable to perturbative techniques, which is not the case for other constructions of physical Hilbert spaces.

Different gauge fixings of the constrained system can in some cases be seen to be related to different kinematical Hilbert space structures. Results will thus be independent of ambiguities that otherwise arise by choosing a particular kinematical Hilbert space as one does in group averaging. Moreover, since Hilbert-space representations are not referred to, no difference in the treatment arises between constraints with zero in their discrete or continuous spectra.

13.2.1 Non-Relativistic Constraints

Non-relativistic constraints are characterized by a linear dependence on one momentum variable which can be considered as an energy. In non-relativistic deparameterizable systems, this linear momentum is conjugate to a global time variable. From the point of view of physical Hilbert spaces as well as effective constraints this case is easier to deal with, and we discuss it first by way of examples.

Example 13.6 (*Linear constraint*) Let us take a constraint $\hat{C} = \hat{p}$. It immediately implies $\langle \hat{p} \rangle = 0$, while $\langle \hat{q} \rangle$ is pure gauge. At second order of the moments, we have $(\Delta p)^2 = 0$ fully constrained, while $C_q = \langle \hat{q}\hat{p} \rangle = 0$ implies the complex-valued kinematical covariance

$$\Delta(qp) = \frac{1}{2}\langle \hat{q}\hat{p} + \hat{p}\hat{q} \rangle - \langle \hat{q} \rangle \langle \hat{p} \rangle = -\frac{1}{2}i\hbar.$$

At least one of the kinematical variables must thus take complex values, which means that we indeed have to redefine our kinematical inner product for the physical Hilbert space. Dealing with complex variables cannot always be avoided; it even has an advantage: the uncertainty relation

$$(\Delta q)^2 (\Delta p)^2 - \Delta(qp)^2 \geq \frac{1}{4}\hbar^2$$

remains respected even though one of the fluctuations, Δp, vanishes. (The other one, Δq, is pure gauge.) With the uncertainty relation at one's disposal, one can analyze dynamical coherent states also for constrained systems.

Example 13.7 (*Two-component linearly constrained system and physical observables* [7]) We now assume a constraint $\hat{C} = \hat{p}_1 - \hat{p}$ which is still linear but defined for a system with two independent pairs of degrees of freedom. We denote the kinematical quantum variables as

$$\Delta(p^a q^b p_1^c q_1^d) = \langle (\hat{p} - p)^a (\hat{q} - q)^b (\hat{p}_1 - p_1)^c (\hat{q}_1 - q_1)^d \rangle_{\text{symm}}.$$

As second-order constraints in addition to $C_Q = \langle \hat{p}_1 \rangle - \langle \hat{p} \rangle$ we have

$$C_q = -\frac{i\hbar}{2} - \Delta(qp) + \Delta(qp_1), \quad C_p = \Delta(pp_1) - (\Delta p)^2$$

$$C_{p_1} = (\Delta p_1)^2 - \Delta(pp_1), \quad C_{q_1} = \frac{1}{2}i\hbar - \Delta(pq_1) + \Delta(q_1 p_1)$$

solved by $\langle \hat{p}_1 \rangle = \langle \hat{p} \rangle$ and

$$\Delta(qp_1) \approx \frac{1}{2}i\hbar + \Delta(qp), \quad \Delta(pp_1) \approx (\Delta p)^2$$

$$(\Delta p_1)^2 \approx \Delta(pp_1) \approx (\Delta p)^2, \quad \Delta(pq_1) \approx \frac{1}{2}i\hbar + \Delta(q_1 p_1).$$

To derive observables, we also need the gauge flows which, for instance for C_p, are of the form

$$\delta\Delta(qp) = \Delta(pp_1) - 2(\Delta p)^2 \approx -(\Delta p)^2$$
$$\delta(\Delta q)^2 = 2\Delta(qp_1) - 4\Delta(qp) \approx i\hbar - 2\Delta(qp)$$
$$\delta\Delta(q_1 p_1) = \Delta(pp_1) \approx (\Delta p)^2$$
$$\delta\Delta(qq_1) = \Delta(q_1 p_1) + \Delta(qp) - 2\Delta(pq_1) \approx \Delta(qp) - \Delta(q_1 p_1) - i\hbar$$
$$\delta(\Delta q_1)^2 = \Delta(pq_1) \approx i\hbar + 2\Delta(q_1 p_1).$$

It is easy to check that all gauge flows satisfy $\delta\Delta(qp) = -\delta\Delta(q_1 p_1)$, $\delta\Delta(qq_1) = -\frac{1}{2}\left(\delta(\Delta q)^2 + \delta(\Delta q_1)^2\right)$. Thus, the functions

$$(\mathscr{D}q)^2 := (\Delta q)^2 + 2\Delta(qq_1) + (\Delta q_1)^2 \tag{13.25}$$

$$(\mathscr{D}p)^2 := (\Delta p)^2 \tag{13.26}$$

$$\mathscr{D}(qp) := \Delta(qp) + \Delta(q_1 p_1) + \frac{1}{2}i\hbar \tag{13.27}$$

on the kinematical state space are gauge invariant. They also satisfy the correct algebra

$$\{(\mathscr{D}q)^2, \mathscr{D}(qp)\} = 2(\mathscr{D}q)^2 \tag{13.28}$$

$$\{(\mathscr{D}q)^2, (\mathscr{D}p)^2\} = 4\mathscr{D}(qp) \tag{13.29}$$

$$\{\mathscr{D}(qp), (\mathscr{D}p)^2\} = 2(\mathscr{D}p)^2 \tag{13.30}$$

expected for second-order quantum variables on a reduced state space. For this, the imaginary contribution in (13.27) is required, which need not be added just to make $\mathscr{D}(qp)$ gauge-invariant. At this stage, all moments involving the pair (q_1, p_1) are either expressed in terms of (q, p)-moments by solving constraints, or subject to gauge. (One could certainly as well choose to express the (q, p)-moments in terms of the (q_1, p_1) ones.)

We finally impose reality conditions: $\mathcal{D}(q^a p^b) \in \mathbb{R}$. The observable moments can then be seen as those computed in a physical Hilbert space. Expressed in terms of kinematical quantum variables, this again requires an imaginary part $\mathrm{Im}(\Delta(qp) + \Delta(q_1 p_1)) = -\frac{1}{2}i\hbar$ for $\mathcal{D}(qp) := \Delta(qp) + \Delta(q_1 p_1) + \frac{1}{2}i\hbar$ to be real. Since kinematical quantum variables must be complex to provide correct physical reality (although they appear symmetrically ordered), we have an explicit example in which the kinematical Hilbert-space structure must be changed when deriving the physical Hilbert space. We can also see explicitly that different kinematical choices are possible, such as $\Delta(qp)$ real or $\Delta(q_1 p_1)$ real. They imply different kinematical reality conditions, all resulting in the same physical observables, and thus correspond to different kinematical Hilbert space structures.

This simple example can be used to illustrate one aspect of the problem of time and how it can be overcome by effective techniques; see also Sect. 13.2.3. In the preceding example we could easily choose q as time instead of q_1; the choice is simply a question of gauge fixing of the effective constraints. If one were to implement the constraints for physical states in a Hilbert space, on the other hand, the situation would be more subtle. The constraint equations could certainly still be solved easily, and physical inner products be constructed, for instance by group averaging. However, the different choices of time are difficult to compare at the Hilbert-space level, where no obvious mapping between the two deparameterized Hilbert spaces exists. In this simple model one could easily postulate a natural mapping thanks to the symmetric way in which q and q_1 and their momenta appear. For general constrained systems, however, no such mapping would be available, while effective constraints still provide comparisons between different choices of time via the corresponding gauge fixings.

Explicit derivations as in the case of a linear constraint are more complicated when constraints are non-linear since different orders of moments are mixed in expectation values. But similar structures for solutions to constraints and observables remain realized.

Example 13.8 (*Free particle*) A parameterized free particle has the constraint $\hat{C} = \hat{p}_t + \hat{p}^2/2M$, and thus $C_Q = \langle \hat{p}_t \rangle + \langle \hat{p} \rangle^2/2M + (\Delta p)^2/2M$, $C_q = \Delta(qp_t) + i\hbar \langle \hat{p} \rangle/2M + \langle \hat{p} \rangle \Delta(qp)/M$ and so on [7]. Solving the constraints is more involved than in the linear case, but the procedure is the same. Now, second-order observables are $\mathcal{P} = \langle \hat{p} \rangle$, $\mathcal{Q} = \langle \hat{q} \rangle - \langle \hat{t} \rangle \langle \hat{p} \rangle/M - \Delta(pt)/M$ and

$$(\mathcal{D}p)^2 = (\Delta p)^2$$

$$\mathcal{D}(qp) = \Delta(qp) + \Delta(tp_t) - \frac{\langle \hat{t} \rangle}{M}(\Delta p)^2 + \frac{i\hbar}{2}$$

$$= (\Delta q)^2 - 2\frac{\langle\hat{p}\rangle}{M}\Delta(qt) + \frac{\langle\hat{p}\rangle^2}{M^2}(\Delta t)^2$$
$$- \frac{2\langle\hat{t}\rangle}{M}\left(\Delta(qp) + \Delta(tp_t) + \frac{i\hbar}{2}\right) + \frac{\langle\hat{t}\rangle^2}{M^2}(\Delta p)^2$$

with explicit couplings between kinematical expectation values and moments. Also here, second-order moments satisfy the correct algebra. For a free particle, we do not expect quantum back-reaction in the kinematical quantities, even though physical expectation values depend on some moments. The relationship can be understood by deparameterizing: We invert our equations for the observables to obtain

$$\langle\hat{q}\rangle(\langle\hat{t}\rangle) = \mathscr{Q} + \frac{\langle\hat{t}\rangle}{M}\mathscr{P} + \frac{1}{M}\Delta(pt)$$
$$\approx \mathscr{Q} + \frac{\langle\hat{t}\rangle}{M}\mathscr{P} - \frac{1}{\langle\hat{p}\rangle}\left(\Delta(tp_t) + \frac{i\hbar}{2}\right)$$
$$= \mathscr{Q} + \frac{\langle\hat{t}\rangle}{M}\mathscr{P} - \frac{1}{\mathscr{P}}\left(\mathscr{D}(qp) + \frac{\langle\hat{t}\rangle}{M}(\mathscr{D}p)^2 - \Delta(qp)\right)$$

and now interpret the observable coefficients as integration constants. For the gauge choice $\Delta(tp_t) = -\frac{1}{2}i\hbar$ identical to what we used before, we have correct deparameterized solutions

$$\langle\hat{q}\rangle(\langle\hat{t}\rangle) = \mathscr{Q} + \frac{\langle\hat{t}\rangle}{M}\mathscr{P}, \quad \Delta(qp)(\langle\hat{t}\rangle) = \mathscr{D}(qp) + \frac{\langle\hat{t}\rangle}{M}(\mathscr{D}p)^2 \qquad (13.31)$$

in terms of initial values. These equations can also be interpreted as relational observables between the remaining degree of freedom and time. (The gauge choice used is not unique, but is the only one good for all $\mathscr{P}$. Moreover, it implies real $\Delta(qp)$, but imaginary $\Delta(tp_t)$ as it is realized in the deparameterized, time-independent kinematical Hilbert space. This gauge choice is the one corresponding to the usual Schrödinger representation of the deparameterized system, whose solutions for $\langle\hat{q}\rangle(t)$ and $\Delta(qp)(t)$ are the same as (13.31). Other gauge choices provide different deparameterized relationships, and are more difficult to formulate in a Hilbert-space setting.)

The presence of quantum variables such as the covariance opens up new possibilities for internal times and deparameterization, compared to a classical system. In fact, the squeezing of a state has often been brought in contact with entropy and the second law of thermodynamics, implying monotonic behavior [8–14]. Thus, a quantum variable such as $\Delta(qp)$ may provide a good internal time even in regions where all classical variables would behave in an oscillatory manner [15, 16]. It would also imply that a natural concept of time would in fact include a strong quantum component, showing a possible quantum origin of time.

13.2.2 Relativistic Systems

Relativistic systems can be treated in much the same way as non-relativistic ones. A major advantage of effective techniques in this context is the fact that square roots need not be taken at the operator level (for which one would have to know the complete spectral decomposition of the Hamiltonian) but simply for numbers such as expectation values. We demonstrate these features using the previous Example 12.5 of the relativistic harmonic oscillator [17]. Up to second order in moments, the effective constraints are

$$C = \langle \hat{p}_\varphi \rangle^2 - \langle \hat{p} \rangle^2 - \langle \hat{q} \rangle^2 + (\Delta p_\varphi)^2 - (\Delta p)^2 - (\Delta q)^2 \tag{13.32}$$

$$C_\varphi = 2\langle \hat{p}_\varphi \rangle \Delta(\varphi p_\varphi) + i\hbar \langle \hat{p}_\varphi \rangle - 2\langle \hat{p} \rangle \Delta(\varphi p) - 2\langle \hat{q} \rangle \Delta(\varphi q) \tag{13.33}$$

$$C_{p_\varphi} = 2\langle \hat{p}_\varphi \rangle (\Delta p_\varphi)^2 - 2\langle \hat{p} \rangle \Delta(p_\varphi p) - 2\langle \hat{q} \rangle \Delta(p_\varphi q) \tag{13.34}$$

$$C_q = 2\langle \hat{p}_\varphi \rangle \Delta(p_\varphi q) - 2\langle \hat{p} \rangle \Delta(qp) - i\hbar \langle \hat{p} \rangle - 2\langle \hat{q} \rangle (\Delta q)^2 \tag{13.35}$$

$$C_p = 2\langle \hat{p}_\varphi \rangle \Delta(p_\varphi p) - 2\langle \hat{p} \rangle (\Delta p)^2 - 2\langle \hat{q} \rangle \Delta(qp) + i\hbar \langle \hat{q} \rangle. \tag{13.36}$$

Thus, some kinematical moments must be complex. Going through the solution procedure for first-class constraints applied to this system [18], solving (13.32) for $\langle \hat{p}_\varphi \rangle$ and eliminating all p_φ- and φ-moments using the other effective constraints, one can see that the system is deparameterizable in φ with quantum Hamiltonian $\langle \hat{p}_\varphi \rangle = \pm H_Q$,

$$H_Q = \sqrt{\langle \hat{p} \rangle^2 + \langle \hat{q} \rangle^2} \left(1 + \frac{\langle \hat{q} \rangle^2 (\Delta p)^2 - 2\langle \hat{q} \rangle \langle \hat{p} \rangle \Delta(qp) + \langle \hat{p} \rangle^2 (\Delta q)^2}{2(\langle \hat{p} \rangle^2 + \langle \hat{q} \rangle^2)^2} \right).$$

Reality is then imposed on Dirac observables $\langle \hat{q} \rangle (\langle \hat{\varphi} \rangle)$, $\langle \hat{p} \rangle (\langle \hat{\varphi} \rangle)$, $\Delta(\cdots)(\langle \hat{\varphi} \rangle)$ obtained by solving the Hamiltonian equations.

13.2.3 Problem of Time

Dirac observables of the effective constrained system provide the observable information just as one could compute it on the physical Hilbert space when a particular representation is known. But now, in addition to the difficult task of computing a complete set of obervables for the reduced phase space, we have a second option which does not have an analog at the Hilbert-space level: we can treat the constrained system by gauge fixing. An analysis of the constraints and the gauge flow they generate shows that one can fix the gauge (to second order in moments) by requiring the moments $(\Delta \varphi)^2$, $\Delta(\varphi q)$ and $\Delta(\varphi p)$ to vanish when one decides to use $\langle \hat{\varphi} \rangle$ as time. Other moments involving time are then fixed by solving the constraints, such that no φ-moments remain free. This outcome is just as expected if one were to choose φ

as time from the outset and then quantize the system with φ as a parameter, not an operator. Now, however, the procedure easily allows the use of $\langle \hat{\varphi} \rangle$ as a local internal time: one may use different gauge fixings to describe different parts of the phase space, and one may easily transfer to a different choice of time by applying a gauge transformation. Following [19], the gauge fixing that implements a specific choice of time is called a Zeitgeist.

Once a Zeitgeist is specified, we gain access to properties of $\langle \hat{\varphi} \rangle$, which are interesting in the context of time even though this parameter is not an observable on the reduced phase space. In particular, the value changes when a different time is used; it depends on the Zeitgeist. As the following example illustrates, a general consequence is that time is complex.

Example 13.9 (*Time-dependent potential*) We consider a constraint operator $\hat{C} = \hat{p}_\varphi^2 - \hat{p}^2 + V(\hat{\varphi})$ for a relativistic particle in an arbitrary φ-dependent potential $V(\varphi)$. The notation indicates that we are going to choose φ as internal time, but this choice is not required from the outset. In particular, φ may not serve as global internal time classically; there may be turning points where p_φ vanishes and φ fails to be a well-defined parameter along solutions. The constraint operator gives rise to the effective constraints

$$C_Q = \langle \hat{p}_\varphi \rangle^2 - \langle \hat{p} \rangle^2 + (\Delta p_\varphi)^2 - (\Delta p)^2 + V(\langle \hat{\varphi} \rangle) + \frac{1}{2} V''(\langle \hat{\varphi} \rangle)(\Delta \varphi)^2 \quad (13.37)$$

$$C_\varphi = 2\langle \hat{p}_\varphi \rangle \Delta(\varphi p_\varphi) + i\hbar \langle \hat{p}_\varphi \rangle - 2p\Delta(\varphi p) + V'(\langle \hat{\varphi} \rangle)(\Delta \varphi)^2 \quad (13.38)$$

$$C_{p_\varphi} = 2\langle \hat{p}_\varphi \rangle (\Delta p_\varphi)^2 - 2\langle \hat{p} \rangle \Delta(p_\varphi p) + V'(\langle \hat{\varphi} \rangle)\left(\Delta(\varphi p_\varphi) - \frac{1}{2}i\hbar\right) \quad (13.39)$$

to second order in the moments. We now implement $\langle \hat{\varphi} \rangle$ as local internal time via the Zeitgeist

$$(\Delta \varphi)^2 = \Delta(\varphi q) = \Delta(\varphi p) = 0. \quad (13.40)$$

We see that $\Delta(\varphi p_\varphi) = -\frac{1}{2}i\hbar$ from $C_\varphi = 0$, which then implies

$$(\Delta p_\varphi)^2 = \frac{\langle \hat{p} \rangle^2}{\langle \hat{p}_\varphi \rangle^2}(\Delta p)^2 + \frac{1}{2}i\frac{V'(\langle \hat{\varphi} \rangle)\hbar}{\langle \hat{p}_\varphi \rangle}$$

because of $C_{p_\varphi} = 0$. With this, we arrive at the expression

$$C = \langle \hat{p}_\varphi \rangle^2 - \langle \hat{p} \rangle^2 + \frac{\langle \hat{p} \rangle^2 - \langle \hat{p}_\varphi \rangle^2}{\langle \hat{p}_\varphi \rangle^2}(\Delta p)^2 + \frac{1}{2}i\frac{V'(\langle \hat{\varphi} \rangle)\hbar}{\langle \hat{p}_\varphi \rangle} + V(\langle \hat{\varphi} \rangle) \quad (13.41)$$

for the constraint $C_Q = \langle \hat{C} \rangle$ on the space on which C_φ and C_{p_φ} are solved.

In (13.41), all terms except the last two are expected to be real-valued: $\langle \hat{p} \rangle$ and Δp are physical observables for the class of systems considered, and $\langle \hat{p}_\varphi \rangle$ can be interpreted physically as the local energy value. When the constraint is satisfied, we obtain the imaginary part of $\langle\hat{\varphi}\rangle$ from

$$\frac{1}{2}i\frac{V'(\langle\hat{\varphi}\rangle)\hbar}{\langle\hat{p}_\varphi\rangle} + V(\langle\hat{\varphi}\rangle) = 0. \qquad (13.42)$$

For semiclassical states, to which this approximation of effective constraints refers, we Taylor expand the potential

$$V(\langle\hat{\varphi}\rangle) = V(\mathrm{Re}\langle\hat{\varphi}\rangle + i\,\mathrm{Im}\langle\hat{\varphi}\rangle) = V(\mathrm{Re}\langle\hat{\varphi}\rangle) + i\,\mathrm{Im}\langle\hat{\varphi}\rangle\,V'(\mathrm{Re}\langle\hat{\varphi}\rangle) + O((\mathrm{Im}\langle\hat{\varphi}\rangle)^2)$$

in the imaginary term, expected to be of the order $\hbar$. To order $\hbar$, the imaginary contribution to C is given by

$$\frac{1}{2}i\frac{V'(\mathrm{Re}\langle\hat{\varphi}\rangle)\hbar}{\langle\hat{p}_\varphi\rangle} + iV'(\mathrm{Re}\langle\hat{\varphi}\rangle)\mathrm{Im}\langle\hat{\varphi}\rangle + O(\hbar^{3/2}) = 0.$$

Thus,

$$\mathrm{Im}\langle\hat{\varphi}\rangle = -\frac{\hbar}{2\langle\hat{p}_\varphi\rangle}. \qquad (13.43)$$

As a consistency result, one can check that changing the Zeitgeist by a gauge transformation transfers the imaginary contribution from $\langle\hat{\varphi}\rangle$ to the new variable used as local internal time [20].

At this stage, one may note that the possibility of time as an operator has often been considered in quantum mechanics, with the conclusion that it could not be self-adjoint. Otherwise, it would generate unitary transformations changing the energy by arbitrary amounts, in conflict with the expectation that energy should be bounded from below for stability. A non-self-adjoint time operator would not be guaranteed to have a real-valued expectation value, a consequence which seems in agreement with what we have found here in explicit form. However, the result (13.43) is of a more general nature: it applies whenever there is a time-dependent potential, irrespective of whether it has a lower bound. Even if the potential is bounded neither from above nor from below do our results hold, but arguments using the energy-translation generated by time would no longer apply to tell us whether time could be real.

13.3 Applications of Effective Constraints in Quantum Gravity and Cosmology

Canonical quantum gravity is plagued by several long-standing problems which have not been resolved so far. While loop quantum gravity has introduced many new ingredients in this field, it has barely been able to touch these difficult issues. Central among them are the problem of time [21] and the anomaly problem, and they must both be solved before a consistent framework to derive physical predictions can be obtained. Effective-constraint techniques allow one to address these problems in a new way, and to solve them in semiclassical regimes where effective constraints can be truncated and analyzed. This procedure does not eliminate the problems altogether, but it considerably tames them for most practical purposes.

13.3.1 Problem of Time

The effective procedure to deal with non-deparameterizable constraints has several advantages. Regarding the problem of time, local internal times and local relational observables can be consistently implemented in quantum systems [19, 20]. Changing the internal time simply amounts to a gauge transformation; the equivalence of different choices of time easily follows. As illustrated for instance by the fact that the imaginary part of time is transferred when the time is changed, the procedure is consistent. In particular, one can require that all physical observables be real even in the presence of complex time. In fact, the complexity of time is crucial for the consistency of the whole framework; requiring time expectation values to be real would lead to contradictions with physical reality conditions. Complex time is thus an important part of a consistent procedure to solve quantum constrained systems.

13.3.2 Anomaly Problem

Effective constraints are also useful to analyze the anomaly problem, as we have already done for perturbation equations subject to inverse-triad corrections. For an anomaly-free quantization of several constraints $\hat{C}_I$, effective constraints $C_{f,I} = \langle f(\hat{q}, \hat{p}) \hat{C}_I \rangle$ satisfy a first-class system. But it is not always easy to quantize in an anomaly-free way, such that constraint operators of a classical first-class system would indeed be first-class. The first-class nature required for this is rather strong since it is off-shell: it must be realized even off the solution space of the constraints since anomaly-freedom would have to be checked before solving the constraints. (The importance of properties of the off-shell algebra has been emphasized in [22].)

Effective techniques allow more direct ways of implementing anomaly-freedom: One first formulates effective constraints for a possible quantization parameterized by different choices (such as factor ordering or other ambiguities). At this stage no inconsistency arises even if the corresponding constraint operators are not anomaly-free; inconsistencies would result only when one tries to solve the constraints. Before doing so, one can compute the Poisson algebra of the effective constraints and check under which conditions, or if at all, it can be first-class. As always, this can be done order by order and is much more feasible than a calculation of the full algebra of operators. If an anomaly-free version exists, one can pick this first-class version of effective constraints for further calculations. This is the procedure we followed in Sect. 10.3 for inverse-triad corrections in loop quantum gravity.

All this is more tractable than full quantum commutators, and yet it provides consistent constraints incorporating quantum corrections. As a side product, it shows how strongly kinematical quantization ambiguities are reduced by dynamical consistency.

References

1. Bojowald, M., Brizuela, D., Hernandez, H.H., Koop, M.J., Morales-Técotl H.A.: arXiv:1011.3022
2. Bojowald, M., Strobl, T.: Rev. Math. Phys. **15**, 663 (2003). hep-th/0112074
3. Taveras, V.: Phys. Rev. D **78**, 064072 (2008). arXiv:0807.3325
4. Bojowald, M., Skirzewski, A.: Rev. Math. Phys. **18**, 713 (2006). math-ph/0511043
5. Cametti, F., Jona-Lasinio, G., Presilla, C., Toninelli, F. In: Proceedings of the International School of Physics "Enrico Fermi", Course CXLIII, pp. 431–448. IOS Press, Amsterdam (2000). quant-ph/9910065
6. Simon, J.Z.: Phys. Rev. D **41**, 3720 (1990)
7. Bojowald, M., Sandhöfer, B., Skirzewski, A., Tsobanjan, A.: Rev. Math. Phys. **21**, 111 (2009). arXiv:0804.3365
8. Gasperini, M., Giovannini, M.: Class. Quantum Grav. **10**, L133 (1993)
9. Kruczenski, M., Oxman, L.E., Zaldarriaga, M.: Class. Quantum Grav. **11**, 2317 (1994)
10. Koks, D., Matacz, A., Hu, B.L.: Phys. Rev. D **55**, 5917 (1997). Erratum: Koks, D., Matacz, A., Hu, B.L.: Phys. Rev. D **56**, 5281 (1997)
11. Kim, S.P., Kim, S.W.: Phys. Rev. D **49**, R1679 (1994)
12. Kim, S.P., Kim, S.W.: Phys. Rev. D **51**, 4254 (1995)
13. Kim, S.P., Kim, S.W.: Nuovo Cim. B **115**, 1039 (2000)
14. Kiefer, C., Polarski, D., Starobinsky, A.: Phys. Rev. D **62**, 043518 (2000)
15. Bojowald, M., Tavakol, R.: Phys. Rev. D **78**, 023515 (2008). arXiv:0803.4484
16. Bojowald, M. The Arrow of Time. Springer, Berlin (2011). arxiv: 0910.3200
17. Bojowald, M., Tsobanjan, A.: Class. Quantum Grav. **27**, 145004 (2010). arXiv:0911.4950
18. Bojowald, M., Tsobanjan, A.: Phys. Rev. D **80**, 125008 (2009). arXiv:0906.1772
19. Bojowald, M., Höhn, P.A., Tsobanjan, A.: Class. Quantum Grav. **28**, 035006 (2011). arXiv:1009.5953
20. Bojowald, M., Höhn, P.A., Tsobanjan, A.: Phys. Rev. D. arXiv:1011.3040
21. Kuchař, K.V.: General Relativity and Relativistic Astrophysics. In: Kunstatter, G., Vincent, D.E., Williams, J.G. (eds.), Proceedings of the 4th Canadian Conference, World Scientific, Singapore (1992)
22. Nicolai, H., Peeters, K., Zamaklar, M.: Class. Quantum Grav. **22**, R193 (2005). hep-th/0501114

Chapter 14
Outlook

Quantum cosmology can be seen as the executive arm of quantum gravity. It provides the best chance for observations and tests of the whole theory, thereby holding enormous power to decide about its ultimate fate. This power is often undercut by the weakly built relationship with the legislative branch of constructing the fundamental theory, making it easily evadable. Fortunately, the ties are being strengthened by ongoing research, as is the observational judiciary. With a long-term perspective, quantum cosmology and quantum gravity thus seem on a promising and scientifically sustainable way.

The current phase, however, is a critical one. Quantum cosmology has to fight with severe conceptual and technical problems which it inherits from full quantum gravity. The hope always is that the reduced setting of quantum cosmology may show solutions to such problems more easily, and that solutions can then slowly be extended to the full setting. Unfortunately, too often quantum cosmology contents itself with the shiny arraignment of special effects which do not have a large influence on the whole system. A recent example is bouncing activities which have been reproduced in various guises, without so far providing lessons for general implementations or consistent perturbations around them. There is always a danger that comparatively simple derivations in models are done for their own sake, without paying attention to lessons learned and challenges confirmed by more complicated considerations of how those models may lie in a full framework. Nevertheless, as perturbations around homogeneous models are being brought under better control, one can hope to enhance our understanding of specific candidates of full theories such as loop quantum gravity—or possibly to rule them out by showing irreconcilable conflicts between predictions and observations.

M. Bojowald, *Quantum Cosmology*, Lecture Notes in Physics 835,
DOI: 10.1007/978-1-4419-8276-6_14, © Springer Science+Business Media, LLC 2011

Index

6-valent vertex, 139

A
Abelianization, 134, 139, 140, 189, 226, 230
Acceleration, 106
Adiabatic approximation, 96, 282, 285, 288
Affine quantum gravity
Algebra, 275
Almost-periodic function, 30, 38, 57, 80, 134,
    139, 143, 147, 152, 154, 160
Anharmonicity, 281, 283, 286
Anisotropic metric, 136
Anisotropy, 150, 156
    potential, 137
Annihilation operator, 21
Anomaly, 225, 240–241
    freedom, 52, 180, 193, 199, 224, 229, 232,
    236, 298
Anti-friction, 239
Arrow of time, 117
Ashtekar–Barbero connection, 26, 141, 190
Asymptotic
    expansion, 12, 282
    properties, 121
Asymptotically almost-periodic function, 135
Atomic nature, 20
averaging, 203, 279
    problem, 150

B
Back-reaction, 237
Background
    independence, 21
    state expansion, 85, 107, 121, 220–221,
    230

Barbero–Immirzi parameter, 27
Bardeen variables, 226
Basic operators, 276
Belinsky–Khalatnikov–Lifshitz (BKL)
    conjecture, 122
Bessel function, 262
Bianchi
    I model, 137, 254
    models, 136
    type A, 139
Big bang, 99, 122
    nucleosynthesis, 238
Black hole, 157, 193, 194, 258, 260
    collapse, 162
Bohmian viewpoint, 121, 125
Bohr compactification, 30, 38, 40, 52, 61, 134,
    139, 149, 159, 173
Bounce, 21, 104, 107, 109, 112, 114, 120,
    126–127, 145, 235, 239, 241–242
Boundary, 9, 191
    condition, 14, 37
Boundedness, 215, 261

C
Canonical
    quantization, 22, 275
    quantum gravity, 297
    variables, 9
Casimir condition, 101, 280
Causality, 236
Classical limit, 215
Closed model, 10, 23, 89
Co-normal, 28
Coarse graining, 200
Coherent state, 272, 281
Commutator, 89, 127, 206, 209

M. Bojowald, *Quantum Cosmology*, Lecture Notes in Physics 835,
DOI: 10.1007/978-1-4419-8276-6, © Springer Science+Business Media, LLC 2011